Prospects of Science, Technology and Applications

A Compendium of Symposium

Prospects of Science, Technology and Applications

A Compendium of Symposium

Editors

Prof. Renu Sharma

Siksha 'O' Anusandhan Deemed to be University,
Bhubaneswar, Odisha, India

Prof. Dilip Kumar Mishra

Siksha 'O' Anusandhan Deemed to be University,
Bhubaneswar, Odisha, India

Prof. Satyanarayan Bhuyan

Siksha 'O' Anusandhan Deemed to be University,
Bhubaneswar, Odisha, India

CRC Press is an imprint of the
Taylor & Francis Group, an **informa** business

First edition published 2024
by CRC Press
4 Park Square, Milton Park, Abingdon, Oxon, OX14 4RN

and by CRC Press
2385 NW Executive Center Drive, Suite 320, Boca Raton FL 33431

© 2024 selection and editorial matter, Renu Sharma, Dilip Kumar Mishra and Satyanarayan Bhuyan; individual chapters, the contributors

CRC Press is an imprint of Informa UK Limited

The right of Renu Sharma, Dilip Kumar Mishra and Satyanarayan Bhuyan to be identified as the authors of the editorial material, and of the authors for their individual chapters, has been asserted in accordance with sections 77 and 78 of the Copyright, Designs and Patents Act 1988.

All rights reserved. No part of this book may be reprinted or reproduced or utilised in any form or by any electronic, mechanical, or other means, now known or hereafter invented, including photocopying and recording, or in any information storage or retrieval system, without permission in writing from the publishers.

For permission to photocopy or use material electronically from this work, access www.copyright.com or contact the Copyright Clearance Center, Inc. (CCC), 222 Rosewood Drive, Danvers, MA 01923, 978-750-8400. For works that are not available on CCC please contact mpkbookspermissions@tandf.co.uk

Trademark notice: Product or corporate names may be trademarks or registered trademarks, and are used only for identification and explanation without intent to infringe.

British Library Cataloguing-in-Publication Data
A catalogue record for this book is available from the British Library

ISBN: 978-1-032-78833-3 (pbk)
ISBN: 978-1-003-48944-3 (ebk)

DOI: 10.1201/9781003489443

Typeset in Times LT Std
by Aditiinfosystems

Contents

List of Figures ix
List of Tables xv
Forward xvii
About Editors xix
Preface and Acknowledgements xxi

1. **Brain Tumor Segmentation and Classification from MR Images with Feature Extraction** 1
   M. Disha, A. Abhisek Patro, Kaberi Das, Swati Sucharita Roy and Bharat J. R. Sahu

2. **Adsorption of Co (II) by Banana (*Musa paradisiaca*) Peel** 12
   Lipsa Mayee Mishra, Biswaprakash Sarangi, Lopita Priyadarshini Pradhan and Sneha Prabha Mishra

3. **Recovery of Fe, Sc, and V from Tailing Sample Using Acidic Lixiviants** 18
   Milli Agarawala, Smruti Rekha Patra, Sibananda Sahu and Niharbala Devi

4. **Examining Efficiency of Nitrogen Based Extractants for the Solvent Extraction of Acids** 27
   S. P. Moharana, P. K. Khillar, D. Swain, R. L. Bhutia, P. Panda and S. Mishra

5. **Revisiting the Kinetics of Ester Hydrolysis Using ODE, DDE and FDE** 34
   Priyadarshan Garnayak, Sunita Chand, Santoshi Panigrahi and Sujata Mishra

6. **Scope of Acidic Ionic Liquids as Catalyst** 42
   Debasmita Jena, Pravanjan Mishra and Sanghamitra Pradhan

7. **Study of Agro-Waste Biorefineries for Circular Bio-Economy** 52
   Nayak Sanchita and Das Mira

8. **A Review on, Recent Developments in Solid Adsorbents for CO_2 Capture** 61
   Biswal Shaswata and Das Mira

9. **Developing a Machine Learning Model for HAI Risk Prediction Using EHR Data** 72
   Sriya Das Mohapatra, Naveen Kumar, Dishant Yadav, Swati Sucharita Ray and Bharat Jyoti Ranjan Sahu

10. **Interpretable Machine Learning Model for Diabetes Disease Detection using Explainable Artificial Intelligence (xAI)** 81
  Abhilasha Panda and Anukampa Behera
11. **Improvement of LSB Image Steganography Using Feistel Cipher** 89
  Rupesh Kumar Mohapatra, Spandan Udgata and Susmita Panda
12. **Efficient Data Management for Blood Banks: Integrating Data Warehousing, Database Management, and Location Tracking APIs** 102
  Sakshi, Anitesh Raj, Swati sucharita Roy, Bharat J R Sahu
13. **PicoPager: An Opensource Raspberry-Pi Pico W Based IP Paging System** 108
  Manish Kumar Tiruwar and Bharat Jyoti Ranjan Sahu
14. **Multivariate Machine Learning Approaches for Dynamic Prediction of Air Quality and Estimating Heatwave Occurrence** 116
  Abhinandan Roul, Shubhaprasad Padhy, Sambit Kumar Sahoo, Ayush Pattanayak, Manoranjan Parhi and Abhilash Pati
15. **Software Bug Classification Using Machine Learning Approach** 136
  Sandeep Soumya Sekhar Mishra and Swadhin Kumar Barisal
16. **Hindi Image Captioning Using Deep Learning** 149
  Soumyarashmi Panigrahi and Jaydev Sutar
17. **An Integrated Model for Smart Healthcare Solutions With 5G Network Slicing** 156
  Swati Sucharita Roy, Bharat Jyoti Ranjan Sahu and Shatarupa Dash
18. **Elderly Fall Detection Using Machine Learning** 168
  Amisha Sinha, Dezy Jha, Malaya Kumar Swain, Soumya Sagar Rath and Nimisha Ghosh
19. **2D CNN based Pituitary, Meningioma and Glioma Tumor Classification** 177
  Reetichi Pattanaik, Suraj Sahu, Vishwas Kumar and Samrudhi Mohdiwale
20. **Dynamic Long Short-Term Memory Model for Stock Market Price Forecasting** 189
  Indrajit Sahu, Kiran Shankar Paira, Priti Rani Bhoi and Samrudhi Mohdiwale
21. **Health is Wealth: Menu Driven Health Monitoring System for Improved Quality of Life** 199
  Shreeja Mahapatra, Ayushi Pradhan, Amlanjyoti Satapathy and Nimisha Ghosh
22. **Fine-Tuning and Comparing XLSR Models on Odia Speech** 213
  Pragyan Prusty and Ajit K Nayak
23. **Chronic Kidney Disease Detection Using Machine Learning Model** 218
  Rishav Pandit, Amlan Mahapatra, Afraa Jumaa, Sushant Kumar and Suman Sau
24. **WhatsApp Chat Analysis Application** 226
  Sarthak Kumar Behera, Zarit Ahmed, Sahil Saswat Jena, Abhijit Pattanaik and Suman Sau

25. **Characterization of Antimony Chalcogenide-based PIN Photodiode** — 237
   Vedika Pandey and Sumanshu Agarwal

26. **Customer Churn Prediction in Banking Sector Using Machine Learning Techniques** — 244
   Anshul Srivastava and Abhigyan Bhadra Laxmipriya Moharana

27. **Plant Disease Detection** — 250
   Tirthankar Biswas and Swastik Kumar Pati and Sunita Sarangi

28. **Smart Hospital and Healthcare: Hospital Management System** — 257
   Taniya Baral and Anish Mohanty

29. **From Pollution to Power: Harnessing the Potential of Non-Biodegradable Wastes for Electricity Generation and Wireless Air Pollution Monitoring** — 266
   Aditya Kumar Lenka, Amlan Adarsh and Biswaranjan Swain

30. **A Formulation for Maximizing Solar Irradiance Based on Adjustment of Optimum Inclination Angle** — 274
   Bibekananda Jena, Sonali Goel and Renu Sharma

31. **Complimentary Filter-Based Technique for Identification of Unmanned Aerial System Parameters** — 282
   Dhrutidhara Behera, Sobhit Panda, Subhranshu Sekhar Puhan, Sonali Goel and Renu Sharma

32. **Innovative Methods and Greener Technology for Remediation of Microfiber Pollutants** — 295
   Biswanath Naik, Lala Behari Sukla and Aditya Kishore Dash

33. **Micro-Raman and ESR Studies for the confirmation of RTFM in $Zn_{1-x}Cu_xO$ $(0.00 \leq x \leq 0.1)$** — 316
   Urmishree Routray, Jyoshnarani Mohapatra, V. V. Srinivasu and Dilip Kumar Mishra

List of Figures

1.1	Generic proposed model	3
1.2	Segmented MR image	4
1.3	Watershed segmented MR image	4
1.4	Testing with dataset having no tumor	7
1.5	Testing with dataset having tumor	7
1.6	Comparision between different classification algorithms	8
1.7	Comparision between the accuracy of different works	9
1.8	Comparision between the F1 Score of different classification algorithms	10
1.9	Comparision between the Precision of different classification algorithms	10
2.1	Determination of point of zero charge	14
2.2	Effect of contact time on Co(II) adsorption	15
2.3	Effect of pH on Co(II) adsorption	15
2.4	Effect of adsorbent dose on Co(II) adsorption	16
2.5	Effect of temperature on Co(II) adsorption	16
3.1	Impact of different acids on the leaching rates. Leaching Conditions. (Acid concentration = 1 M, temperature = 30°C, S/L = 10 g/L, and leaching time = 60 min.)	22
3.2	Impact of H_2O_2 concentration on the leaching rates. (Acid concentration = 1 M, temperature = 30 °C, S/L = 10 g/L, and leaching time = 60 min.)	22
3.3	Impact of H_2SO_4 concentration on the leaching rates. (Temperature = 30 °C, S/L = 10 g/L, and leaching time = 60 min.)	23
3.4	Impact of Solid-to-Liquid on the leaching rates. (H_2SO_4 concentration = 3 M, temperature = 30 °C, S/L = 10 g/L, and leaching time = 60 min.)	24
3.5	Impact of temperature on the leaching rates. (HCl concentration = 1 M, temperature = 30 °C, S/L = 10 g/L, and leaching time = 60 min.)	24
3.6	Impact of HCl concentration on the leaching rates. (Temperature = 60 °C, S/L = 50 g/L, and leaching time = 60 min.)	25
4.1	Schematic representation of the liquid-liquid extraction procedure	30

4.2	*Plot of %E versus [Acid], M for the extraction of acids using 0.1M TEHA in kerosene; Shaking time = 15 min, O/A = 1; Temperature = 298K*	31
4.3	*Plot of %E versus [Acid], M for the extraction of acids using 0.1M Aliquat 336 in kerosene. Shaking time = 15 min, O/A = 1; Temperature = 298K*	32
5.1	*Graphical representation of the scheme of present investigation*	36
5.2	*Plot of [Ester], mol/L against time for the acid hydrolysis of ethyl acetate using ordinary differential equation*	37
5.3	*Plot of [Ester], mol/L against time for the acid hydrolysis of ethyl acetate using delay differential equation*	38
5.4	*Plot of [Ester], mol/L against time for the acid hydrolysis of ethyl acetate using partial differential equation*	39
6.1	*Representation of types of AILs*	44
6.2	*Scheme of synthetic route of 1-butyl-2,3-dimethylimidazolium chloride/$ZnCl_2$ [BMMI][$ZnCl_2$]*	45
6.3	*Scheme of synthetic route of 1-butyl-3-methylimidazolium tetrachloroferrate [Bmim][$FeCl_4$]*	45
6.4	*Scheme of synthetic route of 1-butyl-3-methylimidazolium hydrogen sulfate ([Bmim]$^+$ [HSO_4]$^-$) and l-butyl-3-methylimidazolium hydrogen phosphate ([Bmim]$^+$[HPO_4]$^-$)*	46
6.5	*Scheme of synthetic route of 1-(3-sulfonic acid)-propyl-3-methylimidazole ferric chloride [MIMPS]$FeCl_4$.$^-$*	46
7.1	*The categories of biorefinery feedstocks, conversion technologies, products and applications*	54
7.2	*Schematic representation of pretreatment of lignocellulosic biomass*	56
8.1	*Classification of adsorbent materials*	63
8.2	*CO_2 adsorption on MOF based mixed matrix membrane*	64
8.3	*Nitrate doped MgO adsorbent for enhancing CO_2 capture*	65
9.1	*Relationship between BP and WBC count in dataset*	77
9.2	*Relation between blood report and infections associated with blood report*	77
9.3	*Accuracy with different algorithms*	78
9.4	*Relationship between algorithm performance and cross validation scores*	78
9.5	*Relationship between accuracy scores and cross validation scores*	79
10.1	*Experimental framework*	85
10.2	*Average impact on model output magnitude*	86
10.3	*Impact on model output by using SHAP value*	86
10.4	*A test case showing the probability of a person having diabetes or not*	87
11.1	*Types of steganography*	91
11.2	*Flowchart depicting entire flow of the proposed mechanism*	92

11.3	Mechanism of LSB stegnography	94
11.4	Flowchart for Fiestel Cipher encryption	95
11.5	Home page for the app	96
11.6	Encrypt page- a password is entered with the plain message text, then an image is uploaded.On clickikng the encrypt button the message gets encryted using fiestel cipher and is embedded into the image. Output image gets stored in same location	96
11.7	Decrypt page- secret password is entered, then stego image is uploaded. On clicking the decrypt button the message gets extracted from the image and then decrypted using Fiestel cipher. Displays the secret message as output	97
11.8	Line chart depicting comparison of different values for MSE	99
11.9	Line chart depicting comparison of different values for PSNR	100
11.10	Line chart depicting comparison of different values for SSIM	100
12.1	Diagram illustrating the lack of centralization in current blood bank management systems	104
12.2	Flowchart for proposed system	105
12.3	ER Diagram for possible database	106
13.1	A top view of our implementation	109
13.2	A top view of 3d printed filed in PLA	110
13.3	A top view of snap fit Raspberry Pi Pico W and SSD1306 oled	110
13.4	A 3-layer ESP-NOW protocol vs full OSI stack	111
13.5	A pictorial block representation of how my implementation. At the core of our implementation is the paging controller. It connects to a given network with password and SSID given. It then collects string from user via terminal window. And broadcasts a encypted string on its IP. This broadcasted string can be viewed in the browser on any device in the network. The nodes have | the SSID, Password, IP of Paging controller, Node id. According to the Node Id, each node concatenates the string, decrypts the string and displays on SSD1306.	112
13.6	A top view of our implementation where random number is sent to each of the nodes. Raspberry Pi 4B is used to create a local network. And the IP paging controller is used to send random number to each of the nodes	113
14.1	Process flow for methodology	123
14.2	Correlation analysis	124
14.3	AQI shows a repeating pattern	124
14.4	Actual vs prediction values of PM10	127
14.5	Actual vs prediction values of NO	128
14.6	Actual vs prediction values of NO2	128
14.7	Actual vs prediction values of NH3	128

14.8	*Actual vs prediction values of SO2*	129
14.9	*Actual vs prediction values of AQI*	129
14.10	*Actual vs prediction values of Max Temp*	129
14.11	*Actual vs prediction values of Min Temp*	130
14.12	*AQI Error comparison*	131
14.13	*Heatwave Error comparison*	131
14.14	*Neural prophet model metrics for Multivariate AQI*	132
14.15	*Neural Prophet Model metrics for Temperature prediction (for Heatwave)*	132
14.16	*Max Temp Prediction using TFT (for future timeline)*	133
14.17	*Encoder variable importance as per TFT model*	133
15.1	*Architecture of suggested method*	141
15.2	*Graphical representation of RMSE of different ML techniques*	145
15.3	*Graphical representation of precision values of different ML techniques*	145
15.4	*Graphical representation of recall score of different ML techniques*	146
15.5	*Graphical representation of F1 score for different ML techniques*	146
16.1	*A model representation of how the IC generator consisting of CNN-LSTM works*	151
16.2	*Comparison of A. Rathi et al.[3] model Vs. our proposed model at 25 epochs BLEU-1(unigram), BLEU-2(bigram), BLEU-3(trigram) and BLEU-4(4-gram)*	153
16.3	*Image caption generated from various input images*	154
17.1	*Application areas of IoT*	157
17.2	*IoT application with requirement specifications*	157
17.3	*Various Dimensions of the Healthcare System with Their Requirements*	158
17.4	*Overall configuration of IoT-based healthcare system*	162
17.5	*5G network slicing with different IoT applications*	164
17.6	*Overall configuration in network slicing*	164
18.1	*Flow chart of elderly fall detection*	170
18.2	*Relationship between every feature*	171
18.3	*Confusion matrix*	174
19.1	*Framework for the proposed model*	181
19.2	*Convolution Neural Network 22 Layers*	182
19.3	*Training and validation accuracy/loss for reference model*	185
19.4	*Training and validation accuracy/loss for proposed Model*	186
19.5	*Confusion matrix for proposed model*	187
20.1	*Schematic layout/model diagram*	194
20.2	*Closing price year wise*	194

20.4	*Predicting the weekly data based on the previous data by trained model*	195
20.3	*Training and testing of the model using stacked LSTM model*	195
20.5	*Training and testing of the model using stacked LSTM model*	196
20.6	*Final image of predicted data weekly along with the previous data*	196
21.1	*Pair-plot for breast cancer dataset*	203
21.2	*Pair-plot for diabetes dataset*	203
21.3	*Pipeline of the work*	204
21.4	*Comparative analysis of ROC curves for the 4 diseases*	208
21.5	*Homepage of the website, where we can see 4 buttons representing 4 diseases viz. heart disease, diabetes, breast cancer and mental health issues*	209
21.6	*Shows the tab that opens when a user clicks on the mental health button. After entering the details in the form, if the result is negative it is displayed in red colour at the bottom of the screen as shown*	210
21.7	*Shows the positive result after entering the details*	210
22.1	*Flow diagram of fine tuning a model*	214
22.2	*The XLSR approach*	214
22.3	*Training Loss of XLSR-53 and XLSR-300-M*	216
22.4	*Validation loss of XLSR-53 and XLS-R-300M*	216
22.5	*WER of XLSR-53 and XLS-R-300M*	217
23.1	*Specified structure of random forest*	221
23.2	*Comparative study in terms of accuracy*	223
23.3	*Comparative study in terms of precision*	223
23.4	*Comparative study in terms of sensitivity*	224
24.1	*Design of system framework diagram*	229
24.2	*Data acquisition stage flow chart*	230
24.3	*Total number of data transfers in the chat*	231
24.4	*Monthly count of messages over a certain period*	232
24.5	*Daily count of messages over a certain period*	232
24.6	*Most active day of the week and most active month of the year*	233
24.7	*Total no of messages on a per-hour basis in the week*	233
24.8	*Most active users and their contribution percentage in the chat*	234
24.9	*Wordcloud of all words used in the chat*	234
24.10	*Frequency of top 20 words of a chat*	235
24.11	*Emoji count and usage percentage*	235
25.1	*Schematic design of antimony chalcogenide ($Sb_2(S_{1-x}Se_x)_3$) based p-i-n photodiode*	239
25.2	*Comparision between (a) simulated and (b) Analytical result of current-voltage characteristics of photodiode under darkness with mole fraction*	241

25.3	Variation of responsivity with wavelength for various mole fractions (x = 0 to 0.5)	242
25.4	Variation of quantum efficiency with wavelength for various mole fractions (x = 0 to 0.5)	242
26.1	Framework for bank churn prediction using machine learning techniques	246
26.2	Projection of learning model to churn rate calculation	247
27.1	The workflow diagram of the model describe in the thesis	252
27.2	The plot between validation and training accuracy and loss	256
28.1	Architectural block diagram of proposed system	262
28.2	Smart hospital flow diagram	263
29.1	Flowchart of the proposed system for electricity generation and wireless air pollution monitoring	269
29.2	Experimental setup for harnessing non-biodegradable wastes for electricity generation and air pollution monitoring using ESP32	270
29.3	Wireless Air Pollution Monitoring on IoT Platform using ESP32 and ThingSpeak	271
29.4	Integration of Blynq as an interface for air pollution data visualization and access	272
30.1	Dimension representation of the 11.2 kW roof top PV system	276
30.2	Daily irradiance of each month	278
30.3	Monthly adjustment of tilt for different months of the year	278
30.4	Seasonal variation of tilt angle	279
30.5	Annual adjustment of tilt angle	280
31.1	Hardware used in the system representation	284
31.2	Connection diagram in arduino	285
31.3	(i) Basic control diagram of UAV, (ii) Schematic diagram	287
31.4	Basics of complementary filter design	288
31.5	Roll estimate from raw acceleration data	290
31.6	Roll estimation from filtered acceleration data	291
31.7	Gyro drift present in gyroscope set of IMU	291
31.8	Complementary filter output for roll angle in degrees ($\alpha = 0.4, 0.1$)	292
31.9	Kalman vs. complementary filter	292
32.1	Sources of microfiber pollution	299
32.2	Microfiber degradation by the action of microbes	304
33.1	Micro-Raman spectra of ball milled $Zn_{1-x}Cu_xO$ ($0 \leq x \leq 0.1$) powders	318
33.2	Magnetic field dependent magnetization of ball milled $Zn_{1-x}Cu_xO$ ($0 \leq x \leq 0.1$) powders	320
33.3	ESR spectrum of 1h and 4h ball milled ZnO powders	322
33.4	ESR spectra of ball milled $Zn1-xCuxO$ ($0 \leq x \leq 0.1$) powders	322

List of Tables

1.1	Dataset description	6
1.2	Comparison between different classification algorithms used in the paper	8
1.3	Comparison with other works	9
2.1	Physico-chemical parameters of banana peel	14
3.1	Literature survey on the use of leaching agents and extractants for the recovery of Sc, V and Fe from different ores.	19
3.2	Chemical Composition of the tailing sample	21
4.1	Detail representations of organic extractants and specifications	29
4.2	Detail representations of acids and their specifications	29
4.3	The distribution ratio values obtained in the extraction of acids by using 0.1 M TEHA and Aliquat 336 in kerosene	31
5.1	Particulars of the chemicals used	35
5.2	Ethyl acetate concentration after time t	40
7.1	Four generations of biorefineries [9]	54
7.2	Pretreatments of lignocellulosic wastes used for bio-hydrogen production [24]	57
8.1	An overview of the CO_2 capture by carbon-based materials	63
8.2	CO_2 capture efficiency of different MOFs	64
8.3	CO_2 capture tendencies of various synthetic metal oxides	66
8.4	CO_2 capture tendencies of various amine functionalized mesoporous alumina	66
8.5	CO_2 capture by amine silica composites	67
8.6	CO_2 capture by amine functionalized MOFs	67
8.7	CO_2 capture by amine functionalized zeolites and carbon materials	67
10.1	A brief summary of the results obtained	85
14.1	Literature survey summary of papers	121
14.2	Comparison of results from various models	130
15.1	Evaluation metrics for decision tree model	144
15.2	Evaluation metrics for random forest model	144
15.3	Evaluation metrics for multi-layer perceptron	144

16.1	Bleu score in various epoch values	152
16.2	Comparison of a. Rathi et al.[3] Model with our proposed model	152
18.1	The outcomes of several machine learning algorithms, namely LDA (Linear Discriminant Analysis), CART (Classification and Regression Trees), RF (Random Forest), LR (Logistic Regression), and SVM (Support Vector Machine), were evaluated. The average value across these algorithms for various metrics was found to be 100%, indicating their excellent performance in fall detection	174
18.2	The evaluation of fall detection models involved assessing their accuracy, precision, recall, and F1-score	175
19.1	Performance matrix for reference model	184
19.2	Performance matrix for proposed model	184
21.1	Results of applied models along with hyper-parameter tuning for heart disease	206
21.2	Results of applied models along with hyper-parameter tuning for diabetes	206
21.3	Results of applied models along with hyper-parameter tuning for mental health issues	207
21.4	Results of applied models along with hyper-parameter tuning for breast cancer	207
21.5	Results of best models after hyperparameter tuning for each dataset	208
22.1	Parameters table for XLSR-53 and XLS-R-300M	215
25.1	Typical Parameters for antimony chalcogenide based photodiode	240
25.2	Typical Parameters for antimony chalcogenide based photodiode	241
26.1	Prediction results for the predictors age, gender and area of residence	248
26.2	Prediction results for the age as predictor	248
26.3	Prediction results for the gender as predictor	248
26.4	Prediction results for the area of residence as predictor	248
27.1	Model summary	254
30.1	Specification of PV module at STC ($1000W/m^2$ and $25°C$)	275
30.2	Summary of seasonal variation of optimum tilt angle from MATLAB Simulation	279
31.1	MPU 6050 sensitivities	285
32.1	Microorganisms (Fungal and Bacterial species) having degradation capacity	305
32.2	Algal species having microfiber degradation capacity	307
33.1	Determination of the values of remnant magnetization (M_r), coercivity (H_c) and saturation magnetization (M_s) of $Zn_{1-x}Cu_xO$ compounds ball milled for various hours	321

Forward

In the dynamic landscape of human progress, the intertwining realms of science and technology have emerged as the driving forces propelling us into the future. The "Prospects of Science, Technology, and Applications: A Compendium of Symposium" serves as an intellectual beacon, illuminating the vast possibilities and frontiers that lie ahead in these transformative fields.

This compendium is a testament to the collaborative spirit that characterizes the symposium, bringing together brilliant minds from diverse disciplines to explore, discuss, and envision the future of science and technology. As we stand at the crossroads of innovation and discovery, the need for a comprehensive understanding of the prospects that await us has never been more crucial.

The chapters within this compendium delve into a rich tapestry of topics, ranging from cutting-edge advancements in artificial intelligence, biotechnology, and nanotechnology to the ethical considerations that accompany these scientific breakthroughs. The symposium acts as a melting pot of ideas, fostering interdisciplinary dialogues that transcend traditional boundaries and stimulate novel perspectives.

As readers embark on this intellectual journey, they will encounter a mosaic of insights, analyses, and predictions from leading experts in their respective fields. From the microscopic world of quantum computing to the macroscopic implications of space exploration, the symposium covers a broad spectrum of subjects that collectively shape the trajectory of our technological future. Each contribution is a brushstroke, contributing to the canvas of human knowledge and progress.

As we navigate the complexities of an ever-changing world, "Prospects of Science, Technology, and Applications" invites readers to engage with the profound questions and limitless possibilities that lie ahead. It is a resource for scholars, students, and enthusiasts alike, offering a glimpse into the fascinating world of tomorrow that is being shaped today.

We extend our heartfelt gratitude to the organizers, contributors, and readers who have embarked on this intellectual voyage. May this compendium inspire curiosity, spark creativity, and pave the way for a future where the frontiers of science and technology continue to expand, enriching the human experience.

Renu Sharma
Dilip Kumar Mishra
Satyanarayan Bhuyan

About Editors

Renu Sharma holds the position of Additional Dean of Student Affairs and is a Professor in the Department of Electrical Engineering at Siksha 'O' Anusandhan Deemed to be University, Bhubaneswar, Odisha, India. She is recognized as a Senior Member of IEEE, a Life Member of IE (India), a Member of IET, a Life Member of ISTE, and a Life Member of ISSE. Dr. Sharma also serves as the Chair of the Newsletter for the IEEE Bhubaneswar Subsection.In 2023, Dr. Sharma earned a notable distinction by being featured in Stanford University's list of the top 2 percent of world scientists. Her research expertise spans a diverse range of areas, including Smart Grid, Soft Computing, Solar Photovoltaic systems, Power System Scheduling, Evolutionary Algorithms, and Wireless Sensor Networks. With a keen focus on academic mentorship, she has successfully supervised seven Ph.D. students and contributed significantly to the scholarly landscape with the publication of 158 articles of international repute.Dr. Sharma has taken on the role of Guest Editor for Special Issues in prestigious journals such as the International Journal of Power Electronics (Inderscience) and the International Journal of Innovative Computing and Applications (Inderscience). Her commitment to academic leadership is further evident through her roles as General Chair for several conferences, including IEEE ODICON 2021, ODICON 2022, the flagship conference IEEE WIECON-ECE 2020, Springer conference GTSCS-2020, and IEPCCT-2019 and IEPCCT 2021. Currently, she serves as the Publication Chair for INDISCON-2022 and as the General Chair for the 2023 IEEE 3rd International Conference on SeFeT 2023.

Dilip Kumar Mishra is a Professor of Department of Physics, Faculty of Engineering and Technology (ITER), Siksha 'O' Anusandhan Deemed to be University, Bhubaneswar, Odisha, India. He served as Research Associate at IUAC, New Delhi; fellow scientist at CSIR-IMMT, Bhubaneswar and Visiting Researcher at UNISA-South Africa. The research area of interest of Dr. Mishra includes dilute magnetic semiconductor, defect and interface properties including magnetism, novel methods of materials processing, and studies of nano materials. He has published more than 150 research articles in the journal of international repute with five Indian patents. He has also acted as guest editor for special issue in Advanced Science Letters and International Journal of Nano and Bio Materials.

Satyanarayan Bhuyan is a Professor in the Department of Electronics &Communication Engineering, ITER, Siksha 'O' Anusandhan Deemed to be University, Bhubaneswar, Odisha, India and obtained his Ph.D. degree from School of Electrical and Electronic Engineering,

Nanyang Technological University, Singapore in 2011. He was a Postdoctoral Research Fellow at National University of Singapore (NUS) and was a Scientist-I in RF and Optical Department of Institute for Infocomm Research, A*STAR, Singapore. He has supervised 12 PhD students and published 120 number of research articles. His current research involves fabrications and characterizations of solid state electronic materials, devices, advanced smart ceramics sensor design, piezoelectric actuators, Wave Electronics, Electromagnetism, wireless Energization of electrical and Electronic devices, Electric vehicle charging.

Preface and Acknowledgements

In the rapidly evolving landscape of scientific and technological advancements, the "Prospects of Science, Technology, and Applications: A Compendium of Symposium" endeavors to explore the dynamic future that awaits us. As we stand at the crossroads of innovation and discovery, the need for a comprehensive understanding of the potential trajectories and applications in science and technology has never been more crucial.

This compilation brings together insights from esteemed contributors who are experts in their respective fields, ranging from fundamental sciences to cutting-edge technologies. The diverse perspectives offered within these pages aim to shed light on the exciting possibilities and challenges that lie ahead. Our intention is to inspire curiosity, spark intellectual dialogue, and foster a sense of anticipation for what the future holds.

The book is structured to encompass a wide spectrum of topics, from groundbreaking research in various scientific disciplines to the transformative applications of emerging technologies. Each chapter delves into the nuances of its subject matter, providing readers with a nuanced understanding of the current landscape and potential trajectories.

We hope that this exploration into the "Prospects of Science, Technology, and Applications" will not only inform but also ignite the imagination of readers. Whether you are a seasoned professional, a student embarking on a scientific journey, or an enthusiast eager to grasp the unfolding developments, this A Compendium aims to be a valuable resource.

As we embark on this intellectual journey together, let us delve into the promising prospects that the realms of science and technology hold for our collective future.

The SYMPOSIUM featured a diverse range of activities, including Poster presentations showcasing Research and Innovation, live demonstrations of projects, presentations of research articles through PowerPoint, and an Ideathon evaluated by an esteemed panel of judges.

We would like to express our sincere appreciation and gratitude to Prof. (Dr.) Manojranjan Nayak, Founder President of Siksha 'O' Anusandhan Deemed to be University, Bhubaneswar, Odisha, Prof. (Dr.) Pradipta Kumar Nanda, Vice-Chancellor and Prof. (Dr.) J. K. Nath, Dean of Research & Development.

Acknowledgments are also extended to the faculty at the Faculty of Engineering and Technology, ITER, Siksha 'O' Anusandhan Deemed to be University, including Prof. P. K. Sahoo, the Dean, Prof. Manas Kumar Mallick, the Director, Prof. D. N. Thatoi, Addl. Dean

(Academics), Prof. J. K. Nath, Dean (R&D), Prof. Debahuti Mishra, HOD of the Department of Computer Science Engineering, Prof. Ajit K. Nayak, HOD of Computer Science & Information Technology, Prof. Bidyadhar Basa, HOD of the Department of Civil Engineering, Prof. Saroj Acharya, HOD of the Department of Mechanical Engineering, Prof. Binod Kumar Sahu, HOD of the Department of Electrical Engineering, Prof. Niranjan Nayak, HOD of the Department of Electrical & Electronics Engineering, Prof. Bibhuprasad Mohanty, HOD of the Department of Electrical Engineering, Prof. Kaberi Dash, HOD of the Department of Computer Application, Prof. Sujata Mishra, HOD of the Department of Chemistry, Prof. Jayanta Kumar Dash, HOD of the Department of Mathematics, Prof. Kabita Das, HOD of the Department of Humanities and Social Sciences, Prof. S. K. Kamilla, HOD of Physics along with the Organizing Secretary, Prof. Tanmoy Parida, and Prof. Satish Kumar Samal. We also want to acknowledge Prof. Bharat Jyoti Ranjan Sahu, Prof. Soumojit Shee, Prof. Basant Kumar Panigrahi, Prof. Debabrata Singh, Prof. Kamal Lochan Mohanta, Prof. Naresh Sahoo, Prof. Mitali Madhusmita Nayak, Prof. Nimisha Ghosh, Dr. Prashant Sahoo, Prof. Manoranjan Parhi, Prof. Sanjeet Patra, Prof. Nakul Sahu, Prof. Madhusmita Panda, Prof. Manojranjan Das, Prof. Smrutirekha Pattnaik, Prof. Benazeer Begum, Prof. Abanti Pradhan, Prof. Alok Kumar Mishra, Prof. Anuja Nanda, Prof. Sarada Prasanna Pati, Prof. Aditya Kumar Das, Prof. Sangeet Kumar Patra, Prof. Rajanikanta Parida, Prof. Milu Acharya, and Prof. Bimlesh Nayak.

Our heartfelt gratitude is also extended to Prof. Goutam Rath of the School of Pharmaceutical Sciences, Prof. Kautuk Kumar Sardar of the Institute of Agriculture Sciences, Sri Rashmi Ranjan Sahu of Startup Odisha, and Janmejay Mohapatra, Co-Founder and CEO of Ajatus. Special thanks to all the participants who dedicated their time, expertise, and resources to make this Symposium a reality.

Renu Sharma
Dilip Kumar Mishra
Satyanarayan Bhuyan

Brain Tumor Segmentation and Classification from MR Images with Feature Extraction

M. Disha[1], A. Abhisek Patro[2], Kaberi Das[3], Swati Sucharita Roy, Bharat J. R. Sahu

Department of Computer Applications, ITER, SOA Deemed to be University, Bhubaneswar

Abstract The brain tumor is one of the deadliest diseases, that might prove to be fatal if not detected at an early stage. Early detection by traditional methods is difficult as it is a manual process and might lead to faulty results. To overcome the limitation of the traditional method, many machine learning models have been proposed that lead to early diagnosis as well as more accurate results. In the proposed model, we have taken the MR image dataset from Kaggle having two classes, followed by pre-processing, image enhancement, feature extraction, segmentation, and classification. We have achieved varying accuracies for different techniques used in classification, with Random Forest providing the highest accuracy of 98.56%.

Keywords MR image, Feature extraction, Segmentation, Classification

1. Introduction

The human body is controlled and coordinated by the brain. It helps in the management of different activities that are carried out by a human. But the activities might be affected if the brain is diseased. One such deadly disease is Brain Tumor, which is caused due to the unusual growth of cells. It can be broadly divided into 2 types, that is Benign and Malignant. A benign brain tumor is less harmful as it does not move from one location to another as in the case of a malignant brain tumor. Diagnosis of a brain tumor at an early stage could increase the chances of recovery. Thus, early diagnosis plays a crucial role in the treatment of brain tumors. Many machine-learning techniques have evolved over the years to ease the process of detection and achieve better accuracy than traditional methods. The most popular type of diagnosis is the one using MR (Magnetic Resonance) images. A hybrid model

Corresponding author: [1]mdishabls@gmail.com, [2]abhishekpatrobam@gmail.com, and [3]kaberidas@soa.ac.in

DOI: 10.1201/9781003489443-1

is proposed in [1], that uses the regularized extreme learning machine (RELM) technique. It pre-processes the images using the min-max normalization, Extracts the features using a hybrid approach, and finally classifies them by the use of RELM. Another model uses a method called Generative Adversarial Networks (GANs) [2]. At first, it generates high-resolution MR images using multi-stage noise-to-image GAN, and then Multimodal Unsupervised Image-to-Image Translation (MUNIT), improving the image's shape. Another machine learning-based back propagation neural networks (MLBPNN) technique is analyzed using infrared sensor imaging technology. The fractal dimension is used to extract the features. It improves the exactness of the location of a brain tumor [3]. Another popular method involves the use of data augmentation. A model using data augmentation and a principal-based component (PCA) are proposed [4]. In [5], Machine learning is used with transfer learning. A method called VGG-16 is combined with various classifiers, providing a good accuracy rate. Another method [6], which is a hybrid combination of genetic algorithm (GA), and support vector machine (SVM) is proposed, in which features are extracted using the spatial gray level dependence method (SGLDM). The genetic algorithm helps solve the issue of choosing the optimal features for detecting brain tumors. Another proposed model uses template-based K means and improved fuzzy C means (TKFCM), for the detection of a brain tumor [7]. The proposed method is very fast and achieves a good accuracy rate. Another method is proposed that uses a self-adaptive K-means algorithm that has an edge over the traditional K-means algorithm that required user input. The peaks of the histogram are used for the computation of the number of clusters in the case of self-adaptive K-means clustering [8]. Another methodology uses a set of six features that are extracted using a method called the Cumulative Variance Method (CVM). Then the features are classified using the K-Nearest Neighbour (KNN), Neural Network (NN), and multi-class Support Vector Machine (mSVM) [9]. Another process uses Threshold Based Region Optimization (TBRO) for segmentation [10]. It enhances the accuracy and helps in easier recognition of segmented features. A proposed approach uses modified fuzzy c-means [11], that have been reformulated and hence provide a boost in the performance of the model. A recent model is proposed that uses a method called an improved sparrow search algorithm (ISSA) [12]. The procedure also uses four classifying algorithms, which are K-nearest neighbor (KNN), Decision tree, Support Vector Machine, and Random Forest. The method proves to be effective in the classification process.

2. Related Work

Numerous machine-learning methods are proposed to detect brain tumors at an early stage. Different approaches have acquired different accuracies, precisions, sensitivities, specificities, etc. In [1], a hybrid approach is proposed that uses the min-max normalization rule for preprocessing, and RELM (Regularised Extreme Learning Machine) for classification. It achieved an accuracy of 94.233%. Another MLBPNN technique [2] is proposed, which achieves a sensitivity of 95.103%, and a specificity of 99.8%. Another model [5], that applies VGG-16 on its dataset, achieves an accuracy of 98.7%, whereas SVM achieved 82.5%, random forest achieved 88.7%, decision tree achieved 68.6%, Naïve Bayes achieved 59.5%, and K-Nearest Neighbour achieved 55.6%. It focuses on improving the performance of Generative Adversarial Networks (GANs) when combined with DA. Another model [6], proposed a hybrid model based on Support Vector Machine (SVM), and Genetic Algorithm (GA) achieved accuracy in the varying range of 94.44% to 98.14%. It extracts a feature set based on the wavelet feature and provides it to the SVM classifier as an input. It focuses on classifying the type of brain tissue. The brain tissue might be normal or might contain benign or malignant tumors. In [7], a template-based K means algorithm along with improved Fuzzy C means is used, which provides an accuracy of 97.5%. It also uses the FCM clustering algorithm to detect the position of the brain tumor. It also detects the tumor in very less time. Another multiclass classification approach is proposed [9], which provides an accuracy of 95.86% using the KNN classifier. It focuses on differentiating the types of malignant brain tumors. It has extracted multiple features in order to dig into the very small details of the image. In [10], an accuracy of 96.57%, specificity of 94.6%, and sensitivity of 97.76%, is achieved by the use of a multi-level threshold-based region optimization technique. A method known as an improved sparrow search algorithm [12], is proposed which provides an accuracy rate of more than 85%. It focuses on eliminating unnecessary features and improving accuracy. It proposed three improving mechanisms namely adaptive crossover operation, novel local search strategy, and tent chaotic initialization.

3. Generic Proposed Model

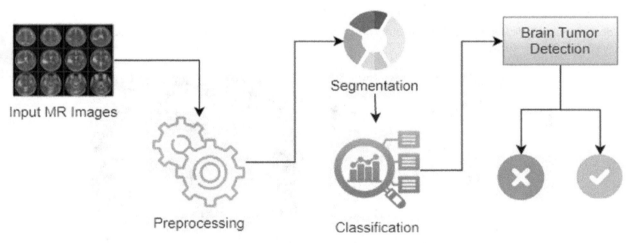

Fig. 1.1 Generic proposed model

3.1 Input MR Images

The initial step of the model begins with the input of MR Images into the model. The input images consist of two sets: the training set and the testing set. 80% of the images are used for training while the rest 20% are used for testing.

3.2 Pre-processing and Image Enhancement

The next step plays a crucial role in detecting brain tumors. The process in which unnecessary details, noise, and insignificant information are removed to extract a clear view of the Region of Interest (ROI) is known as pre-processing. The pre-processed images are further enhanced using the median filter for a better view of the images, which would ensure easy and more

4 Prospects of Science, Technology and Applications

accurate results. Image Enhancement involves multiple techniques such as cropping, using filters, using contrast, greyscale, and many more.

3.3 Segmentation

In this procedure, the pre-processed and enhanced image is now divided into several fragments, which helps in understanding the image at a pixel level. It helps in easier recognition of the region of the brain having a brain tumor. We have used the watershed segmentation technique as well.

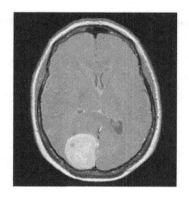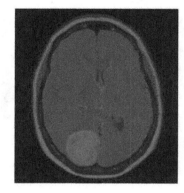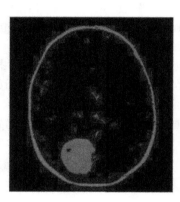

Fig. 1.2   Segmented MR image

Source: Msoud Nickparvar. (2021). Brain Tumor MRI Dataset [Data set]. Kaggle. https://doi.org/10.34740/KAGGLE/DSV/2645886

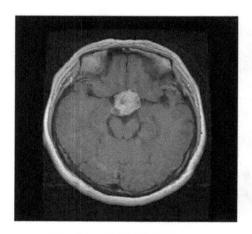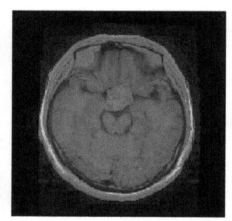

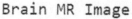

Fig. 1.3   Watershed segmented MR image image

Source: Msoud Nickparvar. (2021). Brain Tumor MRI Dataset [Data set]. Kaggle. https://doi.org/10.34740/KAGGLE/DSV/2645886

3.4 Feature Extraction

Feature Extraction is a process of extraction of important features which is required for the analysis and processing of images. Some important features include Contrast, Covariance, Correlation, Smoothness, Skewness, Homogeneity, Energy, Entropy.

Contrast is the difference between the luminance of an object that makes it distinguishable.

$$\text{con} = (1/N) \sum (P_i - \mu)^2 \qquad (1)$$

where N is the total number of pixels, P_i is the intensity of pixel at i^{th} position, and μ is the mean intensity.

Covariance is used to measure the variance between two variables. Mathematically, it can be represented as:

$$\text{cov}(M, N) = (1/n) \sum (Mi - mean(M))(Ni - mean(N)) \qquad (2)$$

where, n is the number of samples, Mi and Ni are the feature value at ith position.

Correlation is used to measure the direction and strength of the relationship between two variables. Mathematically, it can be represented as:

$$\text{cor}(M, N) = cov(M, N)/(sd(M) X sd(N)) \qquad (3)$$

where, cov(M, N) is the covariance and sd(M), and sd(N) are the standard deviations of M and N respectively.

Smoothness is the degree of change of the model according to the input values. Mathematically, it can be represented as:

$$\text{tv}(m) = \iint \sqrt{((\delta m/\delta x)2 + (\delta m/\delta y)2) dx dy} \qquad (4)$$

where, tv is the total variation and $\delta m/\delta x$, and $\delta m/\delta y$ are the partial derivatives of m with respect to x and y, respectively.

Skewness is used to measure the asymmetric distribution of an object. Mathematically, it can be represented as:

$$\text{sk} = (1/N) \sum [(P_i - \mu)^3/\sigma^3] \qquad (5)$$

where N is the total number of pixels, P_i is the intensity of pixel at i^{th} position, μ is the mean intensity, and σ is the standard deviation.

Homogeneity is the measure of the degree of similarity between samples in a cluster. Mathematically, it can be represented as:

$$\text{hom} = \Sigma m \Sigma n \left[\frac{1}{(1 + (m - n))^2} \right] p(m, n) \qquad (6)$$

where m and n are the pixel intensities, and $p(m, n)$ is the probability of intensity pair occurrence.

Energy is the measure of the intensity of an image or object. Mathematically, it can be represented as:

$$e = \Sigma m \Sigma n p(m, n)^2 \qquad (7)$$

where m and n are the pixel intensities, and $p(m, n)$ is the probability of intensity pair occurrence.

Entropy is the measure of a model's uncertainty or disorder. Mathematically, it can be represented as:

$$\text{ent} = -\Sigma m \Sigma n p(m, n) log^2 p(m, n) \qquad (8)$$

where m and n are the pixel intensities, and p(m,n) is the probability of intensity pair occurrence.

3.5 Classification

And the final step is known as classification. It is a supervised machine learning technique, that helps the model to predict a result based on the given data. Here, after the segmentation of the image is done, it is finally classified using different classification techniques. Different classification techniques used are:

K-Nearest Neighbour (KNN) KNN is a supervised machine learning technique used for regression and classification. It does not make any assumptions about the underlying data and hence it is non-parametric. During the training phase, it stores the dataset and uses it to classify the data based on the similarity of the given data. It uses distance metrics such as Manhattan distance or Euclidean distance, to find the similarity between the training and testing data.

Support Vector Machine (SVM) SVM is a supervised machine learning technique, used for both regression and classification. It is used to create a decision boundary known as the hyperplane which is used to segregate the n-dimensional plane into different classes, which can further make the task of segregating the data into different classes easier. The points or the vectors at the extreme distance help in creating the hyperplane. These points or vectors are called support vectors.

Decision Tree A decision tree is a supervised machine-learning technique that is used for both classification and regression. As the name suggests, it's a tree-structured algorithm with internal nodes, leaf nodes, and branches. Features of a dataset are represented using the internal nodes, while the decision rules are represented using the branches, and the outcome is represented using the leaf nodes.

Random Forest Random Forest is a supervised machine learning algorithm used for classification and regression. It uses the concept of ensemble learning. Ensemble learning is a technique of combining multiple classifiers which are used to solve a complex problem and also improve a model's performance. It consists of several decision trees. The greater the number of trees, the higher the accuracy of the model.

4. Experiments and Result Analysis

4.1 Dataset

This paper uses a Kaggle dataset having two classes that are no tumor and pituitary tumor. A total of 300 images having pituitary tumors are used out of which 240 are used for training and 60 for testing. And a total of 405 images having no tumor are used out of which 324 are used for training and 60 for testing. 80% of the images are used for training and 20% for testing.

Table 1.1 Dataset description

Total MR Images: 705			
Class	Total MR Images	Training Set	Testing Set
No Tumor	405	324	81
Pituitary Tumor	300	240	60

Brain Tumor Segmentation and Classification from MR Images with Feature Extraction 7

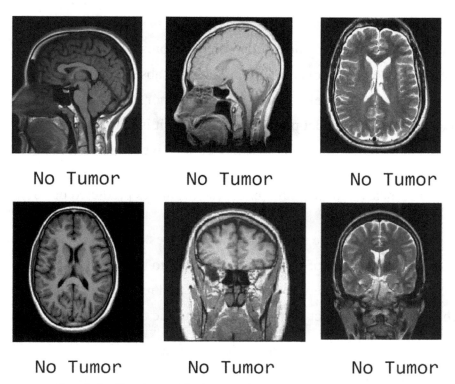

Fig. 1.4 Testing with dataset having no tumor image

Source: Msoud Nickparvar. (2021). Brain Tumor MRI Dataset [Data set]. Kaggle. https://doi.org/10.34740/KAGGLE/DSV/2645886

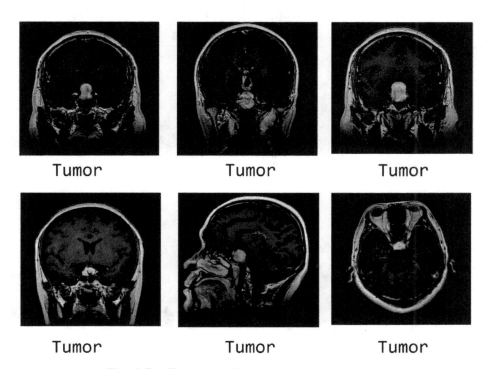

Fig. 1.5 Testing with dataset having tumor

Source: Msoud Nickparvar. (2021). Brain Tumor MRI Dataset [Data set]. Kaggle. https://doi.org/10.34740/KAGGLE/DSV/2645886

4.2 Evaluation Matrices

True Positive(T_{pt}): When a model correctly predicts the presence of a tumor.

True Negative(T_{nt}): When a model correctly predicts the absence of a tumor.

False Positive(F_{pt}): When a model does not predict a tumor but it is present.

False Negative(F_{nt}): When a model predicts a tumor but it is not present.

Sensitivity: It is the measure of the number of true positives predicted out of all the positives present in the dataset.

Specificity: It is the measure of the number of true negatives predicted out of all the negatives present in the dataset.

Precision: It is the measure of the number of true positives out of predicted positives.

Recall: The ratio of the number of positives in a dataset to the number of predicted positives.

F1 Score: The harmonic mean between the precision and recall.

Accuracy: The ratio of the correct predictions made to the total number of predictions made.

Table 1.2 Comparison between different classification algorithms used in the paper

Model	Sensitivity	Specificity	Precision	F1-score	Accuracy
DT	0.9500	0.9512	0.9344	0.9421	0.9507
SVM	0.9667	0.9753	0.9667	0.9667	0.9716
KNN	0.9831	0.9756	0.9667	0.9748	0.9787
RF	0.9828	0.9877	0.9828	0.9828	0.9856

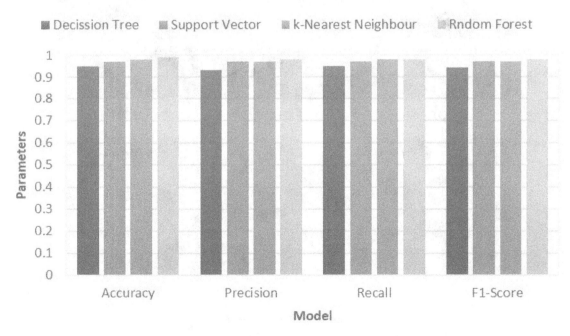

Fig. 1.6 Comparision between different classification algorithms

Table 1.3 Comparison with other works

Reference	Model Used	Accuracy
[1]	SVM-RBF	91.51%
	DT	84.33%
[5]	SVM	82.5%
	RF	88.7%
	DT	68.6%
	KNN	79.6%
[9]	mSVM	92.5%
	KNN	88.43%
Generic Proposed Model	DT	95.07%
	SVM	97.16%
	KNN	97.87%
	RF	98.56%

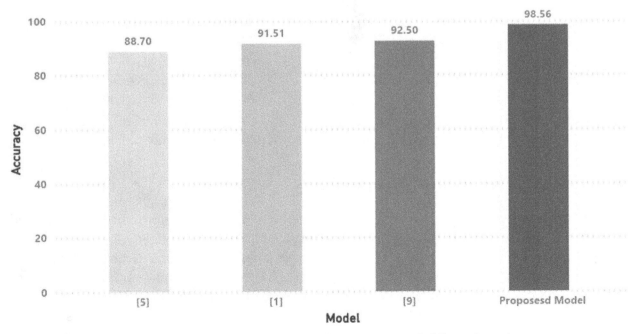

Fig. 1.7 Comparision between the accuracy of different works

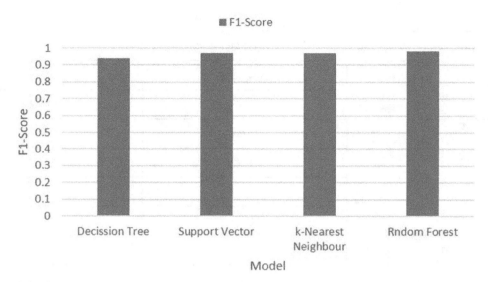

Fig. 1.8 Comparision between the F1 Score of different classification algorithms

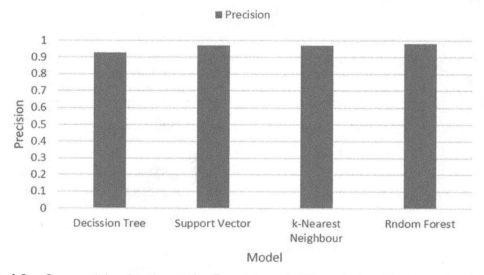

Fig. 1.9 Comparision between the Precision of different classification algorithms

5. Conclusion and Future Scope

In the paper, we have used two classes of images, i.e. images having pituitary tumors and images with no tumors. We have filtered and enhanced the images, followed by segmentation of the image and extraction of features. Then we applied four classification algorithms: Decision Tree, Support Vector Machine, K-Nearest Neighbour, and Random Forest. Different algorithms provided different accuracies with Random Forest providing the highest accuracy of 98.56%. The accuracy achieved in the paper surpasses the accuracy of other works referred to.

In the future, we would use a larger dataset as well as multiclass algorithms for the detection of brain tumors. We would also apply optimization techniques as well as use deep learning methods for detection.

REFERENCES

1. Gumaei, Abdu, et al. "A hybrid feature extraction method with regularized extreme learning machine for brain tumor classification." IEEE Access 7 (2019): 36266–36273.
2. Han, Changhee, et al. "Combining noise-to-image and image-to-image GANs: Brain MR image augmentation for tumor detection." Ieee Access 7 (2019): 156966–156977.
3. Shakeel, P. Mohamed, et al. "Neural network based brain tumor detection using wireless infrared imaging sensor." IEEE Access 7 (2019): 5577–5588.
4. Anaya-Isaza, Andrés, and Leonel Mera-Jiménez. "Data augmentation and transfer learning for brain tumor detection in magnetic resonance imaging." IEEE Access 10 (2022): 23217–23233.
5. Iliass Zine-dine, Jamal Riffi, Khalid El Fazazy, Mohamed Adnane Mahraz, Hamid Tairi."Brain Tumor Classification using Machine and Transfer Learning " Science and Technology Publications.
6. Kharrat, Ahmed, et al. "A hybrid approach for automatic classification of brain MRI using genetic algorithm and support vector machine." Leonardo journal of sciences 17.1 (2010): 71–82.
7. Alam, Md Shahariar, et al. "Automatic human brain tumor detection in MRI image using template-based K means and improved fuzzy C means clustering algorithm." Big Data and Cognitive Computing 3.2 (2019): 27.
8. Kaur, Navpreet, and Manvinder Sharma. "Brain tumor detection using self-adaptive K-means clustering." 2017 International Conference on Energy, Communication, Data Analytics and Soft Computing (ICECDS). IEEE, 2017.
9. Vidyarthi, Ankit, et al. "Machine Learning Assisted Methodology for Multiclass Classification of Malignant Brain Tumors." IEEE Access 10 (2022): 50624–50640. 10.Kanmani, P., and P. Marikkannu. "MRI brain images classification: a multilevel threshold based region optimization technique." Journal of medical systems 42 (2018): 1–12.
10. El-Melegy, Moumen T., and Hashim M. Mokhtar. "Tumor segmentation in brain MRI using a fuzzy approach with class center priors." EURASIP Journal on Image and Video Processing 2014.1 (2014): 1–14.
11. Yu, Wenyu, et al. "Bio-Inspired Feature Selection in Brain Disease Detection via an Improved Sparrow Search Algorithm." IEEE Transactions on Instrumentation and Measurement 72 (2022): 1–15.
12. Sharma, Komal, Akwinder Kaur, and Shruti Gujral. "Brain tumor detection based on machine learning algorithms." International Journal of Computer Applications 103.1 (2014).
13. Siar, Masoumeh, and Mohammad Teshnehlab. "Brain tumor detection using deep neural network and machine learning algorithm." 2019 9th international conference on computer and knowledge engineering (ICCKE). IEEE, 2019.
14. George, Dena Nadir, Hashem B. Jehlol, and Anwer Subhi Abdulhussein Oleiwi. "Brain tumor detection using shape features and machine learning algorithms." International Journal of Advanced Research in Computer Science and Software Engineering 5.10 (2015): 454–459.

Adsorption of Co (II) by Banana (*Musa paradisiaca*) Peel

**Lipsa Mayee Mishra, Biswaprakash Sarangi,
Lopita Priyadarshini Pradhan, Sneha Prabha Mishra**[*]

Department of Chemistry, Institute of Technical Education and Research,
Siksha 'O' Anusandhan (Deemed to be University), Bhubaneswar-751030, Odisha, India

Abstract Biosorption is a newly developed low-cost technology used for removing heavy metals from waste water. In biosorption technology different natural, low cost, eco-friendly materials such as bacteria, fungus, algae, different waste materials, fruits and vegetable wastes etc. can be used as biosorbents. Co(II) is a toxic pollutant present in aqueous system and it causes various diseases. That's why in the present investigation a low-cost material i.e., banana (*Musa paradisiaca*) peel has been utilized for the removal of Co(II) from the aqueous solutions. In order to know the surface properties of the adsorbent, characterization of banana peel has been carried out in terms of pH, point of zero charge (pzc), mass loss on ignition, moisture content, density and surface area. Its pH, density, moisture content, loss of mass on ignition, pzc and surface area were found to be 6.26, 1.68 g/mL, 66%, 82.03%, 5.1, 99.8m^2/g respectively. Adsorption of Co(II) was carried out on banana peel varying different experimental variables such as contact time, pH, temperature, initial Co(II) concentration and adsorbent concentration. The most favourable contact time, pH, temperatures were found to be 30 min, 6 and 323K respectively. The Co(II) uptake increased with increase of Co(II) concentration. From these experiments we found that banana peels, which is a trash product, have good capacity as an adsorbent to extract toxic metals such as Co(II) from aqueous system.

Keywords Banana peel, Adsorption, Adsorbent, Heavy metal, Cobalt

1. Introduction

Water plays a significant role in our life. It is also beneficial to all living plants and animals. But due to industry and many chemical resources it is getting contaminated day

---
[*]Corresponding author: snehamishra@soa.ac.in

DOI: 10.1201/9781003489443-2

by day. Industrialization such as plating, fertilizers, pulp and paper mills, steel works, organic chemicals release heavy metals into the water ways. Heavy metals are gradually accumulated in human tissues and are non-biodegradable [1]. These heavy metals exist in cationic form (in most of the cases) in aqueous solutions. Heavy metal ions are very much toxic to the existing flora and fauna of the earth. In case of human being, these may cause severe health issues like cancer, nervous system failure, organ damage etc. For this reason, the industry waste water must be treated suitably before being cleared to the environment [2]. A number of selections have been evolved in order to treat this waste water, which is coagulation, ion-exchange, complexation, chemical precipitation, solvent extraction, electrochemical deposition [3]. The use of adsorption technology can be a good alternative to all these methods because of its cost effectiveness, high selectivity and high efficiency [4]. Biosorption is a newly developed technology for removing heavy metal ions from waste water in which various low cost, eco-friendly bio and waste materials are used as the biosorbents [5].

Cobalt is present in the waste water of nuclear plant, metallurgical, electroplating, paint, pigment and electronic industries [6]. As it is bio-nondegradable it causes severe diseases to human beings like diarrhoea, paralysis, high blood pressure, lung irritation etc. [1] So, in this study we choose banana (*Musa paradisiaca*) peel, a cheap biomaterial which can be used to remove the cobalt ion from water. Banana peel waste has also been utilized by several researchers for the removal of different heavy metal ions like Cu(II), Zn(II), Ni(II), Cd(II), Pb(II) etc. [7]. Characterization of the peel was conducted to study the surface properties and effect of different experimental parameters on Co (II) adsorption was also evaluated to know the optimum condition of Co (II) adsorption on banana peel.

2. Experimental

2.1 Adsorbent/adsorbate Preparation

Banana peel was collected from the local market. It was cleaned properly with running water to eliminate any dirt or foreign matters. Again, it was kept under sun for 5-6 days to remove the moisture. Thereafter it was dried at 70°C for 2 h in an oven and the dried banana peel was blended properly, ground to powder with a kitchen grinder. The banana peel powder was kept in air tight bottles. Standard solution of cobalt was made by dissolving necessary amount of sulphate salt of cobalt ($CoSO_4.5H_2O$) in deionized water. This stock solution was diluted successively to prepare the required concentration of the experimental solutions.

2.2 Adsorption Experiments

The adsorption experiments were carried out in batch equilibrium method on temperature controlled magnetic stirrer with a purpose to study the effect of various experimental variables like contact time, pH, adsorbent/adsorbate doses and temperature. The experiments were conducted in 250 mL conical flasks taking a pre-weighed amount of adsorbent with 50 mL of Co(II) solution. The pH of the solutions was adjusted by 0.1 M HNO_3 or 0.1 M NaOH. Co(II) ion concentrations of the solutions before and after adsorptions were analyzed by Atomic

Absorption Spectrophotometer. The % elimination of Co(II) on banana peel was determined using equation (1).

$$\text{Removal (\%)} = 100\,(C_o - C_e)/C_o \tag{1}$$

Where, C_o and C_e are the initial and final concentrations of the Co (II) ion in the solution. All the experiments were repeated thrice and the mean value was used in the analysis of data.

3. Results and Discussion

3.1 Characterization of Banana Peel

Different physico-chemical parameters of the banana peel were determined and are represented in Table 2.1. The point of zero charge was determined from Fig. 2.1.

Table 2.1 Physico-chemical parameters of banana peel

Sl. no.	No of parameters	Observed values
1	pH	6.26
2	Density	1.68 g/mL
3	Moisture content	66 %
4	Loss of mass on ignition	82.03 %
5	Point of zero charge	5.1
6	Surface area	99.8 m²/g

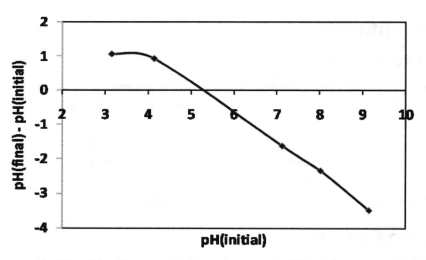

Fig. 2.1 Determination of point of zero charge

3.2. Determination of optimum contact time

Impact of contact time on adsorption of Co(II) was explored. The time was varied from 0–70 min at the Co(II) concentration of 10 mg/L. Most of the adsorption happened within first 30 min of contact time and after that no alteration was observed for Co(II) adsorption, It concludes that adsorption equilibrium is achieved within 30 min and 30 min is the optimum contact time for Co(II) adsorption on banana peel. The result is shown in Fig. 2.2.

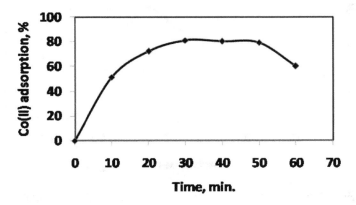

Fig. 2.2 Effect of contact time on Co(II) adsorption

3.3. Determination of optimum pH

Figure 2.3 shows the effect of pH on percentage of Co(II) removal on banana peel. It was found that the percentage of cobalt adsorption increased with increase of pH and reached a maximum at pH 5, which is the optimum pH for cobalt ion adsorption. After pH 6, adsorption decreases due to precipitation and aqueous metal hydroxide formation. Also, in alkaline conditions, the binding site may not be as active as in acidic condition.

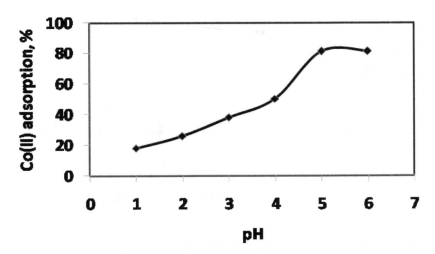

Fig. 2.3 Effect of pH on Co(II) adsorption

3.4. Effect of adsorbent dose

Co(II) adsorption % on banana peel was increased from 55 to 81 % as the adsorbent amount increased from 1 g/L to 3 g/L as shown in Fig. 2.4, This is due to increased surface area at higher adsorbent dose which leads to increase in adsorption site. Once the maximum limit attended, there was no further increase in Co(II) adsorption with increase in adsorbent dose. Because, at high adsorbent doses, partial aggregation of active sites occur, while at low doses active binding sites are less.

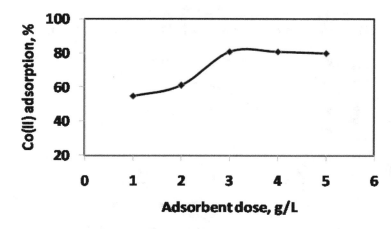

Fig. 2.4   Effect of adsorbent dose on Co(II) adsorption

3.5. Effect of temperature

Figure 2.5 displays the effect of temperature on the adsorption of Co(II) on banana peels. Adsorption % increased from 81 to 86% when the temperature increased from 303 to 323 K. So, the present adsorption study is found to be endothermic in nature.

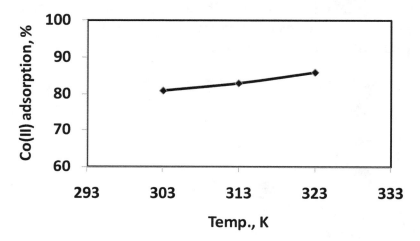

Fig. 2.5   Effect of temperature on Co(II) adsorption

4. Conclusions and future prospective

In this project, the aims are accomplished successfully. From the characterization study it has been observed that banana peel can be effectively used as a bio adsorbent to remove Co(II) ion in waste water and other aqueous solutions. Its pH, density, moisture content, loss of mass on ignition, point zero charge, surface area were found to be 6.26, 1.68g/mL, 66%, 82.03%, 5.1, 99.8m^2/g respectively. It was concluded that the ideal pH, dose of adsorbent, temperature for cobalt adsorption were 5, 3 g/L and 323 K respectively. Based upon the world-wide high consumption of banana, banana peel is a remarkable agricultural waste material and its alternative uses should be explored. As the banana peel has admirable capability of Co(II) removal, it can be utilized successfully for the decontamination of aqueous systems.

REFERENCES

1. J. Briffa, E. Sinagra, R. Blundell, Heavy metal pollution in the environment and their toxicological effects on humans, *Heliyon*, 6 (9) (2020) e04691, https://doi.org/10.1016/j.heliyon.2020.e04691
2. L. Lin, H. Yang, X. Xu, Effects of water pollution on human health and disease heterogeneity: A review, *Front. Environ. Sci.*, 10 (2022) 880246, https://doi.org/10.3389/fenvs.2022.880246
3. N.A.A. Qasem, R.H. Mohammed, D.U. Lawal, Removal of heavy metal ions from wastewater: a comprehensive and critical review, *npj Clean Water*, 4 (2021) 36. https://doi.org/10.1038/s41545-021-00127-0
4. B.S. Rathi, P.S. Kumar, Application of adsorption process for effective removal of emerging contaminants from water and wastewater, *Environmental Pollution*, 280 (2021) 116995, https://doi.org/10.1016/j.envpol.2021.116995.
5. V. Krishnamurthy, R.V. Kavitha, L. Shantha, N. Gopika, G. Megha, C. Niranjana, S. Sasmita, Biosorption: An eco-friendly technique for the removal of heavy metals, *Indian Journal of Applied Research*, 2 (2012) 1-8, https://doi.org/10.15373/2249555X/DEC2012/1.
6. S.S. Al-Shahrani, Treatment of wastewater contaminated with cobalt using Saudi activated bentonite, *Alexandria Engineering Journal*, 53(1) (2014) 205–211, https://doi.org/10.1016/j.aej.2013.10.006.
7. K.G. Akpomie, J. Conradie, Banana peel as a biosorbent for the decontamination of water pollutants. A review. *Environ Chem Lett* 18 (2020) 1085–1112, https://doi.org/10.1007/s10311-020-00995-x

Recovery of Fe, Sc, and V from Tailing Sample Using Acidic Lixiviants

Milli Agarawala[a], Smruti Rekha Patra[a]
Department of Chemistry, Institute of Technical Education and Research,
Siksha 'O' Anusandhan Deemed to be University, Bhubaneswar, Odisha, India

Sibananda Sahu[b]
Biofuels and Bioprocessing Research Center, Institute of Technical Education and Research,
Siksha 'O' Anusandhan Deemed to be University, Bhubaneswar, Odisha, India

Niharbala Devi*
Department of Chemistry, Institute of Technical Education and Research,
Siksha 'O' Anusandhan Deemed to be University, Bhubaneswar, Odisha, India
Biofuels and Bioprocessing Research Center, Institute of Technical Education and Research,
Siksha 'O' Anusandhan Deemed to be University, Bhubaneswar, Odisha, India

Abstract Scandium and Vanadium are rarely distributed across the crust of the planet and are frequently discovered in conjunction with minerals. Due to their distinctive characteristics, these metals have a variety of physical and chemical properties, which leads to their numerous usage. The need for the development of extraction techniques in both the industrial and laboratory sectors is driven by the limited resources in the earth's crust, the complexity of the extraction process, and the high demand for these metals. Leaching was suggested as a unique method for recovering iron, vanadium, and scandium from the tailing sample. Different leaching parameters were investigated such as acid concentration, the function of a reductant, solid-to-liquid ratio, and temperature. Leaching tests using 1 M HCl as lixiviant produced positive results for 10 g/L of the material at 90°C for 60 minutes, with recoveries of roughly 90% for scandium, 55% for vanadium, and 63% for iron. However, total Scandium leaching with a leaching efficiency of over 91% of Vanadium and minimal iron leaching were observed when using the 3 M H_2SO_4 leaching agent at a S/L ratio of 10 g/L, 30°C, and for 60 minutes. It is observed that the difficulty of iron to leach with sulphuric acid, which permits the apparent selective leaching of both Scandium and Vanadium, thus sulphuric acid leaching was used as an efficient selective leaching process.

Keywords Scandium, Vanadium, Leaching, Hydrometallurgy

*Corresponding author: niharbaladevi@soa.ac.in

DOI: 10.1201/9781003489443-3

1. Introduction

The chemical element scandium, with the atomic number 21 and the symbol Sc is a soft, silvery-white metal that belongs in the d-block of the Periodic Table. Scandium is categorized as a critical and rare earth element by the International Union of Pure and Applied Chemists. Vanadium, a soft silvery-grey metal with the chemical symbol V and atomic number 23, is also listed in the d-block of the Periodic Table. With regard to the presence of Scandium in the +3 oxidation state, octahedral coordination (ionic radius of 75 pm) in minerals, and high charge-to-radius ratio (Z/r), it shares traits with rare earth elements (Ln, Y, etc.), transitional elements bearing diagonal relationships (Al). In nature, Vanadium often exists in trivalent and pentavalent forms. V^{3+}, via isomorphism, arises in iron and aluminum minerals because of its similar ionic radius to that of Fe^{3+}, much like Scandium does [1,2]. In lithosphere, they form solid solutions with low concentrations in more than 100 minerals since it has no affinity for the typical anions that form ores [3]. The scarce distribution of these metal classifies them as a critical metal with low individual deposits result in difficulty in extraction process. Scandium oxide (Sc_2O_3) was estimated to be priced at US$ 900/kg and US$ 1400/kg for 99.0% and 99.9% purity, respectively, in 2009. Around 25 to 25 tonnes of scandium are expected to be utilised and consumed yearly on a worldwide basis in 2022, according to estimations. The average Vanadium pentoxide (V_2O_5) prices was reported to be US$ 260/pound for the year 2021. The annual supply and demand for scandium was calculated to be between 15 and 20 tonnes worldwide and that of vanadium was about 120,067 mt [4]. Both Scandium and Vanadium have special qualities with numerous industrial applications. Vanadium is renowned for its strength, hardness, and use in the creation of high-performance steel and energy storage devices, whereas scandium is known for its lightweight and corrosion-resistant alloys. The growing demand and limited resources lead to the extraction and recovery of these elements as difficult and a topic of discussion [5]. The detailed literature survey on the use of leaching agents and extractants for the recovery of Sc, V and Fe from different ores is presented in Table 3.1.

Table 3.1 Literature survey on the use of leaching agents and extractants for the recovery of Sc, V and Fe from different ores.

Sl. No.	Metals	Leaching agent + Extractant	Observation	Reference
1	Sc, V	Conc. HCl + P507(15% P507, 5% TBP, 80% sulfonated kerosene)	99% of vanadium and 96% of scandium were leached by stirring 7 ml of sample per gram with concentrated HCl (30% volume concentration) at 60°C for 90 minutes. Solvent extraction stirred at 300 rpm for 6 minutes was then performed using P507 at pH 1.9. Subsequent stripping of the metals was achieved using H_2SO_4 and NaOH, resulting in a 97% recovery of both metals.	[6]

Sl. No.	Metals	Leaching agent + Extractant	Observation	Reference
2	V, Fe	Oxalic acid/Sodium sulphite	A mixture of 25% oxalic acid and 5% sodium sulfite was utilized for leaching 90.4% and 9.6% of Vanadium and Scandium, respectively at the liquid-to-solid ratio of 5. The leaching process was conducted at 75°C and 300 rpm for 60 minutes. The leaching of vanadium was primarily governed by diffusion at the boundary layer, while the leaching of iron was predominantly influenced by the chemical reactions occurring at different concentrations of oxalic acid.	[7]
3	Sc	Sulphuric acid(H_2SO_4)/ CaF_2 + P507	With the use of 5% CaF_2 in H_2SO_4 (6 M) with an L/S ratio of 5 mL/g at 90°C for 60 min, the leaching efficiency of Sc increased from 74% to 92%, and 98% of Sc was extracted with 10% P507 at pH 0.1 for 4 min.	[8]
4	V	Sulphuric acid(H_2SO_4)/ $FeSO_4$ + P204	Using 200 g/L of H_2SO_4 at an L/S ratio of 1.2 mL/g for 3 hours at 180°C with 6.6 g/L of $FeSO_4$ as an addition, pressure acid leaching of black shale produced a 76% leaching efficiency.	[9]
5	Sc	Conc. HCl	The alkaline KOH sub-molten salt of fergusonite with 0.94% Scandium content was leached with conc. HCl at an L/S of 2.2:1 for 2hr. Scandium leaching increases from 45.2% to 97% from 60°C to 80°C.	[10]
6	V	Sulphuric acid(H_2SO_4)	At a mass ratio of 20% H_2SO_4 to stone coal, a leaching temperature of 95°C, a leaching period of 30 hours, a liquid-to-solid ratio of 1.1 mL/g, and a particle size of 0.1 mm, the two-step vanadium counter current leaching ratio of roasted stone coal may reach 65.1%.	[11]

2. Experimental

2.1 Reagents and Materials

To investigate the leaching behaviour of Sc and V in a different chemical environment, experiments were conducted using tailing samples after processing of vanadium ore. Analytical grade leaching agents such as HCl (≥37%), H_2SO_4 (>98%), HNO_3 (>70%), and H_2O_2 (>30%) were used without any additional purification to study the impact of various parameters on the leaching of these rare metals.

2.2 Analytical method

The initial concentration of the metals in the analyte (sample) was determined by performing an acid digestion process using aqua regia. Complete digestion of the sample involved heating

1g of the sample with 20mL aqua regia (a mixture of HCl and HNO$_3$ in 3:1 ratio) in a beaker at 90°C. The digestion process was marked by the release of dark brown fumes, leaving behind a viscous and concentrated mixture. The sample was then allowed to cool and the concentrated mixture was diluted in a 100 mL volumetric flask to achieve a solid-to-liquid ratio of 10g/L. The diluted mixture was subsequently filtered, and the resulting supernatant containing metal concentrate was analysed to determine the chemical composition of the metals. The sample's chemical composition was analysed using the iCAP 7000 SERIES equipment, which employs Inductively Coupled Plasma – Optical Emission Spectroscopy (ICP-OES). Additionally, Atomic Absorption Spectroscopy (AAS) with the AA S-816 instrument was also utilized to analyze Fe. The composition of Fe, Sc, and V in mg/L in the sample is presented in Table 3.2.

Table 3.2 Chemical Composition of the tailing sample

Metals	Fe	Sc	V
mg/L	319	0.021	1.19

2.3 Procedure

The leaching of rare metals (Scandium and Vanadium) was studied by varying various leaching agents and various leaching parameters involving variation in acid concentration, temperature, solid-to-liquid ratio, etc., to maximize the optimization of the leaching efficiency. The leaching procedure first involved the dissolution and dispersion of the solute particles into the liquid phase using various inorganic lixiviants like hydrochloric acid, sulphuric acid, and nitric acid. The authors examined multiple parameters, including acid concentration, solid-liquid ratio, and temperature, to determine the optimal condition required for the complete recovery of both Sc and V from the sample. All the leaching experiments were conducted by agitating the sample in beakers at 500rpm using a magnetic stirrer, with a constant leaching time of 60 minutes, to facilitate maximum dissolution. The initial step involved selecting an appropriate leaching agent that would provide both metals' highest yield. The experiment was performed by agitating 1g of the sample separately in three beakers containing 1 M solutions of HCl, H$_2$SO$_4$, and HNO$_3$, respectively. For each sampling, the solid-liquid ratio was maintained at 10 g/L, and the temperature of the experiment was fixed at 30°C. Leaching experiments involving H$_2$O$_2$-assisted lixiviant, which involved the simultaneous effect of reductant on leaching using an acid solution, were also conducted. For this, 1g of the tailing sample was added in three beakers, followed by the addition of the three previously chosen leaching agents, then the addition of 5% H$_2$O$_2$ while retaining the molarity of the lixiviant at 1 M.

3. Results and Discussions

3.1 Effect of Different Acids

The effect of different acids such as hydrochloric acid (HCl), sulphuric acid (H$_2$SO$_4$), and nitric acid (HNO$_3$) on the leaching of the ma sample after processing of vanadium ore was well studied, maintaining 1 M acid concentration at an S/L of 10 g/L for 60 minutes at 30°C. The experimental data in Fig. 3.1 depicts iron, scandium, and vanadium recovery using H$_2$SO$_4$

lixiviant to be around 20.2%, 95.2%, and 44.5%, respectively. HCl leaching efficiency of scandium was about 76%, and that of vanadium was about 27%. Thus HCl could also prove beneficiary if the leaching conditions were varied. The nitric acid leaching result did not favour the leaching efficiency.

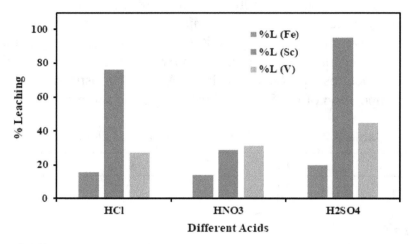

Fig. 3.1 Impact of different acids on the leaching rates. Leaching Conditions. (Acid concentration = 1 M, temperature = 30°C, S/L = 10 g/L, and leaching time = 60 min.)

3.2 H_2O_2-assisted Leaching

The authors studied the H_2O_2-assisted leaching using 5% H_2O_2 in 1 M HCl, H_2SO_4, and HNO_3 at leaching conditions 10 g/L as S/L, temperature of 30°C, and contact time of 60 minutes to understand the effect of reductant on the leaching rate. The effect of reductant (H_2O_2) on leaching efficiency, as depicted in Fig. 3.2 had no such significant impact on the leaching efficiency of the metals (Fe, Sc, and V). The leaching efficiency of scandium was maximum at each leaching condition, followed by vanadium and iron, showing no such positive output in leaching when compared to the standard acid-leaching experiment. Thus, the result denoted that normal acid leaching of metals was sufficient to introduce the metals into the aqueous phase effectively.

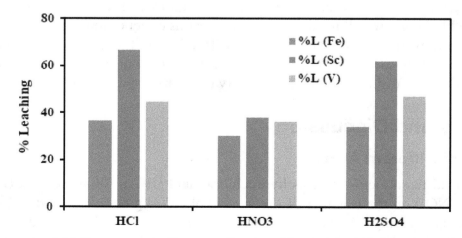

Fig. 3.2 Impact of H_2O_2 concentration on the leaching rates. (Acid concentration = 1 M, temperature = 30 °C, S/L = 10 g/L, and leaching time = 60 min.)

3.3 Impact of H$_2$SO$_4$ Concentration

To investigate the effect of the rate of leaching of iron, scandium, and vanadium on sulphuric acid lixiviant (H$_2$SO$_4$), a series of trials were performed using 1 M, 3 M, and 5 M of H$_2$SO$_4$ with an S/L ratio of 10 g/L using a magnetic stirrer to maintain continuous agitation at 30°C for 60 minutes. The leach recovery of the metals directly depends on acid concentration. The result shows that scandium recovery increased from 95% to 100% with an increase in H$_2$SO$_4$ concentration from 1 M to 3 M, respectively. Vanadium recovery rate increased from 44% to 93% with an increase in concentration from 1 M to 5 M, respectively (Fig. 3.3). Iron recovery was the lowest and had the least increase with acid concentration variation.

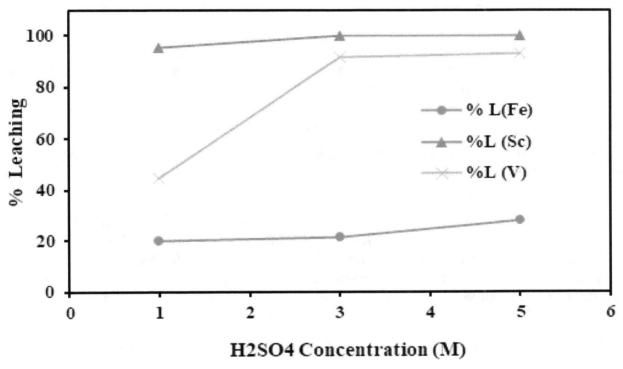

Fig. 3.3 Impact of H$_2$SO$_4$ concentration on the leaching rates. (Temperature = 30 °C, S/L = 10 g/L, and leaching time = 60 min.)

3.4 Solid-to-liquid Ratio Leaching Variation

Leaching trials involving variation in the solid-to-liquid ratio were carried out by maintaining 10 g/L, 50 g/L, 100 g/L, and 300 g/L as S/L ratios and agitating the samples with 3 M H$_2$SO$_4$ at 30°C for 60 minutes. With the increase in the solid-to-liquid ratio, the leaching efficiency decreased for all three metals. The scandium recovery rate dropped from 100% to 53% while the vanadium leaching rate sharply reduced from 91% to 35% with an increase in solid-to-liquid ratio from 10 g/L to 300 g/L, respectively (Fig. 3.4). The recovery and decrease in the leaching rate was not well seen in the case of iron.

24 Prospects of Science, Technology and Applications

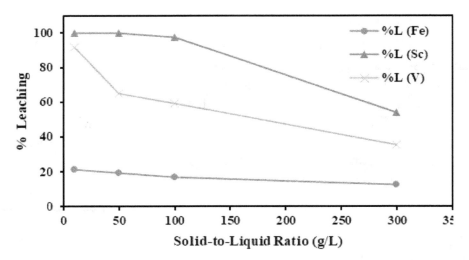

Fig. 3.4 Impact of Solid-to-Liquid on the leaching rates. (H_2SO_4 concentration = 3 M, temperature = 30 °C, S/L = 10 g/L, and leaching time = 60 min.)

3.5 Leaching experiments involving HCl

To investigate the effect of the rate of leaching of Iron, Scandium, and Vanadium on hydrochloric acid lixiviant (HCl), the series of trials performed were:

Temperature variation

A leaching trial for investigating the temperature-variation effect was conducted at 30°C, 60°C, and 90°C using 1 M HCl at S/L ratio 10 g/L and maintaining the agitation time of 60 minutes. With the increase in temperature, the diffusion rate of metals is enhanced, thereby increasing the leaching rate. There was a linear increase in the leaching rate with that of temperature for all the metals. About 90%, 55%, and 63% of scandium, vanadium, and iron were leached at 90°C (Fig. 3.5).

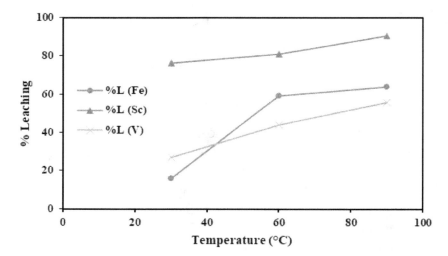

Fig. 3.5 Impact of temperature on the leaching rates. (HCl concentration = 1 M, temperature = 30 °C, S/L = 10 g/L, and leaching time = 60 min.)

Acid Concentration Variation

The leaching efficiency of HCl using 1 M HCl was not sufficient enough for the efficient leaching of metals. The experiments in this series were carried out using 1 M, 3 M, and 5 M of HCl with an S/L ratio of 50 g/L using a magnetic stirrer to maintain continuous agitation at 60°C for 60 minutes. According to the results of the concentration variation for scandium, the leaching efficiency for HCl was notable, increasing first from approximately 33.3% to 59% before declining to 25.7% (Fig. 3.6). This may be explained by the maximum rise in leaching rate with the concentration increase up to the point when the maximum viscosity of the resultant ions in the solution has not been achieved. The abrupt rise significantly influences the pace of ion diffusion in ions that occurred with additional increases in acid content. However, the rate of leaching was linearly increasing for vanadium and iron.

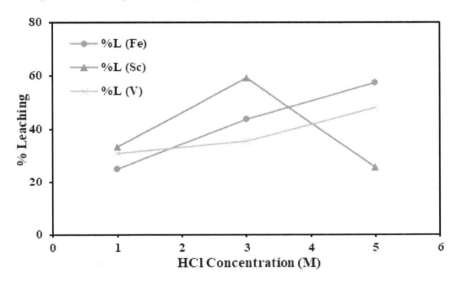

Fig. 3.6 Impact of HCl concentration on the leaching rates. (Temperature = 60 °C, S/L = 50 g/L, and leaching time = 60 min.)

4. Conclusions

Hydrochloric acid and sulphuric acid both produced excellent leaching results under comparable circumstances. Effective leaching rate for all three metals was observed at 90°C by leaching 1 M HCl at an S/L ratio of 10 g/L for 60 minutes. Leaching studies using 50 g/L of the sample at different HCl concentration at 60°C keeping the contact time at 60 minutes shows no significant result in the leaching efficiency of the metals. But complete removal of scandium (100%), more than 91% of vanadium, and over 21% of iron recovered using 3 M H_2SO_4 at an S/L ratio of 10 g/L and temperature of 30°C with a contact time of 60 minutes. The reason for maximum recovery at a 3 M concentration of the lixiviant is attributed to the inability of iron to leach with sulphuric acid medium, thereby allowing apparent selective leaching of both scandium and vanadium. H_2O_2-assisted leaching had no such effect on the leaching rate as compared to that of acid leaching, thereby deliberating no such role of reductant on the leaching experiment for these metals.

Acknowledgments: The authors express their gratitude to S'O'A (Deemed to be University) for providing the research facilities.

Conflicts of Interest: The authors declare that there is no conflicts of interest.

REFERENCES

1. AE, W. J. Vasyukova OV: The economic geology of scandium, the runt of the rare earth element litter. Economic Geology, 113(4), 973-988 (2018). doi:10.5382/econgeo.2018.4579
2. Vanadium mineral resources. In Vanadium (pp. 33–58). Elsevier, (2021). https://doi.org/10.1016/b978-0-12-818898-9.00003-6
3. Wang, W., Pranolo, Y., & Cheng, C. Y.: Metallurgical processes for scandium recovery from various resources: A review. Hydrometallurgy, 108(1-2), 100-108 (2011). https://doi.org/10.1016/j.hydromet.2011.03.001
4. Mineral Commodity Summaries, National Minerals Information Center, (2023). https://www.usgs.gov/centers/national-minerals-information-center/mineral-commodity-summaries
5. Yearbook, Indian Minerals. Part II: Metals & Alloys. 53rd edn. Indian Bureau of Mines (2014).
6. Zhu, X., Niu, Z., Li, W., Zhao, H., & Tang, Q.: A novel process for recovery of aluminum, iron, vanadium, scandium, titanium and silicon from red mud. Journal of Environmental Chemical Engineering, 103528 (2019). doi:10.1016/j.jece.2019.103528
7. Li, W., Yan, X., Niu, Z., & Zhu, X.: Selective recovery of vanadium from red mud by leaching with using oxalic acid and sodium sulfite. Journal of Environmental Chemical Engineering, 9(4), 105669 (2021). doi:10.1016/j.jece.2021.105669
8. Zhu, X., Li, W., Xing, B., & Zhang, Y.: Extraction of scandium from red mud by acid leaching with CaF_2 and solvent extraction with P507. Journal of Rare Earths, 38(9), 1003-1008 (2020). https://doi.org/10.1016/j.jre.2019.12.001
9. Li, M., Wei, C., Fan, G., Li, C., Deng, Z., & Li, X.: Extraction of vanadium from black shale using pressure acid leaching. Hydrometallurgy, 98(3-4), 308–313 (2009). doi:10.1016/j.hydromet.2009.05.005
10. Li, S.-C., Kim, S.-C., & Kang, C.-S.: Recovery of scandium from KOH sub-molten salt leaching cake of fergusonite. Minerals Engineering, 137, 200–206 (2019). doi:10.1016/j.mineng.2018.11.052
11. Wang, M., Xiao, L., Li, Q., Wang, X., & Xiang, X.: Leaching of vanadium from stone coal with sulfuric acid. Rare Metals, 28(1), 1–4 (2009). doi:10.1007/s12598-009-0001-y

Examining Efficiency of Nitrogen Based Extractants for the Solvent Extraction of Acids

S. P. Moharana, P. K. Khillar, D. Swain, R. L. Bhutia, P. Panda, S. Mishra*

Department of Chemistry, Institute of Technical Education and Research (FET), Siksha 'O' Anusandhan Deemed to be University, Khandagiri Square, Bhubaneswar, Odisha, India

Abstract Mineral acids are part of the liquid discharges those usually flow from metal mines as well as coal mines [1]. This also includes active or abandoned mines, coal processing sites, tailings or ponds. In this context, the study of extraction of mineral acids becomes very important. In the current paper, nitrogen based extractants such as TEHA (Tri-n-ethylhexyl amine) and Aliquat 336 (Tricapryl methyl ammonium chloride) diluted in kerosene are tested in terms of extraction efficiency for the extraction of various acids such as HCl, HNO_3, H_2SO_4, $HClO_4$ and CH_3COOH. The percentage of extraction of all acids gradually increases when acid concentration increases from 0.2M to 1M in case of both the extractants of concentration 0.1M. The impact of concentration of extractant variation has been examined. For this purpose, extractants were used of concentration 0.15M and 0.2M and the extraction behavior of 0.5 M of various acids were compared. The distribution ratios have been determined by analyzing the aqueous feed and the raffinate after extraction using volumetric analysis. From the above experimental results, it was concluded that extraction of acids using TEHA showed better performance than Aliquat 336, a quaternary ammonium salt, since it contains nitrogen with lone pair of electrons which forms hydrogen bond with the acid. With TEHA, the extraction of all acids slightly increased with concentration of extractant while in case of Aliquat 336, it does not follow any order and more or less same. From the experimental observations, it has been concluded that more the concentration of acids higher will be the percentage of extraction.

Keywords Liquid-liquid extraction, Acids, TEHA, Aliquat 336

---

* Corresponding author: sujatamishra@soa.ac.in

DOI: 10.1201/9781003489443-4

1. Introduction

Solvent extraction is regarded as most efficient technique for removing mineral acids from effluents or liquid wastes. The extraction of these acids during solvent extraction may result from an interaction between the hydrogen ion of acid and the extractant. Inorganic acids like HCl, HNO_3, H_2SO_4 and H_3PO_4 are generally used in leaching to dissolve metals of interest in ores and various post-consumer products. In the separation of metal ions from the leached solution, the nature of the acids used as leaching agent has great influence in process control. The waste generated from these separation processes comprise these mineral acids and their disposal is of ecological concern [1, 2]. Several schemes are offered for the treatment of effluents having mineral acids, which includes precipitation, diffusion dialysis, neutralization, solvent extraction, ion exchange and membrane processes [3-5]. Tertiary amines such as TIOA (triisooctylamine), TOA (tri-n-octylamine), TEHA (tris 2-ethylhexyl amine), Alamine 308 (tri-isooctyl amine), TDA (tri-n-dodecylamine), and Alamine 336 (tri-C8-10 alkyl amines) were employed for the acid extraction. The extraction mechanism proceeds through protonation reaction [6-8]. Tertiary amines are observed as effective extracting agents for the recovery of HCl as well as H_2SO_4 from aqueous medium [5-8]. S. Brandani et al. [9] have reported the extraction of these strong mineral acids from the aqueous solutions with the use of Amberlite LA-2. With 0.01–0.5 M aqueous solutions of acids and 0.5 M Amberlite in toluene as the organic phase extraction tests were undertaken. They observed that Amberlite LA-2 shows higher selectivity for H_2SO_4 than HCl. Agarwal et al [5] used Alamine 336 for the recovery of H_2SO_4 from aqueous solutions produced through the process of pickling of steel at 30–60°C. Distilled water at 60 °C could effectively stripped off the loaded acid. The long chain amines behave as a complexing agent with the acid, which enables dispersal of the acids into the organic phase.

In this current research work, studies have been undertaken to investigate the extraction of acids like HCl, HNO_3, $HClO_4$, H_2SO_4 as well as CH_3COOH from the aqueous phase using TEHA (Tri-n-ethylhexyl amine) and Aliquat 336(Tricapryl methyl ammonium chloride) diluted in kerosene as the organic phase with the help of solvent extraction technique. Various concentrations of acids and extractants have been chosen for this purpose.

2. Materials and methods

2.1 Materials

The acid solutions were prepared by adding double distilled water to the concentrated mineral acids (Merck life India, Mumbai).The extractants TEHA (Sigma Aldrich), Aliquat 336 (Merck life India, Mumbai) have been used without going for any further purification. The chemical structure, properties, specifications of these extractants and the details of acids have been displayed in Table 4.1 and 4.2, respectively.

2.2. Methods

The liquid-liquid extraction experiments were performed by mixing equal volumes of aqueous phase containing acids (HCl, HNO_3, H_2SO_4, $HClO_4$, CH_3COOH) of desired concentration and

Examining Efficiency of Nitrogen Based Extractants for the Solvent Extraction of Acids

Table 4.1 Detail representations of organic extractants and specifications

Extractant	Structure	Molar mass(g/mol)	Density (g/mL)	Purity(%)	CAS No.
TEHA		353.67	0.817	≥97	1860-26-0
Aliquat336		404.17	0.884	≥88	5137-55-3

Table 4.2 Detail representations of acids and their specifications

Acids	pKa value	Molar mass(g/mol)	Density (g/mL) at 20°C	Purity (%)	Company
HCl	-6.3	36.46	1.16	≥35	Merck
HNO$_3$	-1.3	63.01	1.41	≥68-70	Merck
H$_2$SO$_4$	1.99	98.08	1.84	≥95-98	Merck
HClO$_4$	-15.2±2	100.46	1.768	≥70-72	Merck
CH$_3$COOH	4.8	60.05	1.05	≥99	Merck

the organic phase with required concentration of TEHA/Aliquat 336 diluted in kerosene in a separating funnel for 15 minutes using a mechanical shaker at room temperature. Both the phases were separated after mixing. The concentrations of acid in the aqueous phase before and after the extraction were measured by volumetric analysis [10]. Extraction and titration experiments are performed in triplicate to avoid any error. The schematic representation of the procedure followed is given in Fig. 4.1.

2.3. Data Analysis

The extent of acid extraction is depicted by taking distribution ratio (D_{Acid}) and extraction percentage (%E_{Acid}) and is given as:

$$D_{Acid} = \frac{[Acid]_{org}}{[Acid]_{aq}} \quad (1)$$

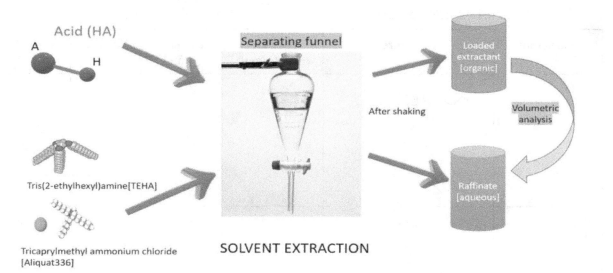

Fig. 4.1 Schematic representation of the liquid-liquid extraction procedure

$$\%E_{Acid} = \frac{100 * D_{Acid}}{D} + \left(\frac{V_{aq}}{V_{org}}\right) \qquad (2)$$

Where V_{aq} and V_{org} stand for the volume of aqueous and that of organic phase, respectively.

3. Results and Discussion

3.1 Effect of Concentration of Acids

For studying the influence of concentrations of various acids mentioned above, the molarity was varied from 0.1M to 1M keeping the concentration of TEHA and Aliquat 336 at 0.1M in kerosene. It was found that in both the cases extraction increases with increase in acid concentration. Amine basicity, steric hindrance and water solubility are relevant parameters and have an influence on extraction performances [11]. The extraction of acids by amines (long chain alkyl) is neutralization of the amine by the extracted acid in the organic phase and also exchange of anions between the aqueous and the organic phases. The amine-acid equilibrium can be well described by knowing the affinity of anion extracted to the organic phase. There are two opposite effects like electrostatic interaction in the organic phase and of ion-water interactions in the aqueous phase [12]. When the aqueous acid concentration is higher, the organic extractant extracts more amount of acid. The extraction of acids by these nitrogen based extractants chosen here (TEHA and Aliquat 336) is based on the following equations:

$$R_3N(org) + HA\ (aq) = R_3NH^+A^-\ (org) \qquad (3)$$

$$RR'_3N^+Cl^-\ (org) + HA\ (aq) = R_3\ NA\ (org) + HCl(aq) \qquad (4)$$

In the process of acid extraction from the diluted aqueous solutions, with high activity of water, water is also extracted to the organic phase [13]. The extent to which acid extraction occurs depends on identity of the anion which is clear from the Hofmeister effect [16]. In this series the ions are seen for the stabilization of the hydrophobic/hydrophilic interface. This includes a strongly hydrated ion group and a group of weakly hydrated ions. Chloride remains in the borderline between the two groups. Kosmotrope ions like suphate,carbonate, dihydrogen phosphate,fluoride, etc. are well hydrated whereas chaotropes ions such as bromide, nitrate, iodide, perchlorate, thiocyanate, etc. are weakly hydrated [14,15]. The extraction data D values (Table 4.1) and Percentage extractions for TEHA (Fig. 4.2)and Aliquat 336 (Fig. 4.3) are represented in the following.

Table 4.3 The distribution ratio values obtained in the extraction of acids by using 0.1 M TEHA and Aliquat 336 in kerosene

Acids	TEHA					Aliquat 336				
	0.2	0.4	0.6	0.8	1.0	0.2	0.4	0.6	0.8	1.0
HCl	0.29	0.40	0.59	0.62	0.65	0.39	0.41	0.44	0.47	0.49
HNO_3	0.40	0.62	0.66	0.68	0.69	0.23	0.27	0.28	0.32	0.35
H_2SO_4	0.43	0.6	0.64	0.66	0.68	0.13	0.18	0.16	0.20	0.25
$HClO_4$	0.39	0.55	0.58	0.63	0.70	0.32	0.51	0.54	0.58	0.68
CH_3COOH	0.32	0.41	0.49	0.61	0.67	0.25	0.31	0.38	0.47	0.58

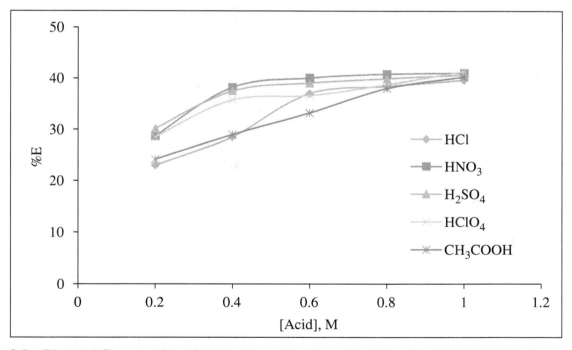

Fig. 4.2 Plot of %E versus [Acid], M for the extraction of acids using 0.1M TEHA in kerosene; Shaking time = 15 min, O/A = 1; Temperature = 298K

As observed from the extraction data in case of TEHA the anions of acids extracted follow the order: nitrate > sulphate > perchlorate > chloride > acetate. With Aliquat 336, the extraction trend follows the order: perchlorate > chloride > acetate > nitrate > sulphate. The observation is not accordance with the Hofmeister series since the molarity of extractant is less and that of acid is higher.

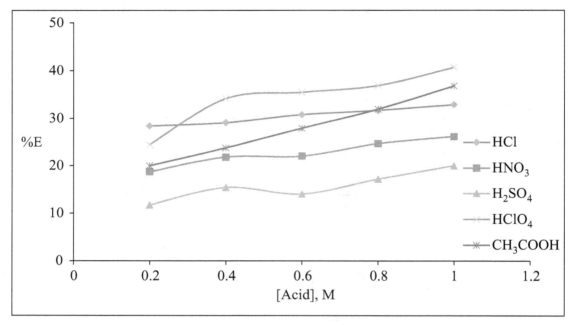

Fig. 4.3 Plot of %E versus [Acid], M for the extraction of acids using 0.1M Aliquat 336 in kerosene. Shaking time = 15 min, O/A = 1; Temperature = 298K

3.2 Effect of Concentration of Extractants

Extractants, TEHA and Aliquat 336 of concentrations 0.15M and 0.2M diluted with kerosene and various acids of 0.5 M concentrations were considered for the extraction purpose. It has been observed that the increase in concentrations of both the extractants has very little effect on the extraction of all acids. The extraction almost remains same with that observed by taking 0.1 M extractant concentrations. This may be due to the saturation of extractants with the acids as well as less concentration of extractants as compared to that of acids.

4. Conclusions

The solvent extraction of acids from aqueous solution was investigated using nitrogen based extractants, TEHA and Aliquat 336 in kerosene. The extractants showed ability to extract acids. TEHA being a tertiary amine exhibits better performance than the quaternary ammonium salt, Aliquat 336 when acid concentration varied from 0.2 to 1M. The study concerning variation of temperature and use of different diluents for the extraction behaviour of acids by these nitrogenous extracting agents may provide more information regarding these extraction systems.

REFERENCES

1. Z. Mohammadpour and H. R. Zare, Met. Mater. Int.2018, 24, 761.
2. M. T. Nguyen, J. H. Kim, J. G. Lee, and J. C. Kim, Met. Mater. Int. 2018, 24, 821.
3. D. F. Haghshenas, D. Darvishi, H. Rafieipour, E. K. Alamdari and A. A. Salardini, Hydrometallurgy, 2009, 97, 173.
4. W. Rudolfs, Ind. Eng. Chem. 1943, 35, 227.
5. A. Agarwal, S. Kumaris, B.C. Ray, K.K. Sahu, Hydrometallurgy. 2007, 88(1–4), 58–66.
6. R. Banda, T. H. Nguyen, and M. S. Lee, Chem. Process Eng.- Inz. 2013, 34, 153.
7. J. Stas and H. Alsawaf, Period. Polytech. Chem. Eng. 2016, 60, 130.
8. A. S. Vieux and N. Rutagengwa, J. Phys. Chem. 1976, 80, 1283.
9. S. Brandani, V. Brandani, Veglio, Industrial & Engineering Chemistry Research. 1998, 37(1), 292–295.
10. A. I. Vogel, Textbook of Quantitative Chemical Analysis, 5th ed., pp. 295–296, Longman Scientific & Technical, New York (1989).
11. G. Kyuchoukov, D. Yankov, Industrial & Engineering Chemistry Research. 2012, 51, 9117–9122.
12. G. Scibona, F. Orlandini, P. R. Danesi, Journal of Inorganic and Nuclear Chemistry, 1996, 28(8), 1701–1706.
13. B. Parbhoo, O. Nagy, J. Chem.soc.Farad.Trans., 1986, 82, 1789–93.
14. T. Janc, M. Luksič, V. Vlachy, B. Rigaud, A. Rollet, A. Korb, J. Phys. Chem. Chem. Phys. 2018, 20, 30340–30350.
15. B.Deyerle, Y. Zhang, Langmuir 2011, 27, 9203–9210.

Revisiting the Kinetics of Ester Hydrolysis Using ODE, DDE and FDE

Priyadarshan Garnayak

Department of Chemistry, Institute of Technical Education and Research (FET), Siksha 'O' Anusandhan Deemed to be University, Khandagiri Square, Bhubaneswar-751030, Odisha, India

Sunita Chand

Centre for Data Science, Institute of Technical Education and Research (FET), Siksha 'O' Anusandhan Deemed to be University, Khandagiri Square, Bhubaneswar-751030, Odisha, India

Santoshi Panigrahi

Department of Mathematics Institute of Technical Education and Research (FET), Siksha 'O' Anusandhan Deemed to be University, Khandagiri Square, Bhubaneswar-751030, Odisha, India

Sujata Mishra*

Department of Chemistry, Institute of Technical Education and Research (FET), Siksha 'O' Anusandhan Deemed to be University, Khandagiri Square, Bhubaneswar-751030, Odisha, India

Abstract Ester hydrolysis has been studied across-the-board ascribable to its importance in chemical, environmental, industrial and biological processes [1]. As important constituents, ccarboxylic acids and their derivatives like amides and esters are broadly used in several drugs and natural molecules. The hydrolysis reaction of an ester in pure water is a slow reaction and in the presence of a mineral acid like HCl, the rate of the reaction gets increased [2]. The kinetics of acid hydrolysis of ethyl acetate has been investigated experimentally in the present study where the concentration of ester has been determined at different time intervals during the course of hydrolysis. The rate constant has been evaluated. Using the models based on ordinary differential equation (ODE), delay differential equation (DDE) and fractional differential equation (FDE), the concentration of ethyl acetate at different time interval has been evaluated. The programs for these calculations has been formulated using MATLAB.

---

*Corresponding author: sujatamishra@soa.ac.in

DOI: 10.1201/9781003489443-5

The advantages of these models are that these need only the initial concentration value and time lag for prediction of the amount reacted at time *t*. The values have been compared with the experimental values to find out the best model. These approaches demonstrated a possible interpretation of the rate of ester hydrolysis. The analysis gives the predicted concentration data at various time intervals.

Keywords   Kinetics, Ester hydrolysis, Rate constant, Differential equations

1. Introduction

Chemical kinetics or reaction kinetics deals with the study of rates of chemical reactions. The chemical kinetics is the quantifiable determination of rate of chemical reactions as well as the parameters influencing the rate. Kinetics is significant because it provides evidence for the mechanism of the chemical reactions. Ethyl acetate is used primarily as coatings solvent, extraction solvent, process solvent and carrier solvent in the paint, metal and pharmaceutical industries. Ethyl acetate gets specific application as a flavour enhancer in foods and pharmaceuticals [1]. Processes like hydrolysis or evaporation are used to recover ethyl acetate found in dilute aqueous process waste streams generally associated with the fine chemicals and also in the pharmaceutical industries. The acetic acid produced due to hydrolysis can again be used in this process. There are reports on the study of rate of hydrolysis of ethyl acetate using ordinary differential equation [2-4].But there are several factors like the dependence of instantaneous rate on the previous state of reaction system, time lag in the experiment, etc. which cannot be determined by ordinary differential equations. The present study aims at analyzing the kinetics of acid hydrolysis of ethyl acetate by determining the concentration of ester at various time intervals using ordinary, delay and fractional differential equations.

2. Materials and Methods

2.1 Chemicals

The details of the chemicals used in this experiment are given below in Table 5.1.

Table 5.1   Particulars of the chemicals used

Chemicals	Company	CAS No.	Molar mass(g/mol)	Density(g/cm³)	Purity
Ethyl acetate	Merck	141-78-6	88.18	0.9 (20°C)	≥ 99.5 %
Hydrochloric acid	Merck	7647-01-0	36.46	1.2 (25°C)	≥ 35 %
Sodium hydroxide	Merck	1310-73-2	40.00	2.13 (20°C)	≥ 97.0 %

2.2 Method

In a glass stoppered bottle 95 mL of M/2 HCl and 5 mL of ethyl acetate was taken. After interval of 10 minutes, 5 mL from the the reaction mixture was transferred to a conical flask containing ice cold water and then volumetric analysis was carried out by titrating against M/5 NaOH to find out the amount of acid formed as a result of acid hydrolysis of ethyl acetate. Therefrom the amount of ester reacted was calculated. The total amount of HCl used was found out by titrating with NaOH without adding ester. The reaction mixture was kept overnight and the acid formed after completion of reaction was found out by volumetric analysis. Concentration of ester at each time interval was calculated experimentally and also by using ordinary differential equation, delay differential equation and fractional differential equation the ester concentration with the help of MATLAB. The detail scheme of the present work is demonstrated in Fig. 5.1.

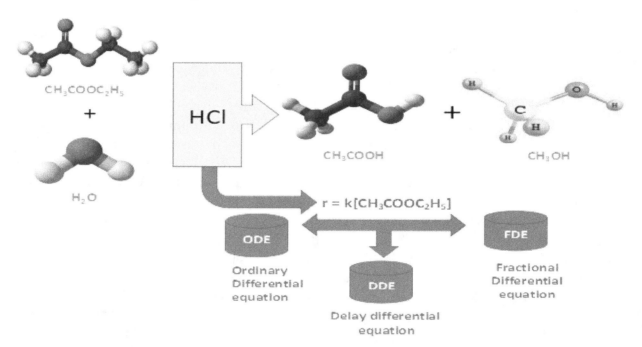

Fig. 5.1 Graphical representation of the scheme of present investigation

3. Results and Discussion

3.1 General Representation

The rate study of hydrolysis of ethyl acetate has been investigated here at room temperature. The reaction is represented as

$$CH_3COOC_2H_5 (l) + H_2O(l) \leftrightarrow CH_3COOH(l) + C_2H_5OH(l) \qquad (1)$$

The rate expression is

$$-\frac{d[CH_3COOC_2H_5]}{dt} = k[CH_3COOC_2H_5] \qquad (2)$$

Where k represents the rate constant. This has been calculated from experimentally determine ethyl acetate concentrations at $t = 0$, $t = t$ and at the end of reaction. By putting the values of rate constant, time lag and initial ethyl acetate concentration as inputs, the concentrations of ethyl acetate at different time have been found out using ordinary (ODE), delay(DDE) and fractional(FDE) differential equations.

3.2 MATLAB command for ODE

ODEs are a kind of distinct and very common group of differential equations. These differential equations are not possible to be determined specifically without some exterior condition, normally either an initial or a boundary value [4]. In this equation, dependent variable is function of one independent variable, for example:

$$-dy/dt = ky \qquad (3)$$

Here y represents the $[CH_3COOC_2H_5]$, mol/L.

Command:

Clear all;

Clc;

$k = 0.0023$;

$f = @ (t, y)[-k * y]$;

$[t, y] = ode45 (f, [0\ 60], 0.51185)$

Plot (t, y)

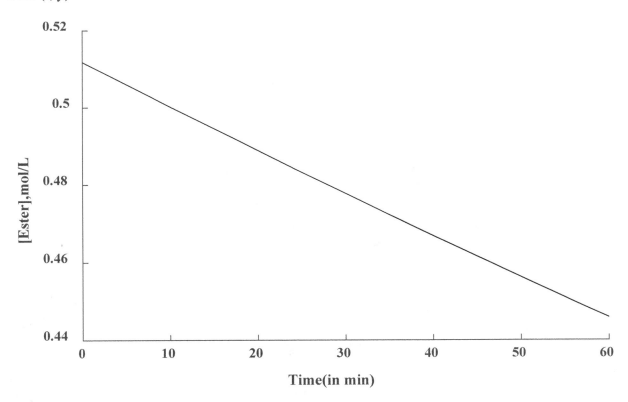

Fig. 5.2 Plot of [Ester], mol/L against time for the acid hydrolysis of ethyl acetate using ordinary differential equation

3.2 MATLAB Command for DDE

DDEs are types of differential equations where the derivative of the unidentified function at a definite time is known in terms of the function at previous times. These are equations which express some derivative of a variable (here ethyl acetate concentration) at time t and its lower order derivatives if any at t and at earlier instants. Every system concerning a response control certainly comprises time delays. A time delay arises since a finite time is needed to sense information and subsequently react to it. A time – dependent solution of DDE is not exclusively regulated by its initial state at a given time but in its place the result summary in an interval with length equivalent to the time lag or delay [5].General equation for delay differential equation:

$$-dy/dt = ky(t - \alpha) \qquad (4)$$

Command:

function DDEN1

sol = dde23(@DDEN1de,10,@DDEN1hist,[0, 60]);

figure;

plot(sol . x, sol . y)

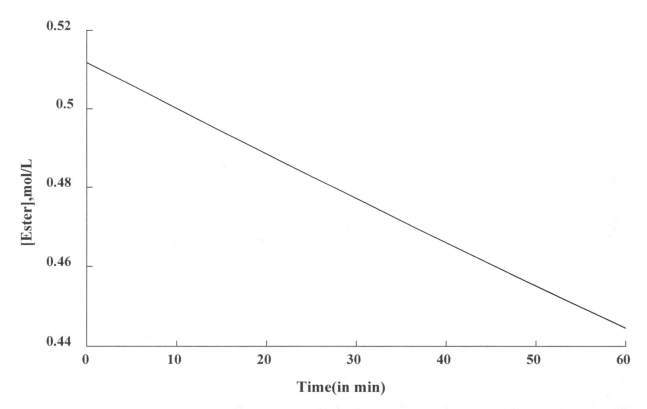

Fig. 5.3 Plot of [Ester], mol /L against time for the acid hydrolysis of ethyl acetate using delay differential equation

xlabel ('time t');

ylabel ('solution y');

function s = DDEN1hist(t)

s = 0.51185;

function dydt = DDEN1de (t, y, z)

ylag1 = z (:, 1);

dydt = [−0.0023*ylag1(1)];

3.3 MATLAB Command for FDE

Fractional differential equation (FDE) involves fractional derivatives of the form d^α/dx^α, and are defined for $\alpha > 0$, where α is not essentially an integer. These are simplifications of the ODEs to a random (non integer) order. They have enticed substantial attention because of their capability to model intricate phenomena [6]. The fractional-order is serving for better understanding of the description of real-world problems than the ordinary integer order.

$$-d^\alpha y/dt^\alpha = ky \qquad (5)$$

Command:

Format compact

k = 0.0023;

myfun = @(t,y)[-k*y];

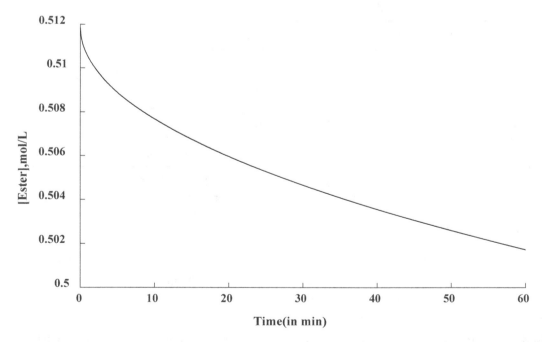

Fig. 5.4 Plot of [Ester], mol/L against time for the acid hydrolysis of ethyl acetate using partial differential equation

t0=0;

tfinal=60;

h=0.1;

alpha=0.5;

tic;

[t, qsol]=fde12(alpha,myfun,t0,tfinal,0.51185,h);

Toc;

Figure;

Plot(t,qsol(1,:));

3.4 Comparison of Performance

The above discussions on prediction of ethyl acetate concentrations at various time intervals during the process of acid hydrolysis of ethyl acetate with the help of three differential equations, ODE, DDE and FDE. From the results it has been observed that FDE gives the best prediction.

Table 5.2 Ethyl acetate concentration after time t

Time (in min)	[Ethyl acetate], mol/L			
	Experimental	Using ODE	Using DDE	Using FDE
0	0.511	0.5118	0.5118	0.5118
10	0.510	0.4996	0.5	0.5077
20	0.510	0.4894	0.4884	0.5060
30	0.510	0.4777	0.4771	0.5047
40	0.509	0.4663	0.465	0.5036
50	0.509	0.4568	0.4551	0.5026
60	0.509	0.4459	0.4446	0.5017

4. Conclusion

The models based on ODE, DDE and FDE to evaluate the concentration of ethyl acetate at different time interval in the course of acid hydrolysis reaction have been demonstrated in this work. A possible approach interpreting the kinetics has been described. The concentration values have been calculated and the graphs show that fractional differential equation model predicts the concentration of ethyl acetate very well and which values seems to be in good agreement with those obtained from the experiments. Therefore, the acid hydrolysis of ethyl acetate kinetics can be well described by following fractional order kinetics. Further studies with varying ester concentration and temperature will certainly throw more light on the mechanism as well as equilibrium kinetics.

Acknowledgements

The authors express their thankfulness to the authorities of Siksha 'O' Anusandhan University for the help provided to carry out this work.

REFERENCES

1. Dale J. Marino, in Encyclopedia of Toxicology (Second Edition), 2005
2. F.Jabeen, Q. Rahman, S.Zafar, International Journal of Applied Chemistry, 6 (2019) 18-23.
3. R.A.Y. Jones, Physical and Mechanistic Organic Chemistry, Cambridge University Press, Cambridge (1979) p. 227.
4. G.Scholz, F. Scholz, First-order differential equations in chemistry. Chem Texts , 1(2015)1.
5. R. D. Driver, Ordinary and Delay Differential Equations, New York Springer–Verlag. (1997).
6. S.Rathinasamy and R. Yong. Approximate controllability of fractional differential equations with state-dependent delay. Results in Mathematics, 63(3-4)(2013)949-963.

Scope of Acidic Ionic Liquids as Catalyst

Debasmita Jena, Pravanjan Mishra and Sanghamitra Pradhan*

Department of Chemistry, Institute of Technical Education and Research (FET), Siksha 'O' Anusandhan, deemed to be University, Khandagiri Square, Bhubaneswar, 751030, Odisha, India

Abstract Acidic ionic liquids (AILs) have a pH of less than 7. They have been the research subject in recent years due to their exceptional features, making them attractive for various chemical reactions. AILs are used as catalysts as they can dissolve organic and inorganic compounds and be used under mild conditions. The acidic property can be of the Brönsted or Lewis acid types or a mixture of both. Lewis acidic ILs, which can act as electron pair acceptors, are generally created by the reaction between a neutral ionic liquid and a Lewis acidic metal halide in anhydrous condition. Brønsted acidic ILs, have the potential to act as proton donors and whose ability to do so can be attributed to either the cation or the anion. The findings illustrate that AILs are used as catalysts in various organic synthesis processes. Lewis acidic catalysts such as choline chloride and benzyltrimethylammonium chloride have been used to protect carbonyls from 1, 3-dioxolanes and 1,3-dioxanes at room temperature without needing solvents. The oxidative removal of sulphur compounds (S-compounds) from diesel fuels is investigated using a series of Lewis acidic ionic liquids (ILs). The paper's primary focus is the synthesis of several forms of AILs and their application as catalysts in the synthesis of organic compounds.

Keywords Acidic ionic liquids, Lewis acidic ionic liquids, Bronsted acidic ionic liquids, Catalyst

1. Introduction

Ionic liquids (ILs) are named as designer solvents as they can be adjusted and tailored for specific applications. These solvents are convenient to use because their ionic properties can

---

*Corresponding author: sanghamitrapradhan@soa.ac.in

be altered, and modification of physicochemical properties like viscosity, density, and melting point is possible in planning novel functional materials [1-3]. ILs are typically understood as liquids entirely made up of ions and have melting point lower than 100°C. They exist in the liquid state below 100°C since their ions are weakly coordinated. A stable crystal lattice cannot be formed because at least one organic component has a delocalized charge [4]. Paul Walden, in 1914, marked the beginning of the field of ILs by the discovery of ethyl ammonium nitrate, which is produced by neutralizing ethylamine with nitric solid acid [5]. ILs are composed of distinct cationic and anionic species, just like all salts, however, unlike conventional salts, they have a lower affinity to crystallize because of their bulky and asymmetrical cation structure. The anions are substantially smaller in volume than the cations and have an inorganic structure, whereas the cation is typically represented by a big organic complex [6]. Imidazolium, pyridinium, ammonium, phosphonium, sulfonium, thiazolium, pyrazolium, and oxazolium cations are the main building blocks of most ILs. Significant variances exist between ILs and various anions because the type of anion dramatically influences the characteristics of an IL. The hydrophobic and physicochemical properties of IL system can be altered by changing the anion. The melting point of an IL serves as the primary parameter for evaluation. The structure and chemical makeup of an IL and its melting point are significantly correlated [7,8]. Thermogravimetric research shows that many ILs have great thermal stability, typically >350°C. Anion's characteristics mainly govern thermal decomposition. Increase in the hydrophilicity of anions also leads to a reduction in further thermal decomposition.[9] Many ILs have viscosities that are one to three orders of magnitude higher than those of conventional solvents, making them relatively viscous. At room temperature, viscosity has been found to range from 10 to 500 mPas-1 for several ILs. [10-12]. The type of cation and anion determine the density of an IL. As the bulkiness of the organic cation increased, the density of similar ILs decreased. The total density of ILs is substantially influenced by the molar mass of the anion [13,14]. It would be predicted that ILs have high conductivities, given that they are made entirely of ions. The conductivity of any solution also depends on physicochemical parameters like viscosity, density, ion size, and ionic mobility in addition to the quantity of charge carriers [15,16]. These environmentally friendly solvents are ideal replacements for volatile organic solvents [17,18]. However, one of the obstacles in their utilization is their increased viscosities compared with other organic solvents. This can be overcome by altering the structure of IL so that ions involved do not pack well. Based on their properties, ILs can be distinguished as acidic, essential, and neutral. Among these, acidic ionic liquids (AILs) have been the subject of much research in recent years as catalysts, making them attractive for use in various chemical reactions. Therefore, this review has been framed to give inclusive coverage of several forms of AILs and their application as catalysts in synthesizing organic compounds.

2. Types of AILs

AILs have been categorized as Bronsted and Lewis type. Metal and organic halides are vital in preparing Lewis-type AILs (LAILs). Due to some disadvantages like instability and toxicity, later research focused on Bronsted type AILs (BAILs) in 2002. There has been rising interest in BAILs as an environment-friendly catalyst. The IL performance has been reported to be improved due to the synergistic effect between Bronsted and Lewis acid [19-22]. Ionic salts with low melting point and acidic properties are known as AILs. AILs can be categorized as Bronsted, Lewis, or a combination of both, and acidic functional groups can be in the cationic, anionic, or both constituents [23]. LAILs display acidity due to a deficit in electrons, whereas BAILs exhibit acidity due to ionizable protons. According to their cation/anion ratios, they can be categorized as either Lewis acidic ILs, which can act as acceptors of electrons (the anion is typically responsible for this type of acidity), or BAILs, which have the potential to act as proton donors and whose ability to do so can be attributed to either the cation or the anion. Usually, BAILs contain imidazolium, ammonium, guanidinium as cations. LAILs, on the other hand, are created by adding the proper quantity of Lewis acids to neutral ILs and typically contain acidic anions [24]. Three types of AILs with their structure are shown in Fig. 6.1.[25].

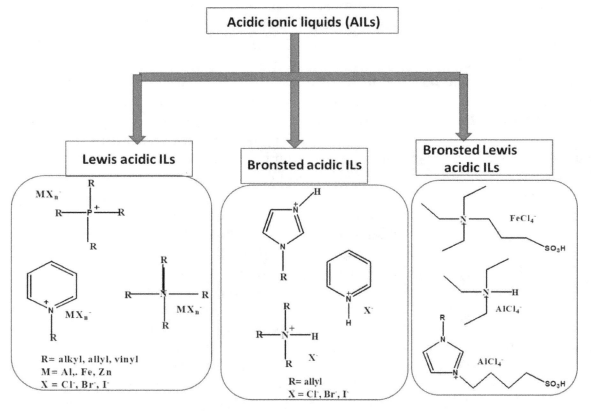

Fig. 6.1 Representation of types of AILs

3. Synthesis strategies for AILs

According to the Lewis theory, the acid takes two electrons while the base gives two electrons. In LAILs, the anion or cation may have this electron-accepting capacity. However, most of the known LAILs are in the anion's electron-accepting ability [26]. Novel BAIL s [C4SCnim] Cl/AlCl$_3$ has also been synthesized [27, 28]. Jiang et al. have synthesized BAILs, which can act as effective recyclable catalysts for converting chitosan into 5-Hydroxymethylfurfural. Both BAILs and LAILs can effectively catalyze different reactions [29]. The ionic liquids were

Scope of Acidic Ionic Liquids as Catalyst **45**

created by adding solid, anhydrous $ZnCl_2$ over time to melted [BMMI][Cl] where X represents the number of equivalents of $ZnCl_2$ to [BMMI][Cl] and R represents the mole ratio of $ZnCl_2$. Synthesis of 1-butyl-2,3-dimethylimidazolium chloride/$ZnCl_2$ [BMMI][ZnCl3] is shown in Fig. 6.2.[30]

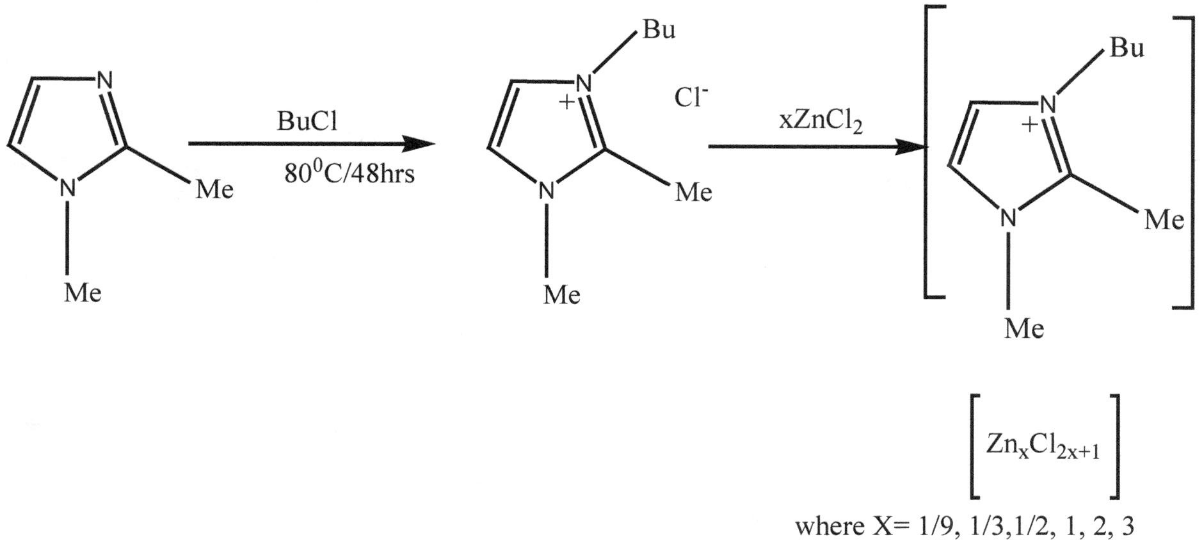

Fig. 6.2 Scheme of synthetic route of 1-butyl-2,3-dimethylimidazolium chloride/$ZnCl_2$ [BMMI][$ZnCl_2$]

By combining equal amounts of 1-butyl-3-methylimidazolium chloride [Bmim][Cl] and $FeCl_3.6H_2O$, [Bmim][$FeCl_4$] was generated. Synthesis of 1-butyl-3-methylimidazolium tetrachloroferrate [Bmim][$FeCl_4$] is shown in Fig. 6.3.[31]

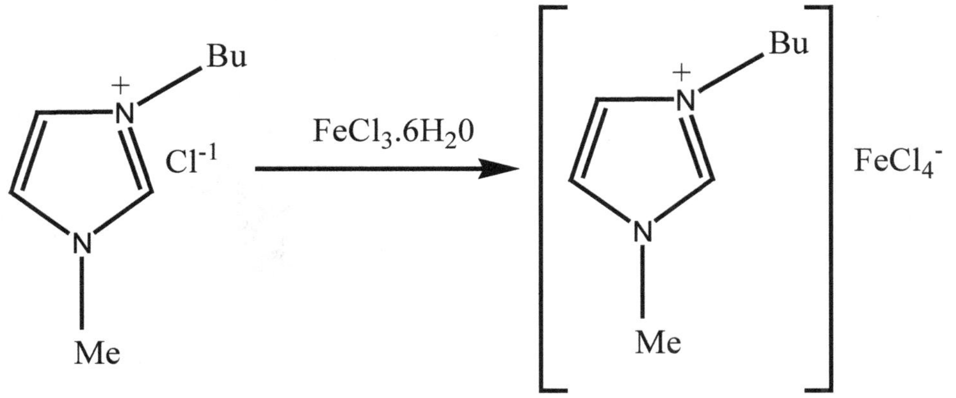

Fig. 6.3 Scheme of synthetic route of 1-butyl-3-methylimidazolium tetrachloroferrate [Bmim][$FeCl_4$] [31]

Addition of equivalent amount of sulphuric and phosphoric acid to 1-butyl-3-methylimidazolium chloride in anhydrous dichloromethane leads to the formation of 1-butyl-3-methylimidazolium hydrogen phosphate and sulphate [Bmim]$^+$[H_2PO_4]- and [Bmim]$^+$[HSO_4]- represented in Fig. 6.4 [32].

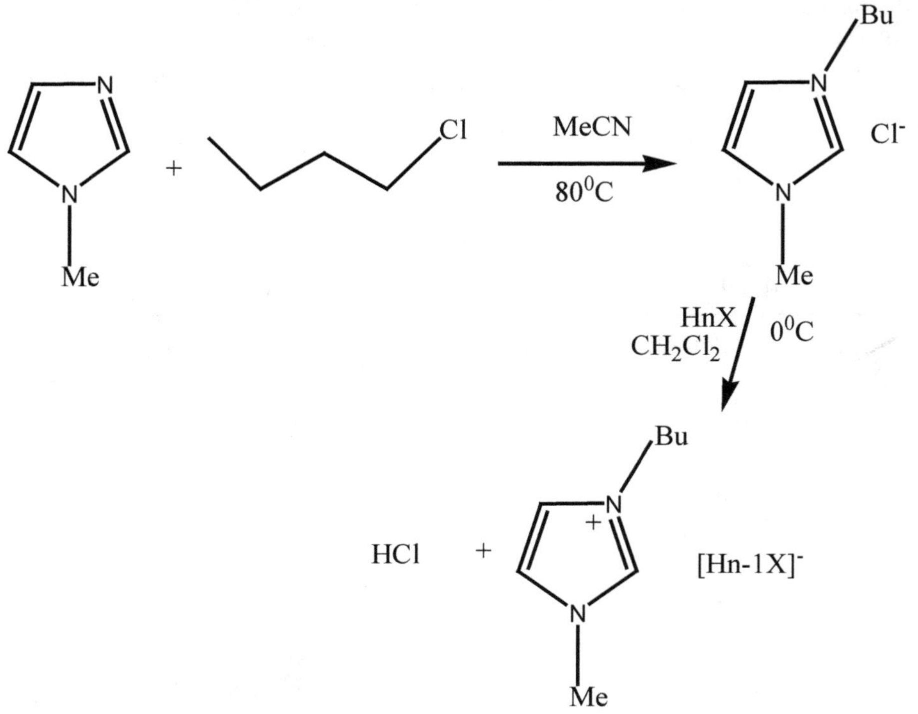

Fig. 6.4 Scheme of synthetic route of 1-butyl-3-methylimidazolium hydrogen sulfate ([Bmim]⁺ [HSO$_4$]⁻) and l-butyl-3-methylimidazolium hydrogen phosphate ([Bmim]⁺[HPO$_4$]⁻) [32]

Ananda et al. have synthesized acid-functionalized IL by first nucleophilic substitution of chlorine and then by acidification using HCl. An acid catalyst was found to be effective for the hydrolysis of cellulose [16]. 1,3-propane sultone was mixed with ethyl acetate and added to methylimidazole to produce ILs. Synthesis of 1-(3-sulfonic acid)-propyl-3-methylimidazole ferric chloride [MIMPS]FeCl$_4$⁻ is shown in the Fig. 6.5[33].

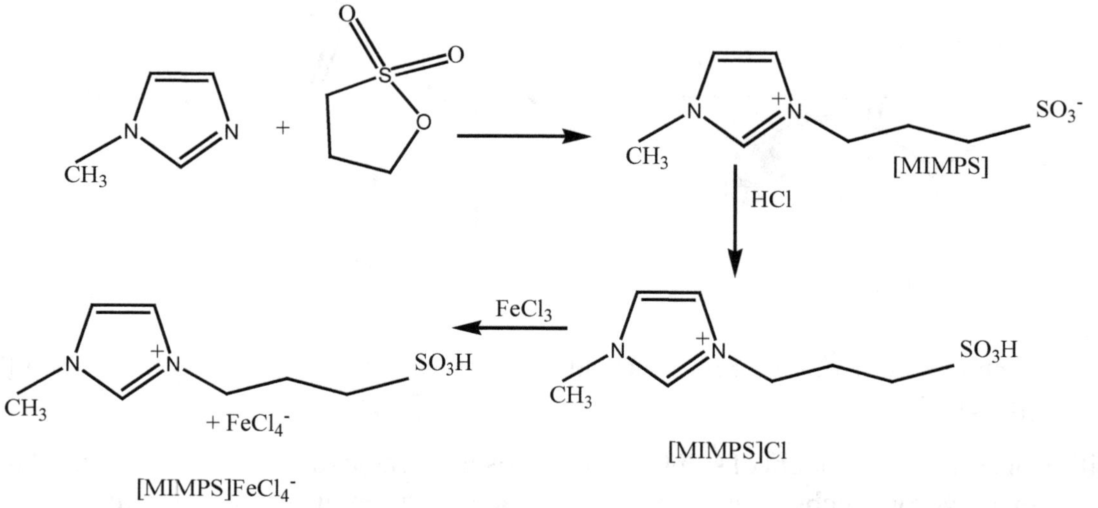

Fig. 6.5 Scheme of synthetic route of 1-(3-sulfonic acid)-propyl-3-methylimidazole ferric chloride [MIMPS]FeCl$_4$⁻ [33]

4. Physical Properties of Acidic Ionic Liquids

The solubilities of AILs are influenced mainly by the solvent's polarity and molecular makeup and by the strength of the hydrogen bonds that connect alcohols and anionic groups [34]. It is evident that when temperature rises, ILs become more soluble in solvents.[35]. Muhammad and co-workers demonstrated the viscosity of four typical BAILs. Except for imidazolium-based hydrogen sulphate ILs, all ILs had increased viscosity when either side of the alkyl chain lengthened. Likewise, the alkyl chain length impacted the density value, which tended to decline with lengthening alkyl chains [29]. These variations in viscosities could be brought based on functionalization and the many intermolecular forces present in each ionic liquid molecule, including hydrogen bonds, the interaction, van der Waals interactions, etc. [36]. For functionalized brönsted acidic ionic liquids, thermal breakdown takes place. The loss in thermal stability was in the order triethanolammonium > pyridinium > methylimidazolium. The primary determinants of the thermal stability of imidazolium-type cyano-functionalized Brönsted acidic ionic liquids with HSO_4^- and HPO_4^- anions are related to the structure of the corresponding anions, and the thermal stability falls in the order $HSO_4^- > HPO_4^-$. The length of the cation's alkyl chain has little impact on the onset temperatures of ionic liquids [37]. In a recent study, Wu and colleagues examined the ionic conductivities of several protic ionic liquids based on acetamide and Brönsted acids. They discovered that most of the samples' ionic conductivities range between 10^{-3} and 10^{-1} S/m at room temperature.

5. Applications of Acidic Ionic Liquid

Lewis acidic catalysts such as choline chloride and benzyltrimethylammonium chloride ($ZnCl_2$) are economic and exhibit moisture stability. They can be effectively used as recyclable catalysts to protect carbonyls from 1,3-dioxolanes. Choline chloride with $ZnCl_2$ is a catalytic system that can be recycled up to five times without significantly losing its effectiveness. This innovative catalytic system's straightforward processes and ease of recovery and reuse are anticipated to aid in developing safer acetalization processes for carbonyl compounds [38, 39]. The oxidative removal of sulphur compounds from diesel fuels is investigated using a series of LAILs based on 1-n-butyl-3-methylimidazolium metal chloride, i.e., [C_4mim]Cl/MCl_2(M = Zn, Fe, Cu, Mg, Sn, Co). ILs act as both an extractant and a catalyst [40]. Waste polycarbonate (PC) is selectively alcoholised to recover the monomer bisphenol (BPA), which is catalyzed by [Bmim]Cl_2.$FeCl_3$. Additionally, without noticeably losing catalytic activity, the [Bmim]Cl_2.$FeCl_3$ catalyst may be employed five times [41]. Ionic liquids (ILs) comprising BmimCl-$FeCl_3$, BmimCl-$AlCl_3$, and BmimCl-$ZnCl_2$ were used as a dual catalyst and solvent in Friedel-Crafts acylation processes to produce benzophenone and its derivatives. This process has a high yield, a straightforward process for isolating the product, allows ILs to be reused, and reduces waste discharge, making the catalytic system effective and environmentally acceptable.[42] These LAILs [Hmim]$ZnCl_3$, [Hmim]$CoCl_3$, [Hmim]$FeCl_4$, and [Hmim]$CuCl_3$ operate as catalysts for the glycolysis of poly (ethylene terephthalate) PET, forming and characterizing the product bis(hydroxyethyl) terephthalate BHET and by-products [43]. Numerous organic chemistry processes use BAILs as a catalyst, including the Mannich reaction,[44] Pechmann reaction,[45] synthesis of amidoalkyl naphthols,[46] -caprolactam synthesis,[47] synthesis of benzoxanthenes,[48]. A variety of BAILS were used as catalysts to study the esterification of aliphatic acids with alcohols. The HSO_4^- and $pCH_3(C_6H_4)SO_3^-$ functionalized SO_3H-functionalized ionic liquids likewise demonstrated good activity, but the $H_2PO_4^-$ and $CF_3CO_2^-$ functionalized ionic liquids had low activity, and the ordinary non-functionalized ionic liquids demonstrated almost little activity[16]. With free long-chain fatty acids or their combinations, low-molecular weight alcohols, and Brønsted acidic ionic liquid N-methyl-2-pyrrolidonium methyl sulfonate ([NMP][CH_3SO_3]) as substrates, biodiesel could be produced [49]. Fischer esterification of alcohols with acids uses BAIL, which has been manufactured in good yield. It contains inorganic anions of the type BF_4, PF_6, and PTSA as well as nitrogen-based organic cations 1-methylimidazole and 1-butyl-3-methylimidazolium.[50]. Imidazolium BAILs; 1-(1-propylsulfonic)-3-methylimidazolium chloride and 1-(1-butylsulfonic)-3-methylimidazolium chloride are efficient in dissolving and hydrolysing cellulose in a single operation at atmospheric pressure and without any pre-treatment. [51, 52].

6. Conclusion

Research in the field of IL has expanded in the past decade, and AIL has become an important branch. This review focuses on the synthetic schemes adopted for synthesizing AILs and highlights their different applications. Blending acidic and IL properties into a single molecule has unlocked novel prospects. It has been surveyed that BAILs exhibit superior catalytic properties compared to organic and mineral acids. This may be attributed to associations of ionic species with the cation/anion in BAIL, which facilitates proton or electron transfer with AILs.

Acknowledgment

The authors are highly grateful to the authorities of Siksha 'O' Anusandhan Deemed to be University for their support in carrying out this present research work.

REFERENCES

1. Marsza, M.P; Kaliszan, R. Application of ionic liquids in liquid chromatography. *Crit. Rev. Anal. Chem.* **2007**, *37*, 127–140.
2. Van Rantwijk, F.; Sheldon, R.A. Biocatalysis in ionic liquids. *Chem. Rev.* **2007**, *107*, 2757–2785.
3. Marsza M.P.; Markuszewski M.J.; Kaliszan R. Separation of nicotinic acid and its structural isomers using 1-ethyl-3-methylimidazolium ionic liquid as a buffer additive by capillary electrophoresis. *J. Pharm. Biomed. Anal.* **2006**, *42*, 3
4. Fannin, A. A.; Floreani, D. A.; King, L. A.; Landers, J. S.; Piersma, B. J.; Stech, D. J.; Vaughn, R., Part 2. *J. Phys. Chem.* **1984**, *88*, 2614–2627.
5. Yang, Q.; Dionysiou, D. D. J. Photolytic degradation of chlorinated phenols in room temperature ionic liquids. *Photochem. Photobiol. A: Chem.* **2004**, *165,* 229–240
6. Singh, G.; Kumar, A. Ionic liquids: Physico-chemical, solvent properties and their applications in chemical processes. *Ind J of Chem.* **2008**, *47A*, 495–503

7. Hayes, R.; Warr, G.G.; Atkin, R. Structure and nanostructure in ionic liquids. *Chem. Rev.* 2015, 115, 6357–6426.
8. Kosmulski, M.; Gustafsson, J. Rosenholm J. B. Thermal stability of low temperature ionic liquids revisted. *Thermochim. Acta.* 2004, *412*, 47–53.
9. Huddleston, J. G.; Visser, A. E.; Reichert, W. M.; Willauer, H. D.; Broker, G. A.; Rogers, R. D. Characterization, and comparison of hydrophilic and hydrophobic room temperature ionic liquids. *Green Chem,* 2001, *3*, 156.
10. Dzyuba, S.; Bartsch, R. A. Influence of structural variations in 1-alkyl (aralkyl)-3-methylimidazolium hexaflurophosphates and bis (trifluoromethylsufonyl)imides of physical properties of ionic liquids. *Chem Phys Chem.* 2002, 3, 161–166.
11. Cardo-Broch, S.; Berthold, A.; Armstrong, D. W. Solvent properties of the 1-butyl-3methylimidazolium hexafluorophosphate ionic liquid. *Anal Bioanal. Chem.* 2003, 375, 191.
12. Pringle, J. M.; Golding, J.; Baranyai, K.; Forsyth, G. B.; Deacon, G. B.; Scott, J. L.; McFarlane, D. R. The effect of anion fluorination in ionic liquids physical properties of a range of bis(methanesulfonyl) amide salts. *New J. Chem,* 2003, 27, 1504–1510.
13. Xu, W.; Wang, L. M.; Nieman, R. A. Angell, C. A. Ionic liquids of chelated orthoborates as model ionic glassformers. *J. Phys. Chem. B.* 2003, *107,* 11749.
14. Xu, W.; Cooper, E. I.; Angell, C. Ionic liquids: Ionic mobilities, glass temperatures and fragilities. *J. Phys. Chem. B,* 2003, *107,* 6170.
15. Every, H. A.; Bishop, A. G.; Oradd, G.; Forsyth, M.; McFarlane, D. R. Transport properties in a family of dialkylimidazolium ionic liquids. *Phys Chem Chem Phys,* 2004, 6, 1758.
16. Ananda S. A. Acidic ionic liquids. *Chem. Rev.* 2016, 116, 6133–6183.
17. Wells, A.S.; Coombe, V.T. On the freshwater ecotoxicity and biodegradation properties of some common ionic liquids. *Org. Process Res. Dev.,* 2006, *10.* 794-798.
18. Studzinska, S.; Buszewski, B. Study of toxicity of imidazolium ionic liquids to watercress (Lepidium sativum L.). *Anal. Bioanal. Chem.*, 2009, *393*, 983–990.
19. Amarasekara, A.S.; Hasan, M.A. 1-(1-Alkylsulfonic)-3-methylimidazolium chloride Bronsted acidic ionic liquid catalyzed skraup synthesis of quinolines under microwave heating. *Tetrahedron. Lett.* 2014, *55*, 3319.
20. Zhao, P.; Cui, H.; Zhang, Y.; Zhang, Y.; Wang, Y.; Zhang, Y.; Xie, Y.; Yi, W. Synergistic effect of Bronsted/Lewis acid sites and water on the catalytic dehydration of glucose to 5- hydroxymethylfurfural by heteropolyacid based ionic hybrids. *Chem. Open.* 2018, *7*, 824–832.
21. Patra, T.; Alfreen, G.; Parveen, F.; Kumar, K.; Bahri, S.; Upadhya, S. Synergistic Bronsted – Lewis acidity effect on upgrading biomass- derived phenolic compounds: Statistical optimization of process parameters, kinetic investigations and DFT study. Chemistry Select. 2018.
22. Jiang, X.; Ye, W.; Song, X.; Ma, W.; Lao, X.; Shen, R. Novel ionic liquid with both Lewis and Bronsted acid sites for Michael addition. *Int. J. Mol. Sci.* 2011, 12, 7438.
23. Chiappe, C.; Rajamani, S.; Structural effects on the physico-chemical and catalytic properties of acidic ionic liquids. An overview. *Eur. J. Org. Chem.* 2011, 5517–5539.
24. Muazzam, R.; Asim, A. M.; Uroos, M.; Muhammad, N.; Hallett, J.P. Evaluating the potential of a novel hardwood biomass using a superbase ionic liquid.*J Mol. Liq.* 2019, 287, 110943
25. Estager, J.; Holbrey, J. D.; Swadźba-Kwaśny, M. Halometallate ionic liquids. *Chem. Soc. Rev.* 2014, *43 (3)*, 847–886.
26. Lecocq, V.; Graille, A.; Baudouin, A.; Chauvin, Y.; Basset, J. M. Synthesis and characterisation of ionic liquids based upon 1-butyl-2,3-dimethylimidazolium chloride/ZnCl$_2$ *New J. Chem.* 2005, 2,700–706.

27. Wang, X.H.; Tao, G-H.; Zhang, Z.Y.; Yama, K. Synthesis and Characterization of Dual Acidic Ionic Liquids. *Chin. Chem. Lett.* **2005**, *16,* 1563–1565.
28. Wei, Y.; Keke, C.; Xiaofang Z.; Yingying K.; Xiujuan, T.; Xiaoxiang, H. Synthesis of novel Brønsted–Lewis acidic ionic liquid catalysts and their catalytic activities in acetalization. *J. Ind. Eng. Chem.* **2015**, *29,* 185–193.
29. Jiang, Y., Zang, H., Han, S., Yan, B., Yu, S. and Cheng, B. Direct conversion of chitosan to 5-hydroxymethylfurfural in water using Brønsted–Lewis acidic ionic liquids as catalysts. RSC Adv. 2016, 6, 103774–103781.
30. Xie, Z.L.; Jeličić, A.; Wang F.P.; Rabu, P.; Friedrich A. Transparent, flexible, and paramagnetic ionogels based on PMMA and the iron-based ionic liquid 1-butyl-3-methylimidazolium tetrachloroferrate (III) [Bmim][FeCl$_4$]. *J. Mater. Chem.* 2010, 20, 9543–9549.
31. Johnson, K.E.; Pagni R. M.; Bartmess, J. Bronsted acids in ionic liquids: Fundamentals, organic reactions, and comparisons. *Monatsh. Chem.* **2007**, 138, 1077.
32. Lunagariya, J.; Dhar, A.; Vekariya, R. L. Dependency of anion and chain length of imidazolium based ionic liquid on micellization of the block copolymer F127 in aqueous solution: An experimental deep insight. *RSC Adv.*, **2017**,*7,* 5412–5420
33. Liu, S.; Xie, C.; Yu, S.; Liu, F. Dimerization of rosin using Bronsted-lewis acidic ionic liquid as catalyst. *Catal. Commun.* **2008**, 9, 2030–2034.
34. Dyson, P.J.; Ellis, D.J.; Welton, T. Parker, D.G. Arene hydrogenation in a room temperature ionic liquid using a ruthenium cluster catalyst. *Chem. Commun.* **1999**, 25-26.
35. Muhammad, N.; Man, Z.; Elsheikh, Y.A.; Bustam, M.A.; Abdul Mutalib, M.I. Synthesis and thermophysical properties of imidazolium based bronsted acidic ionic liquids. *J. Chem. Eng. Data.* **2014**, *59,* 579–584.
36. Liang, W.D.; Li, H.F.; Gou, G.J.; Wang, A.Q. *Asian J. Chem.* **2013,** *25,* 4779–4782.
37. Bull. Chem. Soc. Jpn. Vol. 80, No. 12, 2365–2374 (2007)
38. Chen, X.; Song, D.; Asumana, C.; Yu, G.; Chen, X. Deep oxidative desulfurization of diseal fuels by Lewis acidic ionic liquids based on 1-n-butyl-3-methylimidazolium metal chloride. *J. Mol. Catal. Chemical.* **2012**, *359,* 8–13.
39. Liu, M.; Guo, J.; Gu, Y.; Gao, J.; Liu, F.; Yu, S. Pushing the limits in alcoholysis of waste polycarbonate with DBU-based ionic liquids under metal and solvent free conditions. *ACS Sustainable Chem. Eng*, **2018**, *6,* 13114–13121.
40. Zhao, G.; Jiang, T.; Gao, H.; Han, B.; Huang, J.; Sun, D. Mannich reaction using acidic ionic liquids as catalysts and solvents. *Green. Chem.* **2004**, *6,* 75–77.
41. Li, C.; Liu, W.; Zhao, Z. Efficient synthesis of benzophenone derivatives in Lewis acidic ionic liquids. *Catal. Commun.* **2007**, *8,* 1834–1837.
42. Shuangiun, C.; Weihe, S.; Haidong, C.; Hao, Z, Zhenwei, Z.; Chaonan, F. J. Therm. Anal. Calorim. 2020.
43. Vafaeezadeh, M.; Alinezhad, H. Bronsted acidic ionc liquids: Green catalysts for essential organic reactions. J. *Mol. Liq.* **2016**, *218,* 95–105.
44. Gu, Y.; Zhang, J.; Duan, Z.; Deng, Y. Adv. Synth. Catal. 2005, 347, 512–516.
45. Zhang, Q.; Luo, J.; Wei, Y. A silica gel supported dual acidic ionic liquid: an efficient and recyclable heterogenous catalyst for the onw-pot synthesis of amidoalkyl naphthols. *Green. Chem.* **2010**, *12,* 2246–2254.
46. Safaei, S.; Mohammadpoor-Baltork, I.; Khosropour, A.R.; Moghadam, M.; Tangestaninejad, S.; Mirkhani, V. *Catal. Sci. Technol.* **2013**, *3,* 2717–2722.

47. Safari, J.; Zarnegar, Z. Brosnted acidic ionic liquid based magnetic nanoparticles: a new promoter for the Biginelli synthesis of 3,4-dihydropyrimidin-2(1H)-ones/thiones. *New. J. Chem.* **2014**, *38*, 358–365.
48. Zhao, Y.; Long, J.; Deng, F.; Liu, X.; Li, Z.; Xia, C.; Peng, J. Catalytic amounts of Brosnted acidic ionic liquids promoted esterification: study of acidity-activity relationship. *Catal. Commun.* **2009**, *10*, 732–736.
49. Zhang, L.; Xian, M.; He, Y.; Li, L.; Yang, J.; Yu, S.; Xu, X. A Bronsted acidic ionic liquid as an efficient and environmentally benign catalyst for biodiesel synthesis from free fatty acids and alcohols. *Bioresour. Technol.* **2009**, *100*, 4368–4373.
50. Joseph, T.; Sahoo, S.; Halligudi, S.B. Bronsted acidic ionic liquids: A green, efficient and reusable catalyst- system and reaction medium for Fischer esterification. *J. Mol. Catal.* **2005**, *234*, 107–110.
51. Lunagariya J.; Dhar, A.; Vekariya R.L. Efficient esterification of n-butanol with acetic acid catalyzed by the Bronsted acidic ionic liquids: Influence of acidity. *RSC. Adv.* **2017**, *7*, 5412–5420.
52. Liu, S.; Chen, C.; Yu, F.; Li, L.; Liu, Z.; Yu, S.; Xie, C.; Liu, F. Alkylation of isobutane.isobutene using Bronsted Lewis acidic ionic liquids as catalyst. *Fuel.* **2015**, *159*, 803–809.

Study of Agro-Waste Biorefineries for Circular Bio-Economy

Nayak Sanchita*, Das Mira

Department of Chemistry, ITER, SOA Deemed to be University

Abstract The primary source of concern for the world's energy is traditional fossil fuels. The combustion of fossil fuels contributes to greenhouse gas emissions, which pose the greatest threat to the environment and consequently cause changes in the planet's climate. Regarding sustainability, fuels and products made from organic or plant wastes solve this drawback, providing a solution to the shortage of fossil fuels. It turns out that using agricultural waste as a feedstock for the biorefinery approach is an environmentally beneficial procedure for the creation of biofuel, strengthening energy security. Therefore, using this renewable biomass in the future to create green fuel and other green biochemicals will have a positive influence on both cost and sustainability. The biorefinery strategically develops bioenergy that eventually fits in a circular bioeconomy by utilising various agricultural biomass and experimenting with various biomass conversion strategies. This paper presents a critical explanation of the sources and production of agricultural waste and paves the way for future value addition using diverse technologies.

Keywords Agro-waste, Biorefineries, Circular bio-economy, Lignocellulosic biomass, Pre-treatment methods, Hydrodynamic cavitation

1. Introduction

Two of the most significant issues facing modern society are lowering the amount of solid biowaste and limiting reliance on fossil fuel resources. In terms of biowaste, lignocellulosic biomasses include mash and paper, horticulture, nutrition, forestry, and municipal solid waste. Biowaste is usually made up of waste from botanical gardens, decomposable plants, industrial biowaste, food-related biowaste from family cafeterias, and waste from industrial facilities

---

*Corresponding author: nayaksanchita712@gmail.com

[1,2]. The transition to post-carbon fossil social regimes on a worldwide scale may be facilitated by both advances and improvements in cleaner biomass production and processing [3]. Since industrial and agricultural biowaste is a cost-effective renewable resource, researchers have recently focused their efforts on creating bioproducts and bioenergy from it [4]. Researchers' interest has been drawn to microalgae as an alternative source of feedstock for the manufacture of several bioenergy products, including biodiesel, biohydrogen, and bioethanol [5]. The development of the circular economy places a high value on the life cycles, designs, and generation strategies of products as well as on the efficient use of resources and the production of biowaste across a bioproduct's entire life cycle. With regard to aims for resource efficiency and maintainability, the circular and bioeconomy architectures are complementary. The creation of biorefineries as well as improved sustainability measures will be necessary for such a circular bioeconomy [6]. The life cycle assessment (LCA) is an approach that is simple for anybody to apply to assess the effects of the energy used, hazardous substances emitted, and natural resources used throughout the life cycle of a product or activity. In order to characterise the programme and predict future trends, system dynamics (SD) can be applied to biowaste management arrangements. The management of biowaste is one area where SD has recently shown to be quite helpful [7]. In order to produce energy, vitality, and chemicals, coordinated biorefineries use the entire biomass waste component. These facilities form the foundation of the CE. Moreover, the life cycle assessment (LCA) methodology will be used in this study to assess various natural waste management scenarios.

2. Progress on the Biorefinery Studies

Table 7.1 illustrates the growth of the biorefinery over four generations. Using a single feedstock and technology, the initial generation of biorefineries could only generate a few products. Typically, high-sugar plants, maize stovers, or wheat straw made up the feedstock [8]. Even if some of these feedstocks have the potential to yield biofuels, the production process will likely take a lot of grains, which could increase food costs and have a negative impact on economic growth. A single feedstock and technique were employed by the second generation of biorefineries. Lignocellulosic feedstocks were also used; they were not just food crops. In comparison to the first and second generations, the third generation of biorefineries constituted a significant advancement.

3. Feedstock Categories for Biorefineries

A number of bioprocesses can be employed to manufacture biofuels and biobased compounds from a wide range of wastes, whether they are solid, liquid, or gaseous [10]. Aquatic creatures, woody residues, herbs, manures, industrial residues, municipal garbage, and a wide variety of agricultural and industrial waste have all been utilised as feedstocks in biorefineries. The three biomass feedstocks that are most commonly discussed in the literature are urban garbage, industrial waste, and agricultural crops and residues, with frequencies of 30, 27, and 19, respectively (Fig. 7.2). There are significant financial and environmental advantages to using these top three feedstocks to make gasoline or other high-value biobased goods. A significant problem, though, is how consistently available these feedstocks are.

Table 7.1 Four generations of biorefineries [9]

1st Generation	2nd Generation	3rd Generation	4th Generation
Feedstocks: • High-sugar crops (sugarcane, sugar beet) • Maize Stovers • Wheat straw • Corn **Products:** • Bio-oils • Biofuels • Biogases **Flaws:** • A great amount of cereals will be consumed in the production process	**Feedstocks:** • Lignocellulosic feedstocks • Agriculture • Industry • Woody Residues (Forestry) • Herbs **Products:** • Biofuels • Biogases • Bio-oils • Heat • Power • Electricity • Biochemicals **Flaws:** • Although feedstocks are not limited to cereals, the production is inefficient	**Feedstocks:** • Aquatic organisms (Microalgae, Macroalgae) **Products:** • Biofuels • Bio-oils • Biogases • Heat • Power • Electricity • Fertilizer • Biochemicals **Flaws:** • The growth of micro-algae requires the extraction of nutrients from the fossil resources, resulting in the depletion of non-biofossil resources	**Feedstocks:** • Aquatic organisms based on high-solar-efficiency cultivation • (Microalgae, Macroalgae) **Products:** • Biofuels • Bio-oils • Biogases • Heat • Power • Electricity • Fertilizer • Biochemicals **Flaws:** • The large-scale cultivation of algae is limited by the associated high costs

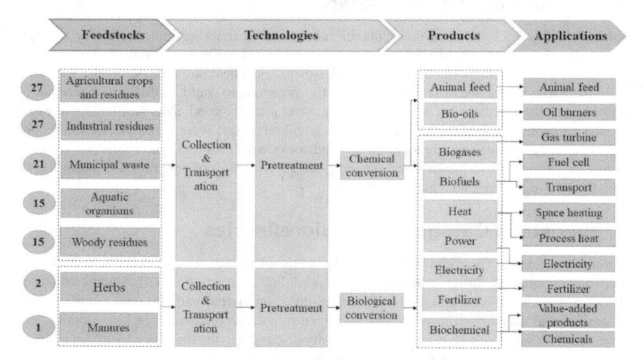

Fig. 7.1 The categories of biorefinery feedstocks, conversion technologies, products and applications.[9]

4. Agro-industrial Waste's Causes and Impacts

The top-ranking agro-industrial waste is sugarcane bagasse, a dry, fibrous trash that is produced after the juice from the plant has been removed. Orange, pomegranate, banana, apple, and other fruit peels are examples of agro-industrial wastes. Crop remnants are items that are discarded after being taken straight from an agricultural field. The most accessible and affordable waste that can be used to create products with value-added is this one [11]. Wheat straw, corn straw, and rice straw are the most widely available crop leftovers for the generation of bioethanol. The biomass produced in huge quantities worldwide with the most potential value added of the three crops is rice straw [12]. The most prevalent type of biomass used in manufacturing of cellulosic ethanol is corn husks. According to agriculture industrial wastes, sorghum, barley, and oat residues are also included. Agro-industrial wastes make up every unprocessed vegetable and fruit, including oranges, jackfruit, tomatoes, and bananas, among others [13]. Markets and the food processing industry are the sources of these trash. Due to the likelihood of contamination, these materials pose a serious hazard to the ecosystem because they are easily and quickly perishable. In most countries, there is a significant amount of waste produced from different units, including processing, packing, distributing, and consumption. These wastes are produced more frequently each year and are disposed of improperly and pollutingly by dumping. [14]. Due to the high nutritional value of these agro-industrial waste materials, inappropriate handling or disposal of these materials may cause environmental contamination, including the discharge of solid waste, air pollution, and waste effluent [15]. For enhancing agricultural output, both developing and developed countries are heavily reliant on and in need of these fertilisers and pesticides. However, using these goods with chemical formulations causes the soil's fertility and properties to decline. Water pollution is a side effect of using these insecticides and fertilisers with chemicals. In order to prevent these environmental issues, it is possible to recycle and transform agro-industrial waste into bioenergy in the form of biofertilizers and biopesticides.[16]

5. Pre-treatment of Lignocellulosic Biomass

Cellulose, lignocellulose, and lignin are the main components of lignocellulosic biomass. Amounts between 35 and 50 percent, 20 to 35 percent, 15-20 percent, and 15-20 percent, respectively, of cellulose, hemicellulose, lignin, and other components can be found in lignocellulose [17]. The pre-treatment process (Fig. 7.2) breaks down complex lignocellulose into simpler elements like cellulose, hemicellulose, and lignin. lignin will eventually be eliminated, whereas hemicellulose will be preserved, cellulose crystallinity will be decreased, and material porosity will be increased. A cost-effective pre-treatment procedure must lessen the degradation of the carbohydrates and the production of inhibitors for hydrolysis and fermentation [18]. It also needs to increase the creation of sugars in the succeeding phase of enzymatic hydrolysis. An efficient pre-treatment procedure's primary goals are: Direct sugar formation through hydrolysis, prevention of loss or degradation of sugars generated, restriction of the creation of inhibitory compounds, reduction of energy requirements and minimization of the cost of biofuel production.

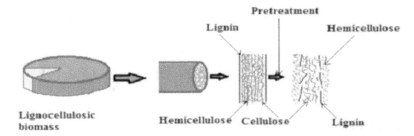

Fig. 7.2 Schematic representation of pretreatment of lignocellulosic biomass [19]

6. Methods of Pre-treatment of Lignocellulosic Biomass

6.1 Physical Pre-treatment

The lignocellulosic waste is pre-treated using a variety of physical pre-treatment techniques, including milling, chipping, grinding, freezing, and radiation. With these techniques, lignocellulosic materials' surface area and particle size are increased. Some of these pre-treatment techniques are combined with others because they are ineffective when used alone. In order to lower the size of lignocellulosic biomass, size reduction is obviously the first stage in the pre-treatment process. However, in order to prevent the production of inhibitors, size reduction should not be carried out to an excessively high or low level. Instead of using a ball mill or a basic mill to reduce the particle size, beating uses more energy [20]. The kind and makeup of the biomass being utilised to produce biofuel completely determines which pre-treatment techniques are chosen. Extrusions(EXT) is hailed as the greatest physical pre-treatment method among those discussed here because it combines mixing, heating, and shearing to liberate a significant amount of simple sugars, which are subsequently subjected to maximal biofuel production. For the pre-treatment of biomass, EXT is a simple and affordable procedure. If carried out at the proper temperature in accordance with the type of biomass, Microwave (MW) would also be superior to conventional heating [21]. Pyrolysis could be helpful for the manufacture of gaseous biofuels, but its application is restricted in the production of liquid biofuels. Pyrolysis has not been widely used because of its high operating costs.

6.2 Chemical Pre-treatment

Due to the high levels of cellulose and hemicellulose solubilization and lignin removal in Acid Pre-treatment (Acd) and Alkaline pre-treatments, lignocellulosic biomass was more frequently treated using these pre-treatment techniques than any other. Due to the usage of expensive ionic and organic liquids and the impossibility of recovering these solvents, Ionic liquids and Organosolv pre-treatment have substantial operational expenses. Due to their high running costs, Ozonisation and Oxidative pre-treatment are likewise only occasionally used. Acd can be stated to work best when biomass is pre-treated with diluted acid at a high temperature, despite the fact that each of these pre-treatment techniques has advantages and disadvantages of their own [22].

6.3 Physico-chemical Pre-treatment

Due to the lack of a chemical or catalyst requirement, pH regulated Liquid hot water (LHW) pre-treatment can be said to be the best. In addition to the physical pre-treatment process, additional pre-treatment procedures also included chemicals including NH_3, CO_2, acid, alkali, and catalysts. Compared to Steam explosion, LHW requires less pressure, and it doesn't operate with expensive corrosion-resistant containers. At neutral pH, LHW exhibits the least amount of inhibitor generation.

6.4 Biological Pre-treatment

Microbial Consortium is the best pre-treatment method out of the three ones discussed here because it uses a variety of microorganisms, lowering the risk of contamination and maintenance costs. Because extra attention should be taken during pre-treatment, Enzymatic pre-treatment have significant operating costs. As part of the pre-treatment process for fungi, great care must be taken to avoid cross-contamination across fungi species, which could reduce the quantity of biofuel produced.

6.5 Combined Pre-treatment

Over the past few years, a variety of combinations have been used to pre-treat lignocellulosic biomass. It was discovered that combining two or more pre-treatment techniques was more effective than using just one. The optimal pre-treatment method is Combined enzyme hydrolysis and superfine grinding with steam explosion (EH-SFG-SExp) because it uses less energy than ordinary grinding while lowering particle size, increasing the yield of reducing sugars, improving reactive surface, and doing so [23]. High volumes of xylose can be produced with little to no inhibitor synthesis using Dilute acid-SExp. At 190 °C, DA-MW was shown to be effective, and by applying Bio-SExp and reducing the amount of helocellulose, pre-treatment time was cut in half. Due to the shorter pre-treatment time and lower operating costs, MW in combination with acid or alkali appears to be more effective than conventional heating processes.

Table 7.2 Pretreatments of lignocellulosic wastes used for bio-hydrogen production [24]

Wastes	Pre-treatment used	Bio-hydrogen yield
Food wastes	Ultra-sonication	23% increase in mol/H_2 hexose
Food wastes	Mesophilic heat pre-treatment	55mL/g
Vegetable wastes	Heat pre-treatment at 50-100 °C	98.8% H2 of total gas
Rice straw	Acid and enzyme pre-treatment	771 mL/L
Corn stover	Co-culture of Clostridium cellulolyticum and Citrobacter amalonaticus	51.9 L H2/kg TS
Food wastes	Ultra sonication (US) and alkaline (Alk.) pretreatment	1192 mL/g, 97 mL/g and 46 mL/g for single US and Alk. & 118 mL/g for combined
Food waste, olive mill wastewater (OMWW)	pH between 4.5 and 7.0 with waste and food to micro-organism (F/M) ratio	60.69.0 mL H2/g VS at pH 4.5 and 50.7±0.8 mL H2/g VS at pH 5.0

7. Conclusion

Growing agricultural waste production, sanitization issues, and hazardous gas emissions from decomposing materials are all results of rising global population, agricultural activity, and food processing. Through bioengineering techniques for waste minimization, conversion, and utilization, this work has developed ecologically acceptable and sustainable ways for efficient agricultural waste management. The creation of bioethanol, biomethane, biohydrogen, biobutanol, biomethanol, and biodiesel from agricultural wastes for use in transportation and energy production has also been demonstrated. Agricultural waste can be converted into biofuels, which has advantages in terms of the environment, health, economy, industry, and technology. Using the biofuels in transportation engines also reduces greenhouse gas emissions, improves engine performance, facilitates smoother engine operation, and extends engine life.

By transforming these waste materials into usable forms, the problems caused by the enormous amount of waste produced by agricultural practices and crop leftovers can be lessened. The creation of biofuels from this trash is still an economically and environmentally sound solution. To raise the conversion rate, lower energy consumption, and produce more goods of higher quality, these wastes must first be processed. To improve the digestibility and biodegradability of the biomass, the pre-treatment procedure breaks down the cellulose, hemicellulose, and lignin components of the biomass.

8. Future Perspectives

Going forward, all levels of government should educate farmers about the advantages of turning their wastes into marketable goods. In addition to lowering soil fertility, burning agricultural trash releases smoke that worsens the environment. Waste management methods that worsen deterioration of soil and water quality, erosion, and environmental pollution as well as harm to human health should be abandoned. To lessen the effect of cultural and religious beliefs that inhibit trash conversion and utilization, more understanding is required. Small and medium-sized businesses that are interested in investing in the waste management and conversion sector should be given incentives and tax holidays. Having access to affordable feedstock for biofuel conversion will be guaranteed by a continuous and sustainable supply of agricultural waste. To ensure environmental sustainability, encourage the use of renewable fuels, and stop the impending environmental catastrophe, more financial and human resources must be put into the conversion of agricultural waste into renewable biofuels.

REFERENCES

1. Ardolino F, Parrillo F, Arena U. Biowaste-to-biomethane or biowaste-to-energy? An LCA study on anaerobic digestion of organic waste. J Clean Prod 2018;174: 462–76.
2. Wijekoon P, Koliyabandara PA, Cooray AT, Lam SS, Athapattu BCL, Vithanage M. Progress and prospects in mitigation of landfill leachate pollution: Risk, pollution potential, treatment and challenges. J Hazard Mater 2022;421:126627. https:// doi.org/10.1016/j.jhazmat.2021.126627.
3. Mihai FC, Ingrao C. Assessment of biowaste losses through unsound Waste Manage practices in rural areas and the role of home composting. J Clean Prod 2018;172:1631–8.

4. Abdelghaffar F. Biosorption of anionic dye using nanocomposite derived from chitosan and silver Nanoparticles synthesized via cellulosic banana peel biowaste. Environ Technol Innov 2021;24:101852. https://doi.org/10.1016/j. eti.2021.101852.
5. Khoo KS, Chia WY, Chew KW, Show PL. Microalgal-bacterial consortia as future prospect in wastewater bioremediation, environmental management and bioenergy production. Indian J Microbiol 2021;61(3):262–9. https://doi.org/ 10.1007/s12088-021-00924-8.
6. Husgafvel R, Linkosalmi L, Hughes M, Kanerva J, Dahl O. Forest sector circular economy development in Finland: a regional study on sustainability driven competitive advantage and an assessment of the potential for cascading recovered solid wood. J Clean Prod 2018;181:483–97.
7. Awasthi MK, Sarsaiya S, Wainaina S, Rajendran K, Awasthi SK, Liu T, et al. Techno-economics and life-cycle assessment of biological and thermochemical treatment of bio-waste. Renew Sust Energ Rev 2021;144:110837. https://doi. org/10.1016/j.rser.2021.110837.
8. Karp A, Shield I. Bioenergy from plants and the sustainable yield challenge. New Phytol 2008;179:15–32.
9. Liu Yang, Lyu Yizheng, Tian Jinping. Review of waste biorefinery development towards a circular economy: From the perspective of a life cycle assessment.
10. Mohan SV, Butti SK, Amulya K, Dahiya S, Modestra JA. Waste biorefinery: a new paradigm for a sustainable bioelectro economy. Trends Biotechnol 2016;34:852–5.
11. Prasad, S., Singh, A., Korres, N.E., Rathore, D., Sevda, S., Pant, D., 2020. Sustainable utilization of crop residues for energy generation: A life cycle assessment (LCA) perspective. Bioresour. Technol. 303, 122964. https://doi.org/10.1016/j. biortech.2020.122964.
12. Tajmirriahi, Mina, Momayez, Forough, Karimi, Keikhosro, 2021. The critical impact of rice straw extractives on biogas and bioethanol production. Bioresour. Technol. 319, 124167. https://doi.org/10.1016/j.biortech.2020.124167.
13. Pandey, A., Soccol, C.R, Nigam, P., Soccol, V.T, 2000. Biotechnological potential of agroindustrial residues. I: sugarcane bagasse. Bioresour. Technol. 74 (1), 69–80. https:// doi.org/10.1016/S0960-8524(99)00142-X.
14. Sridhar, Adithya, Kapoor, Ashish, Senthil Kumar, Ponnusamy, Ponnuchamy, Muthamilselvi, Balasubramanian, Sivasamy, Prabhakar, Sivaraman, 2021. Conversion of food waste to energy: A focus on sustainability and life cycle assessment. Fuel. 302, 121069. https://doi.org/10.1016/j.fuel.2021.121069.
15. Sadh, Pardeep Kumar, Duhan, Surekha, Duhan, Joginder Singh, 2018. Agro-industrial wastes and their utilization using solid state fermentation: a review. Bioresour. Bioprocess. 5 (1) https://doi.org/10.1186/s40643-017-0187-z.
16. Ugwu, S.N., Enweremadu, C.C., 2020. Ranking of energy potentials of agro-industrial wastes: Bioconversion and thermo-conversion approach. Energy Rep. 6, 2794–2802. https://doi.org/10.1016/j.egyr.2020.10.008.
17. Mood SH, Golfeshan AH, Tabatabaei M, Jouzani GS, Najafi GH, Gholami M, et al. Lignocellulosic biomass to bioethanol, a comprehensive review with a focus on pretreatment. Renew Sustain Energy Rev 2013;27:77–93.
18. Chiaramonti D, Prussi M, Ferrero S, Oriani L, Ottonello P, Torre P, et al. Review of pretreatment processes for lignocellulosic ethanol production, and development of an innovative method. Biomass Bioenergy 2012;46:25–35.
19. Dolly Kumari, Radhika Singh, Pretreatment of lignocellulosic wastes for biofuel production: A critical review,Renewable and Sustainable Energy Reviews, Volume 90,2018,Pages 877-891, https://doi.org/10.1016/j.rser.2018.03.111.

20. Montingelli ME, Benyounis KY, Stokes JAG. Pretreatment of macroalgal biomass for biogas production. Energy Convers Manag 2016;108:202–9.
21. Li L, Kong X, Yang F, Li D, Yuan Z, Sun Y. Biogas production potential and kinetics of microwave and conventional thermal pretreatment of grass. Appl Biochem Biotechnol 2012;166:1183–91.
22. Xue B, Paul AL, Paul DJ, Sergi A, Steven P. Enhanced methane production from algal digestion using free nitrous acid pre-treatment. Renew Energy 2016;88:383–90.
23. Jin S, Chen H. Superfine grinding of steam-exploded rice straw and its enzymatic hydrolysis. Biochem Eng J 2006;30:225–30.
24. Dolly Kumari, Radhika Singh. Pretreatment of lignocellulosic wastes for biofuel production: A critical review, Renewable and Sustainable Energy Reviews. Volume 90, 2018, Pages 877–891.

Prospects of Science, Technology and Applications – Prof. Renu Sharma (eds)
© 2024 Taylor & Francis Group, London, ISBN 978-1-032-78833-3

A Review on, Recent Developments in Solid Adsorbents for CO₂ Capture

Biswal Shaswata*, Das Mira
Department of chemistry, ITER, Siksha 'O'Anusandhan (Deemed to be university)

Abstract The steady increase in atmospheric carbon dioxide gas (CO_2) levels is a major global concern. Rapid rises in the concentration of atmospheric carbon dioxide (CO_2) may have an impact on climate condition. In this respect, carbon capture and storage (CCS) is acknowledged as having the potential to significantly contribute to the reduction of CO_2 in industrial sectors. Several companies have already adopted the CO_2 capture method based on chemical separation. However, the procedures require a lot of energy and are highly expensive. As an alternative, solid adsorbents can be employed to efficiently capture CO_2. Since it can achieve the goal of a minimal energy penalty and requires few adjustments to power plants, solid adsorption is now regarded as the most practical and least disruptive method of CO_2 capture. This article provides a substantial number of researches that attempt to uptake or minimize CO_2 using various porous classes of materials in an effort to address this widespread problem. Here the adsorbent materials are categorized into four major classes: (I) carbon based porous materials, (II) metal organic frameworks(MOFs), (III) Synthetic metal oxides and (IV) Amine functionalized adsorbent materials. The study also covers the methods for improving the adsorption capacity and effectiveness of these materials. It also highlights the latest advances, possibilities and future research directions and strategies for these adsorbent materials.

Keywords CO_2 adsorption, Carbon capture, Porous materials, Metal-organic frameworks, Solid adsorbent

1. Introduction

One of the most important concerns we are currently facing, is the significant increase in CO_2 in the atmosphere brought on by various natural and anthropogenic activities. Unquestionably,

---
*Corresponding author: shaswatabiswal7@gmail.com

DOI: 10.1201/9781003489443-8

human activity is the primary cause of the high CO_2 concentration in the atmosphere that results in global warming. According to Rochelle,[1] burning fossil fuels accounts for more than 85% of global energy demand, which causes significant CO_2 emissions into the atmosphere. . US EPA states that different industrial processes contribute 24%, and power plants contribute 25% of the global CO_2 emission. As a result, one of the main areas of scientific and engineering research has been on decarbonization technology. Currently several Carbon dioxide capture and storage (CCS) methodologies are introduced to reduce CO_2 emissions effectively by synthesizing suitable adsorbent and storage materials. CO_2 uptake processes, depending on the type of CO_2 production is divided into three types: (1) pre combustion process, (2) oxy fuel combustion and (3) post combustion process [2]. The most advantageous method of capture is the post combustion process, which is compatible with the infrastructure of current power plants. As of now, most industrial sectors use liquid amine-based absorption technique to remove CO_2 from the flue gas after combustion [3]. One of the most researched amines, monoethanolamine (MEA), has been utilised for more than 60 years in the natural gas industry to absorb CO_2. But these, amine solutions have a number of drawbacks. In terms of heating, the solution is rather unstable. As a result, it restricts the temperature at which they can fully renew CO_2 from amine. The nature of amine solutions is also corrosive. In addition to having significantly lower heat capacities than amine solvents, solid adsorbents are superior to liquid amine adsorbents because they require less energy for regeneration and have lower CO_2 partial pressure [4]. In recent years, solid adsorbent-based CO_2 capture technology has been identified as one of the potential scientific techniques for mitigating greenhouse gas emission [5]. Due to their extraordinarily large surface, high CO_2 uptake capacity, and chemical as well as structural diversity, the new class of next generation solid adsorbents, which includes synthetic carbon-based materials, metal organic frame works, synthetic amine incorporated mesoporous composite materials, and several synthetic solid oxide adsorbents, are considered as the perfect platform for CO_2 capture in industrial scales [6]. To develop porous composite materials, numerous studies have been done in the field of material chemistry [7]. However, to assess how well these materials perform in the industrial scale, there are, only a relatively small number of studies that have been conducted.

The purpose of this research is to review and analyse the main adsorption-based CO_2 capture strategies in order to determine which method is most efficient in terms of cost, energy efficiency, environmental friendliness, selectivity, and may be applied practically on an industrial scale.

2. Materials for CO_2 Capture

To achieve the best efficiency for CO_2 uptake, various materials have been developed based on the adsorption conditions. Some of the most studied materials are reviewed here (Fig. 8.1):

2.1 Carbon-based Materials

Carbon-based compounds are regarded to be one of the most promising CO_2 adsorption materials. It is due to their great surface area, cheap price, high temperature stability, stronger inclination to modify pore structure, and relative ease of functionalization and regeneration. Thy major carbon-based compounds used for CO_2 adsorption are activated carbons (ACs),

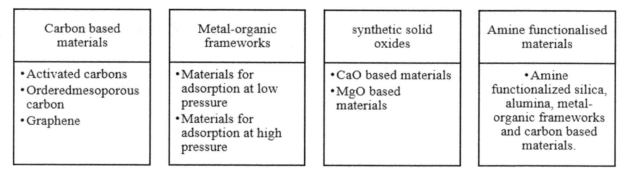

Fig. 8.1 Classification of adsorbent materials [5]

ordered mesoporous compounds (OMCs), and graphene. Following table summarizes the CO_2 capture capacity of various carbon-based materials.

Table 8.1 An overview of the CO_2 capture by carbon-based materials

Composites	Temp. (°C)	Pressure(bar)	CO_2 uptake capacity (mmol/gm)	Reference
AC-MgO	180	1	0.74	[8]
AC-NiO	25	1	2.23	[9]
AC-CaO	25	1	1.79	[10]
AC-CuO	25	1	6.72	[11]
AC-zeolite X	20	1	3.50	[12]
AC-NaUSY	25	1	3.21	[13]
AC-Zeolite 13X	30	0.05	5.31	[14]

2.2 Metal-organic Frameworks

Recently, metal organic frameworks have attracted a lot of interest due to their desirable and distinctive characteristics, such as their flexibility, high surface area, porous structure, functionalizable organic linkers, and metal centres, which have resulted in a wide range of promising applications for CO_2 capture. These properties can be enhanced further by functionalizing the organic linkers, post synthesis modification and incorporating nanomaterials into their frameworks. The heat of adsorption (Q_{st}), a crucial parameter that impacts the adsorption selectivity and the amount of energy required for CO_2 regeneration. The value of Q_{st} ranges from 20-50 KJ/mole. Table 8.2 summarizes the results towards CO_2 adsorption of different MOFs. Another highly effective and popular CCS technique is membrane separation. Membranes can effectively separate CO_2 from $CO_2/CH_4/N_2$ mixture. MOFs are commonly utilized as filler agents, to adding functionality in MMMs, in order to enhance the separation efficiency. Following figure illustrates the CO_2 capture process using a MOF based matrix membrane (MMM). The diagram shows the flow of a feed gas stream containing CO_2 and CH_4 across a MOF based mixed matrix membrane surface with selective permeability to the CO_2 gas over CH_4 gas. As a result of the solution-diffusion mechanism, CO_2 gas molecules diffuse through the membrane's pores to form permeate. Additionally, the retentate is formed by the

CH₄ gas molecule that did not diffuse through the membrane pores. More CO_2 molecules can diffuse through the membrane's holes as a result of the MOF filler's improved selectivity and permeability.

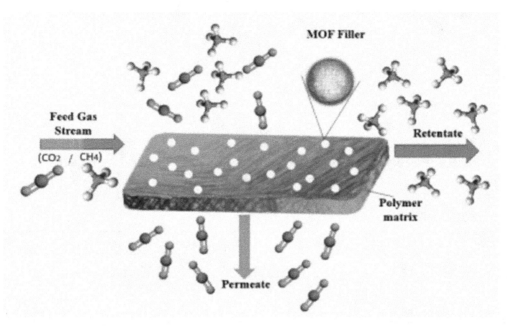

Fig. 8.2 CO_2 adsorption on MOF based mixed matrix membrane[15]

Table 8.2 CO_2 capture efficiency of different MOFs

MOF abbreviation	BET surface area(m/g)	Pressure (bar)	Temperature (K)	CO_2 capture capacity (mmol/g)	-Qst (KJ/mol)	Reference
HNUST-3	2412	20	298	22.4	24.8	[16]
NJU-Bai3	2690	20	273	22.1	36.5	[17]
NOTT-125	2447	20	273	21.2	25.3	[18]
MIL-100	1900	50	304	18	62	[19]
SNU-70	5290	1	195	50	17.2	[20]
NOTT-202a	2220	1	195	20	25	[21]
Mg-MOF-74	1495	1	296	8.0	25	[22]
CPM-33b	808	1	273	7.8	25	[23]

2.3 Synthetic Metal Oxides

Several metal oxides, particularly those made of group 1 and group 2 metals, can be used directly as adsorbents to absorb gaseous CO_2, similar to the various MOFs mentioned above. The metal oxides uptake CO_2 to form corresponding carbonates, and then the carbonates undergo decomposition to reproduce virtually pure CO_2 and the adsorbent. This reversible process can be generalised as follows:

$$M_xO + CO_2 \rightleftarrows M_xCO_3$$

The metal oxides are chosen in such a way that the final carbonised material is thermodynamically stable. In general, a suitable solid oxide material should have the following properties: (1) high uptake and release kinetics; (2) be capable of reversible carbonation at temperatures that closely resemble those of the upstream and downstream processes; (3) display consistent performance across a large number of CO_2 capture cycles; (4) have enough physical hardness to sustain ongoing degradation in circulating fluidized beds; and (5) be able to be produced in large quantities using relatively simple manufacturing processes.

In comparison to other oxides, CaO and MgO both have displayed tremendous action in this regard by generating insoluble carbonates during CO_2 fixation. Operating conditions have an impact on performance of metal oxide adsorbents. Additionally, experimental findings imply that operating in a steam environment increases the sorbents' porosity, which enhances their CO_2 capture efficiency in terms of reaction kinetics. The problem of adsorbent deactivation can be avoided by reactivating spent sorbents. Combining CaO with inert refractory materials (such as Al_2O_3, ZrO_2) to increase sintering resistance and physical hardness can significantly improve the cyclic performance of CaO adsorbents. Similarly, alkali metal nitrates have gained attention recently as a way to boost the CO_2 capture efficiency of MgO-based adsorbents. It is as a result of their melting during the CO_2 capture process, which prevents the solid $MgCO_3$ layer from forming on the surface of MgO. The two most widely used nitrate promoters are $NaNO_3$ and KNO_3. The role of the molten nitrate in promoting MgO carbonation is shown in Fig. 8.3.

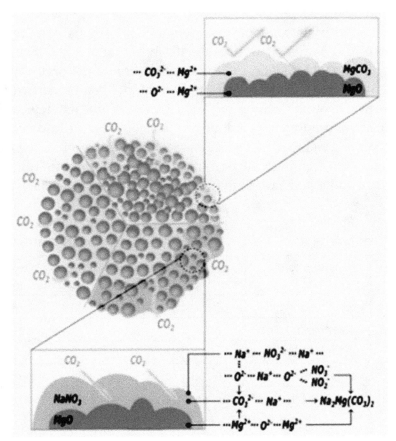

Fig. 8.3 Nitrate doped MgO adsorbent for enhancing CO_2 capture [24]

The CO_2 uptake capacity of various CaO and MgO based materials are listed in the following table.

Table 8.3 CO_2 capture tendencies of various synthetic metal oxides

Name of the adsorbent	Support	Carbonation Temp. [°C]	Calcination Temp. [°C]	Carbonation time[min]	Calcination time[min]	CO_2 uptake capacity [wt%]	Ref.
S20-4x	Zr(NO$_3$)$_2$·6H$_2$O	675	850	20	10	41.8	[25]
CaO-(s-s)	ZrO$_2$	650	780	120	60	57.7	[26]
C74D26	Dolomite	700	850	20	10	54.0	[27]
MgO–NaNO$_3$	NaNO$_3$	325	400	30	20	83.2	[28]
MgO–K$_2$CO$_3$	K$_2$CO$_3$	375	400	20	30	8.6	[29]
MgO–KNO$_3$	KNO$_3$	325	450	20	30	10.2	[30]
5A5M	Al(NO$_3$)$_3$9H2O	200	600	60	60	13.1	[31]

2.4 Amine Functionalized Materials

The CO_2 capture tendency of all the solid adsorbent materials, which were discussed in previous sections, can be enhanced by introducing amine functional groups into their pore structure to form a composite solid material. It is because the basic nature of amine groups facilitates the interaction between weakly acidic CO_2 and the adsorbent material. The incorporation of amine functional groups into the porous adsorbent material can be carried out by employing two methods (1) direct functionalization method (2) post synthesis method. In post synthesis method amine can be incorporated by either physical impregnation process or by chemical grafting. In recent years a number of research works regarding amine-based adsorbents like amine functionalized silica, amine functionalized alumina, amine functionalized MOFs, amine functionalized zeolite and carbon materials were carried out, which are summarized in following tables.

Table 8.4 CO_2 capture tendencies of various amine functionalized mesoporous alumina

Materials	Amine type	Temp. (°C)	Pressure (bar)	CO_2 adsorption capacity (mmol/g)	Ref.
Mesoporous alumina	polyethyleneimine	75	1.0	2.73	[32]
γ-alumina	polyethyleneimine	25	0.1	1.41	[33]
γ-alumina	polyethyleneimine	25	1.0	1.03	[33]
γ-alumina	Diethanolamine	40	0.1	0.68	[34]

Table 8.5 CO$_2$ capture by amine silica composites

Materials	Amine used	Temp. in °C	CO$_2$ partial pressure in atm	CO$_2$ adsorption Capacity in mg/gm	Ref.
MCM-41	3-Aminooropyltriethoxysilane	20	1	47	[35]
MCM-48	3-Aminooropyltriethoxysilane	25	0.05	50	[36]
SBA-15	Poly-(L-lysine)	25	0.0004	26	[37]
KIT-6	Polyethylenimine	75	1	135	[38]
Mesostructured monolith	Tetraethylenepentamine	75	1	171	[39]
AS-synHMS	Dodecyl amine	25	1	34	[40]

Table 8.6 CO$_2$ capture by amine functionalized MOFs

Name of the adsorbent	Amine used	T in °C	CO$_2$ partial pressure in atm	CO$_2$ uptake tendency in mg/g	Ref.
ZTF1	5 aminotetrazole	0	1.0	246	[41]
CPF-13	3,5-diamino-1,2,4-triazole	25	1.0	159	[42]
Bio-MOF-11	Adenine	0	1.0	264	[43]
PPN-6 CH$_2$ DETA	Diethylenetriamine	22	1.0	189	[44]
DETA MIL 101 (Cr)	Diethylenetriamine	0	1.0	79	[45]

Table 8.7 CO$_2$ capture by amine functionalized zeolites and carbon materials

Name of the materials	Amine type	Temp. (°C)	CO$_2$ partial pressure (atm)	CO$_2$ adsorption capacity (mg/g)	Ref.
Zeolite 13x	Monoethanolamine	75	0.15	36	[46]
Zeolite 13x	Isopropanol amine	75	0.15	23	[47]
Y type zeolite	Tetra ethylene pentaamine	60	0.15	113	[48]
B zeolite	Monoethanolamine	30	1.0	61	[49]
Multi wall CNT	3 aminopropyl triethoxysilane	20	0.15	55	[50]

3. Conclusion

An imbalance in the earth's atmosphere and an increase in global temperature are the results of excessive CO$_2$ emissions from anthropogenic sources. This perilous situation has compelled scientists to discover technologies for sequestering or removing CO$_2$ from the atmosphere .The CCS methodology uses a variety of techniques to sequester or trap high levels of greenhouse gases like CO$_2$. Numerous solid materials have been intensively studied for applications in CO_2 collection, including activated carbons, zeolites, metal oxides, silica materials, and metal-organic frameworks. The CO_2 adsorptive properties of the discussed materials can be further improved by modification techniques like amine functionalization and textural property optimization. In recent years amine functionalized carbon materials, metal-organic frameworks, alumina, zeolites and silicates have been extensively studied by a number of researchers at a wide range of temperature and pressure conditions. But unfortunately, as of now not a single adsorbent material can ideally satisfy all the required conditions.

4. Future Perspective

Solid adsorption technology must meet certain criteria including high thermal stability, moisture resistance, and affordability in order to compete with the currently applied amine scrubbing technique. However, the concern of CO_2 capture has no single solution at this time. So, it is worth noting that a numerous studies are still required to investigate new, inexpensive adsorbents and technologies in order to ensure that the capture process operates consistently and that the adsorbents can be recycled after the process. Multiple technological solutions must be integrated to tackle this problem. The future of the process relies on collaboration and interaction between academia and industry from researchers, engineers, and policymakers. Moreover, in order to scale up the operation and deal with the problems related to the issue, engineers and scientists must collaborate and additionally, the engineering economics of the materials must be analysed. It should be tried to create environmentally friendly adsorbent materials from sources like agricultural waste, industrial food and beverage waste, biomass, etc. And to further increase their capacity through effective activation processes, a vast number of research should be carried out.

REFERENCES

1. Rochelle. G.T, Amine scrubbing for CO_2 capture, Science 325, 1652–1654, (2009).
2. Yang, H.; Xu, Z.; Fan, M.; Gupta, R.; Slimane, R. B.; Bland, A. E.; Wright, Progress in Carbon Dioxide Separation and Capture: A Review. J Environ Sci.20(1):14- 27.(2008).
3. E.O. Agbonghae, K.J. Hughes, D.B. Ingham, L. Ma, M. Pourkashanian. Ind. Eng. Chem. Res. 53 (38) 14815–14829 (2014).
4. Kundu N, sarkar S , Porous organic frameworks for carbon dioxide capture and storage. Journal of Environmental Chemical Engineering 9 (2021) .
5. Pardakhti M, Jafari T, Tobin Z, Dutta B, Moharrer Ei, Nikoo S, Suib S, and Srivastava R. Trends in Solid Adsorbent Materials Development for CO_2 Capture. ACS Appl Mater Interfaces 11(38):34533- 34559(2019).
6. K. Sumida, D.L. Rogow, J.A. Mason, T.M. McDonald, E.D. Bloch, Z.R. Herm, T. Bae, J.R. Long, Chem. Rev. 112, 724–781(2012).
7. C. Chen, N. Feng, Q. Guo, Z. Li, X. Li, J. Ding, L. Wang, H. Wan, G. Guan, layer-packed structure and enhanced performance for CO_2 capture, J. Colloid Interface Sci. 513 , 891–902, (2018).
8. Liu Z and Green WH, Experimental investigation of sorbent for warm CO_2 capture by pressure swing adsorption. Ind Eng. Chem Res 52(28):9665–9673 (2013).
9. Jang DI and Park SJ, Influence of nickel oxide on carbon dioxide adsorption behaviours of activated carbons. Fuel 102:439–444 (2012).

10. Song HK and Lee KH, Adsorption of carbon dioxide on chemically modified carbon adsorbents. Sep Sci Techno 33(13):2039–2057 (1998).
11. Boruban C and Esenturk EN, Activated carbon-supported CuO nanoparticles: a hybrid material for carbon dioxide adsorption. J Nanopart Res 20(3): 1–9 (2018).
12. Lee J, Kim J, Kim J, Suh J, Lee J and Lee C, Adsorption equilibria of CO_2 on zeolite 13X and zeolite X/activated carbon composite. J Chem Eng. Data 47(5): 1237–1242 (2002).
13. Zhao Q, Wu F, Xie K, Singh R, Zhao J, Xiao P et al., Synthesis of a novel hybrid adsorbent which combines activated carbon and zeolite NaUSY for CO_2 capture by electric swing adsorption (ESA). Chem Eng. J 336:659–668 (2018).
14. Rostami M, Mofarahi M, Karimzadeh R and Abedi D, Preparation and characterization of activated carbon–zeolite composite for gas adsorption separation of CO_2/N2 system. J Chem Eng. Data 61(7):2638–2646 (2016).
15. Salma Ehab Mohamed Elhenawy, Majeda Khraisheh, Fares AlMomani and Gavin Walker. MetalOrganic Frameworks as a Platform for CO2 Capture and Chemical Processes: Adsorption, Membrane Separation, Catalytic-Conversion, and Electrochemical Reduction of CO2. Catalysis,10(11) (2020).
16. Tan C., Yang S., Champness N.R., Lin X., Blake A.J., Lewis W., and Schröder M.: High-capacity gas storage by a 4,8-connected metal-organic polyhedral framework. Chem. Commune. 47, 4487 (2011).
17. Duan J., Yang Z., Bai J., Zheng B., Li Y., and Li S.: Highly selective CO_2 capture of an agw-type metalorganic framework with inserted amides: Experimental and theoretical studies. Chem. Commun. 48, 3058 (2012).
18. Alsmail N.H., Suyetin M., Yan Y., Cabot R., Krap C.P., Lü J., Easun T.L., Bichoutskaia E., Lewis W., Blake A.J., and Schröder M.: Analysis of high and selective uptake of CO_2 in an oxamide-containing {Cu2(OOCR)4}-based metal-organic framework. Chem. Eur. J. 20, 7317 (2014).
19. Llewellyn P.L., Bourrelly S., Serre C., Vimont A., Daturi M., Hamon L., De Weireld G., Chang J., Hong D., Hwang Y.K., and Jhung S.H.: High uptakes of CO_2 and CH4 in mesoporous metals organic frameworks MIL-100 and MIL-101. Langmuir 24, 7245 (2008).
20. Prasad T.K. and Suh M.P.: Control of interpenetration and gas-sorption properties of metalorganic frameworks by a simple change in ligand design. Chem. Eur. J. 18, 8673 (2012).
21. Lee Y.G., Moon H.R., Cheon Y.E., and Suh M.P.: A comparison of the H2 sorption capacities of isostructural metal-organic frameworks with and without accessible metal sites: [{Zn2(abtc) (dmf)2}3] and [{Cu2(abtc) (dmf)2}3] versus [{Cu2(abtc)}3]. Angew. Chem. Int. Ed. 47, 7741 (2008).
22. Caskey S.R., Wong-Foy A.G., and Matzger A.J.: Dramatic tuning of carbon dioxide uptake via metal substitution in a coordination polymer with cylindrical pores. J. Am. Chem. Soc. 130, 10870 (2008).
23. Zhao X., Bu X., Zhai Q.G., Tran H., and Feng P.: Pore space partition by symmetry-matching regulated ligand insertion and dramatic tuning on carbon dioxide uptake. J. Am. Chem. Soc. 137, 1396 (2015).
24. Chang. R, Wu. X, Cheung. O, and Liu. Wen: Synthetic solid oxide sorbents for CO2 capture: state-of-the art and future perspectives, J. Mater. Chem. A, 10, 1682–1705, (2020).
25. S. M. Hashemi, D. Karami and N. Mahinpey, Solution combustion synthesis of zirconia-stabilized calcium oxide sorbents for CO2 capture, Fuel, 269, 117432,(2020).
26. H. J. Yoon and K. B. Lee, Introduction of chemically bonded zirconium oxide in CaO-based high-temperature CO_2 sorbents for enhanced cyclic sorption, Chem. Eng. J.355, 850–857 (2019).

27. X. Yan, Y. Li, X. Ma, J. Zhao, Z. Wang and H. Liu, CO$_2$ capture by a novel CaO/MgO sorbent fabricated from industrial waste and dolomite under calcium looping conditions, New J. Chem.43, 5116–5125(2019).
28. X. Zhao, G. Ji, W. Liu, X. He, E. J. Anthony and M. Zhao, Mesoporous MgO promoted with NaNO$_3$/NaNO$_2$ for rapid and high-capacity CO$_2$ capture at moderate temperatures, Chem. Eng. J.332, 216–226(2018).
29. G. Xiao, R. Singh, A. Chaffee and P. Webley, Corrigendum to "CO2 capture by adsorption: Materials and process development, Int. J. Greenhouse Gas Control 5, 634–639 (2011).
30. A.-T. Vu, Y. Park, P. R. Jeon and C.-H. Lee, Mesoporous MgO sorbent promoted with KNO$_3$ for CO$_2$ capture at intermediate temperatures, Chem. Eng. J.258, 254–264(2014).
31. K. K. Han, Y. Zhou, Y. Chun and J. H. Zhu, Efficient MgO-based mesoporous CO$_2$ trapper and its performance at high temperature, J. Hazard. Mater.203–204, 341–347(2012).
32. Chen, C., Ahn, W.S.: CO$_2$ capture using mesoporous alumina prepared by a sol-gel process.Chem. Eng. J. 166, 646–651 (2011).
33. Chaikittisilp, W., Kim, H., Jones, C.W. Mesoporous Alumina-Supported Amines as Potential Steam-Stable Adsorbents for Capturing CO$_2$ from Simulated Flue Gas and Ambient Air, Energy Fuels 25, 5528–5537 (2011).
34. Castellazzi, P., Notaro, M., Busca, G., Finocchio, CO$_2$ capture by functionalized alumina sorbents: DiEthano lAmine on γ-alumina, E. Microporous Mesoporous Mater. 226, 444– 453 (2016).
35. X. Xu, C. Song, B. G. Miller and A. W. Scaroni, Influence of Moisture on CO$_2$ Separation from Gas Mixture by a Nano porous Adsorbent Based on Polyethylenimine-Modified Molecular Sieve MCM-41, Ind. Eng. Chem. Res., 44, 8113 (2005).
36. H. Y. Huang, R. T. Yang, D. Chinn and C. L. Munson, Amine-Grafted MCM-48 and Silica Xerogel as Superior Sorbents for Acidic Gas Removal from Natural Gas, Ind. Eng. Chem. Res., 42, 2427 (2003).
37. W. Chaikittisilp, J. D. Lunn, D. F. Shantz and C. W. Jones, Poly(L-lysine) Brush–Mesoporous Silica Hybrid Material as a Biomolecule-Based Adsorbent for CO$_2$ Capture from Simulated Flue Gas and Air, Chem. Eur. J., 17, 10556 (2011).
38. W. J. Son, J. S. Choi and W. S. Ahn, Adsorptive removal of carbon dioxide using polyethyleneimine-loaded mesoporous silica materials, Micropor. Mesopor. Mater., 113, 31 (2008).
39. J.J. Wen, F.N. Gu, F. Wei, Y. Zhou, W.G. Lin, J. Yang, J.Y. Yang, Y. Wang, Z.G. Zou, J.H. Zhu, One-pot synthesis of the amine-modified *meso*-structured monolith CO$_2$ adsorbent, J. Mater. Chem. 20,2840–2846(2010).
40. C. Chen, W. J. Son, K. S. You, J. W. Ahn and W. S. Ahn, Carbon dioxide capture using amine-impregnated HMS having textural mesoporosity, Chem. Eng. J., 161, 46 (2010).
41. T. Panda, P. Pachfule, Y. Chen, J. Jiang and R. Banerjee, Amino functionalized zeolitic tetrazolate framework (ZTF) with high capacity for storage of carbon dioxide, Chem. Commun., 47, 2011 (2011).
42. Q. G. Zhai, Q. Lin, T. Wu, L. Wang, S. T. Zheng, X. Bu and P. Feng, High CO$_2$ and H$_2$ Uptake in an Anionic Porous Framework with Amino-Decorated Polyhedral Cages, Chem. Mater., 24, 2624 (2012).
43. J. An, S. J. Geib and N. L. Rosi, High and Selective CO$_2$ Uptake in a Cobalt Adeninate Metal–Organic Framework Exhibiting Pyrimidine- and Amino-Decorated Pores, J. Am. Chem. Soc., 132, 38 (2010).
44. W. Lu, J. P. Sculley, D. Yuan, R. Krishna, Z. Wei and H. C. Zhou, Polyamine-Tethered Porous Polymer Networks for Carbon Dioxide Capture from Flue Gas, Angew. Chem. Int. Ed., 51, 7480 (2012).

45. S. N. Kim, S. T. Yang, J. Kim, J. E. Park and W. S. Ahn, Post-synthesis functionalization of MIL-101 using diethylenetriamine: a study on adsorption and catalysis, CrystEng.Comm, 14, 4142 (2012).
46. P. D. Jadhav, R. V. Chatti, R. B. Biniwale, N. K. Labhsetwar, S. Devotta and S. S. Rayalu, Monoethanol Amine Modified Zeolite 13X for CO_2 Adsorption at Different Temperatures, Energy Fuels, 21, 3555 (2007).
47. R. Chatti, A. K. Bansiwal, J. A. Thote, V. Kumar, P. Jadhav, S. K. Lokhande, R. B. Biniwale, N. K. Labhsetwar and S. S. Rayalu, Amine loaded zeolites for carbon dioxide capture, Micropor. Mesopor. Mater., 121, 84 (2009).
48. F. Su, C. Lu, S. C. Kuo and W. Zeng, Adsorption of CO_2 on Amine-Functionalized Y-Type Zeolites, Energy Fuels, 24, 1441 (2010).
49. X. Xu, X. Zhao, L. Sun and X. Liu, Adsorption separation of carbon dioxide, methane and nitrogen on monoethanol amine modified β-zeolite, J. Nat. Gas Chem., 18, 167 (2009).
50. F. Su, C. Lu, W. Cnen, H. Bai and J. F. Hwang, Capture of CO_2 from flue gas via multiwalled carbon nanotubes, Sci. Total Environ., 407, 3017 (2009).

Developing a Machine Learning Model for HAI Risk Prediction Using EHR Data

Sriya Das Mohapatra[1]
Department of Computer Science and Engineering, FET-ITER, Siksha 'O' Anusandhan (Deemed to be) University, Bhubaneswar, India

Naveen Kumar[2]
Department of Computer Science and Information Technology, Siksha 'o' Anusandhan (Deemed to be) University, Bhubaneswar, Odisha, India

Dishant Yadav[3], Swati Sucharita Ray[4], Bharat Jyoti Ranjan Sahu[5],
Department of Computer Science and Engineering, FET-ITER, Siksha 'O' Anusandhan (Deemed to be) University, Bhubaneswar, India

Abstract Hospital-Acquired Infections (HAIs) represent a significant public health concern, posing substantial risks to both patients and healthcare systems. To enhance patient safety and address this pressing issue, our study focuses on the development of a machine learning model leveraging Electronic Health Records (EHR) data. The primary objective is to identify patients at elevated risk of developing HAIs and enable early intervention strategies. Our approach utilizes a data-driven methodology, analyzing EHR data and employing the Support Vector Machine (SVM) algorithm. This research aims to create a predictive model for HAI risk, aiming to enhance early detection and intervention for susceptible patients. Our study harnesses EHR data from patients who have experienced HAIs to construct and train a predictive model using the SVM algorithm. The dataset comprises 300 entries encompassing nine variables: age, heart rate, coma scale, blood reports, White Blood Cell (WBC) count, urine reports, types of admission, chronic diseases, and HAI history. Our SVM-based model demonstrates promising results, achieving an accuracy rate of 89% on the test dataset, with precision, recall, and F1-score values of 75%, 80%, and 77%, respectively. Furthermore, the model exhibits a robust Receiver Operating Characteristic (ROC) (Area Under ROC curve) AUC score of 0.86, indicating its proficiency in distinguishing between positive and negative HAI cases. Utilizing a confusion matrix, the model correctly identifies 37 out of 45 instances of HAI, with 8 instances categorized as false negatives. It also accurately predicts 79 out of 90 cases of non-HAI, with 11 instances identified as false positives. Despite potential limitations stemming from the quality and accuracy of EHR data, this study underscores the potential of machine learning in predicting HAI risk. The utilization of EHR data alongside machine learning algorithms offers a promising avenue for healthcare providers to identify high-risk patients and implement preventative measures, ultimately enhancing patient outcomes and reducing healthcare expenditures.

Keywords   Hospital-acquired infections, Risk prediction, Electronic health records, Machine learning, Healthcare

1. Introduction

Hospital-acquired infections (HAIs) continue to be a serious public health concern, affecting millions of individuals worldwide each year [3][4]. The World Health Organization (WHO) estimates that 10% of hospitalize patients in impoverished countries and 7% of hospitalized patients in affluent nations, respectively, contract at least one infection related to healthcare. HAIs cause roughly 1.7 million infections and 99,000 fatalities annually in the US. According to the Centers for Disease Control and Prevention (CDC), approximately 1 in 31 hospitalized patients will develop an HAI, leading to an estimated 99,000 deaths per year in the United States alone [3] HAIs also impose a substantial economic burden, with estimated costs ranging from $28 billion to $45 billion annually [4]. The most typical HAIs are bloodstream infections, pneumonia, urinary tract infections, and surgical site infections.

In [1], the author discussed a Bayesian network model to predict the risk of hospital-acquired CDI. The model achieved good discrimination between patients who did and did not develop CDI, with an area under the ROC curve of 0.82. The model identified several important risk factors for CDI, including prior hospitalization, antibiotic use, and advanced age. However, the model was limited by the small size of the dataset used to develop and validate it, which may limit its generalizability to other patient populations. Solutions to this limitation include validating the model on larger datasets and incorporating additional data sources such as environmental or microbiome data. Additionally, the authors suggest that the model could be used to identify high-risk patients for targeted interventions to prevent CDI, but further refinement and validation are needed before it can be implemented in clinical practice.

In [2], author proposed a solution to automate the detection of hospital-acquired infections using machine learning methods. The proposed solution used supervised learning with data available in electronic patient records, which are largely unstructured free-text. Three different data representations were explored: bag of words, complex symbolic sequences, and simple parameters by information extraction. The classifiers used in the study were SVM and gradient tree boosting. The results showed that both SVM and gradient tree boosting performed similarly, but the focus was on gradient tree boosting due to its visualization capabilities. The best results were obtained by gradient tree boosting, with an F1-score in the range of 0.82-

0.83, recall in the range of 0.88-0.89, and a precision of 0.78 for all three data representations. However, the study was conducted on a relatively small dataset of 300 hospitalizations from a single hospital in Sweden.

This research paper aims to develop a machine-learning model that can predict the risk of spreading HAIs to patients in hospitals with more accuracy. A dataset of patient and hospital information, including age, gender, medical histories, laboratory records, and hospital practices, used to train the model. By leveraging advanced machine learning algorithms like SVM, the model identify the factors that increase the likelihood of HAI infection and generate personalized risk scores for individual patients. The resulting model has the potential to improve patient safety and reduce the incidence of HAIs in hospitals by enabling targeted prevention measures and early interventions.

2. Methodology

In this section, we present the machine learning algorithms considered for the classification of HAI cases. The choice of algorithm is crucial in determining the predictive accuracy and interpretability of the model. We briefly describe three commonly used algorithms before discussing our chosen approach.

1. **Logistic Regression:** Logistic regression is a widely-used algorithm for binary classification tasks. It models the relationship between the dependent variable (HAI occurrence) and independent variables (features) using a logistic function. It is known for its interpretability, making it useful for understanding feature importance.

2. **Random Forest:** Random forests are an ensemble learning technique that combines multiple decision trees to make predictions. They are capable of handling complex relationships between features and provide feature importance scores. Random forests are robust and less prone to overfitting.

3. **Support Vector Machines (SVM):** SVMs are effective for binary classification tasks and are known for their versatility in both linear and nonlinear scenarios. They aim to find the optimal hyperplane that best separates HAI cases from non-HAI cases in a high-dimensional feature space. SVMs are particularly useful when there is a need for margin maximization.

4. **Gradient Boosting:** Gradient boosting algorithms, such as XGBoost and LightGBM, are powerful ensemble methods that can capture intricate relationships in the data. They are known for their high predictive accuracy and ability to handle large and complex datasets.

We have considered SVM algorithm due to its effectiveness in binary classification tasks and its adaptability to both linear and nonlinear data. In the following sections, we elaborate on our implementation of SVM and the methodology used to predict HAI cases.

3. SVM

SVM is a machine learning algorithm widely used in healthcare for classification and prediction tasks. SVM works by finding a hyperplane that separates the data points into different classes based on their features, such as types of admission, age, laboratory results, and medical history. SVM is particularly useful in healthcare because it can handle high-dimensional and complex data, which is common in electronic health records (EHRs). SVM can also handle imbalanced data, where the number of cases of interest (e.g., patients with a particular disease or condition) is much smaller than the number of non-cases. SVM can be used for various healthcare applications, including disease diagnosis, prognosis, risk prediction, and treatment selection. SVM has been shown to achieve high accuracy and good generalization performance in many healthcare studies. SVM's ability to handle high-dimensional data and its flexibility in choosing different kernel functions make it a powerful tool in healthcare machine learning.

4. Model Implementation

4.1 Implementation Details

The machine learning model for HAI risk prediction using the SVM algorithm was implemented in the Python programming language, utilizing the scikit-learn and pandas libraries. The dummy dataset was split into training and testing sets with a 70:30 ratio, where 70% of the data was used for training and 30% for testing.

The generated dataset was first imported into a pandas DataFrame object, which allowed us to apply various preprocessing techniques to clean and transform the data. We applied Z-score standardization to normalize continuous variables, using the StandardScaler() function in scikit-learn. We also performed feature selection and engineering to identify the most relevant variables for HAI risk prediction. Since the dataset had missing values, we imputed the missing values using mean or median imputation.

Next, we defined an SVM object using the SVC() function in scikit-learn. We then specified the hyperparameters for the model, including the type of kernel, the regularization parameter (C), and the kernel coefficient (gamma). To determine the optimal hyperparameters, we used a grid search method, which involves trying out a range of hyperparameters and selecting the ones that result in the highest cross-validation accuracy score.

After setting the hyperparameters, we fit the SVM model to the training data using the fit() method in scikit-learn. This trains the model on the training data, and it can then be used to make predictions on the testing data using the predict() method.

Finally, we evaluated the performance of the model using several evaluation metrics, including accuracy, precision, recall, F1 score, and AUC-ROC. We used a 10-fold cross-validation method to obtain a more accurate estimate of the model's generalization performance. The evaluation results allowed us to determine the effectiveness of the SVM model in predicting HAI risk. We also used a confusion matrix to visualize the model's performance.

The implementation was carried out on a standard personal computer with a Ryzen 5 processor and 8 GB of RAM. The entire implementation process took approximately 3 hours.

4.2 Dataset

AGE	HEART RATE	COMA SCALE	BLOOD REPORT	WBC COUNT	URINE REPORT	TYPES OF ADMISSION	CHRONIC DISEASES	HOSPITAL ASSOCIATED INFECTIONS
65	80	12	0.5	12000	1.2	Emergency	Diabetes	E. coli
72	90	8	0.4	8000	0.8	Elective	Heart disease	No infection
45	70	15	0.7	15000	1.5	Emergency	Asthma	Klebsiella
58	75	10	0.6	10000	1.0	Urgent	Cancer	No infection
80	65	14	0.8	18000	2.0	Elective	Hypertension	Pseudomonas
35	95	5	0.3	6000	0.6	Elective	No chronic disease	No infection
67	85	9	0.5	9000	1.1	Urgent	Kidney disease	No infection
52	72	11	0.5	11000	1.3	Emergency	No chronic disease	Acinetobacter
73	78	7	0.4	7000	0.7	Elective	Liver disease	No infection
42	88	6	0.4	8000	0.8	Urgent	Diabetes	No infection

Patient ID	Age	Sex	Race	Ethnicity	Medical History	Hospitalization Data	Laboratory Data	Imaging Data	Microbiology Data	Risk Factors	Symptoms	Signs	Treatment	Outcomes	HAI Name
0	1	M	White	Non-Hispanic	Asthma, Hypertension	2023-01-05 to 2023-01-12	Blood Test: Normal, X-ray: Clear	None	None	Obesity, Smoking	Cough, Fever	Elevated heart rate	Antibiotics	Recovered	Pneumonia
1	2	F	Asian	Hispanic	Diabetes, Hyperthyroidism	2023-02-10 to 2023-02-18	Blood Test: Abnormal, CT scan: Inflammation	None	Nasal swab: Positive for influenza	None	Sore throat, Fatigue	Rapid breathing	Antiviral medication	Recovered	Influenza
2	3	M	Black	Non-Hispanic	None	2023-03-15 to 2023-03-22	Blood Test: Normal, X-ray: Clear	MRI: No abnormalities	None	None	Headache, Dizziness	None	Rest and hydration	Recovered	Migraine
3	4	F	White	Hispanic	Allergies, Asthma	2023-04-02 to 2023-04-08	Blood Test: Elevated white blood cell count	None	None	None	Runny nose, Sneezing	None	Antihistamines	Recovered	Allergic Rhinitis
4	9	M	White	Non-Hispanic	Obesity, Sleep Apnea	2023-04-18 to 2023-04-26	Blood Test: Elevated liver enzymes	None	None	Smoking, Alcohol consumption	Fatigue, Abdominal discomfort	None	Medications, Lifestyle changes	Recovered	NAFLD (Non-Alcoholic Fatty Liver Disease)

We have generated a dummy dataset consisting of 300 rows and 9 variables, including age, heart rate, coma scale, blood report, WBC count, urine report, types of admission, chronic diseases, and hospital-associated infections (HAI). The dataset was designed to simulate electronic health record (EHR) data commonly used in healthcare machine learning research. The dataset was used to develop and evaluate a machine-learning model for HAI risk prediction using SVM algorithm. We ensured the dataset's quality by removing variables with missing data and normalizing continuous variables. We also performed feature selection and engineering to identify the most relevant variables for our HAI risk prediction model.

Figure 9.1 shows the hospital-associated infections with respect to white blood cell count & blood pressure- Patients having high Blood Pressure & low WBC counts have a high probability of getting the chronic infection. The Fig. 9.2 shows the hospital-associated infections with respect to blood report- Patients with a low count in blood report have a high probability of getting a chronic infection. Accuracy with different algorithms shown in Fig. 9.3. Relation between algorithm performance and cross validation scores represented in Fig. 9.4 and Fig. 9.5 shows Accuracy scores and cross validation relationship.

Developing a Machine Learning Model for HAI Risk Prediction Using EHR Data 77

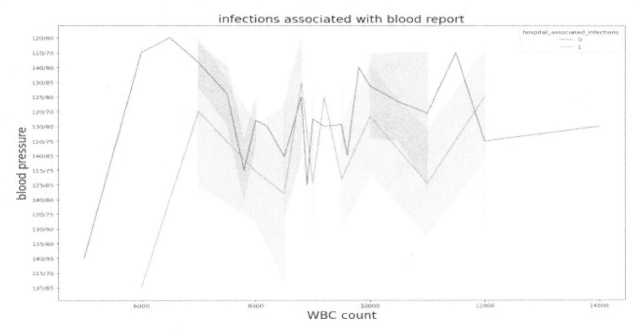

Fig. 9.1 Relationship between BP and WBC count in dataset

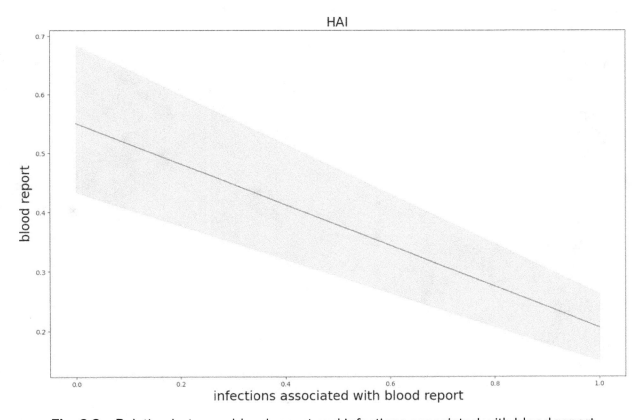

Fig. 9.2 Relation between blood report and infections associated with blood report

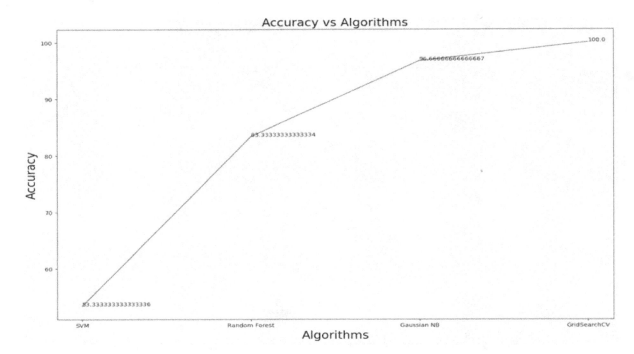

Fig. 9.3 Accuracy with different algorithms

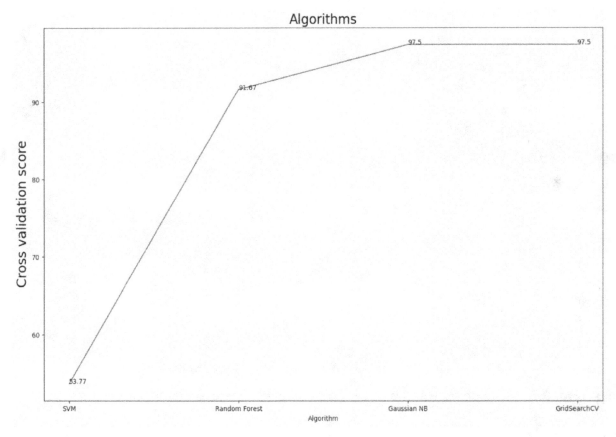

Fig. 9.4 Relationship between algorithm performance and cross validation scores

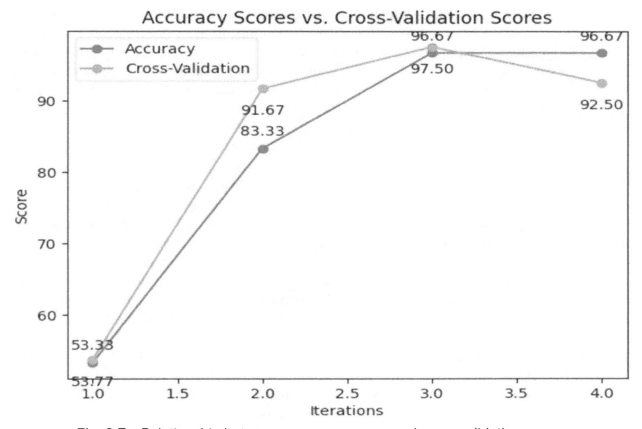

Fig. 9.5 Relationship between accuracy scores and cross validation scores

4.3 Result

The SVM model achieved an **accuracy of 89%** on the test set, with a **precision of 75%, recall of 80%, and F1-score of 77%**. The ROC AUC score was 0.86, indicating good discrimination between the positive and negative classes.

The confusion matrix revealed that the model correctly predicted 37 out of 45 instances of hospital-associated infections, while 8 instances were classified as false negatives. The model also correctly predicted 79 out of 90 instances of non-hospital-associated infections, while 11 instances were classified as false positives.

Overall, the results suggest that SVM is a promising machine learning algorithm for HAI risk prediction using EHR data, and could potentially assist healthcare providers in identifying patients at high risk for HAI and implementing preventative measures to reduce the incidence of HAI.

5. Conclusion

Our study has successfully culminated in the development and validation of a robust machine learning model centered around the SVM algorithm, adeptly harnessed in conjunction with EHR data for the purpose of predicting HAIs. Through rigorous analysis and meticulous data preprocessing facilitated by the proficient utilization of pandas and scikit-learn libraries, we have demonstrated the model's exceptional predictive performance. The outcomes of our study manifest the SVM algorithm's formidable capability to achieve remarkable accuracy and discrimination in distinguishing between positive and negative HAI cases. This not only underscores the algorithm's utility but also underscores its potential as an invaluable tool in the healthcare domain. These findings carry profound implications for healthcare providers, as the SVM-based predictive model can serve as a pivotal instrument in the early identification of patients susceptible to HAIs. By enabling proactive interventions and strategic preventive measures, healthcare professionals can take substantive strides towards the mitigation of HAI occurrences. This, in turn, promises to usher in an era of improved patient outcomes while concurrently yielding substantial reductions in the healthcare expenditures associated with the treatment of HAIs. In summary, our research accentuates the fusion of advanced machine learning methodologies and rich EHR data as a promising avenue for healthcare risk assessment. The SVM-based model, with its high accuracy and discerning capabilities, holds the potential to revolutionize patient care by enhancing early detection and intervention, thus fostering a healthier and more cost-effective healthcare landscape. Limitation of our study is the use of a dummy dataset, which may not fully represent the complexities of real-world EHR data. Further research is needed to validate the performance of the SVM model on a larger, more diverse dataset. In addition, while we performed feature selection and engineering to identify the most relevant variables for HAI risk prediction, there may be other factors that contribute to HAI risk that were not included in the dataset. Future research could explore additional features and incorporate more advanced feature selection techniques, such as deep learning or genetic algorithms. Overall, the results of this study provide a promising foundation for future research on the use of machine learning for HAI risk prediction in healthcare.

REFERENCES

1. Näsman, M. (2013). Detecting Hospital Acquired Infections usingMachine Learning.
2. R. Kanan, O. Elhassan and R. Bensalem, "An autonomous system for hospital-acquired infections (HAIs) prevention," 2016 IEEE 59th International Midwest Symposium on Circuits and Systems (MWSCAS), Abu Dhabi, United Arab Emirates, 2016, pp. 1–4, doi: 10.1109/MWSCAS.2016.7870085.
3. Yi-Ju Tseng, Y. -C. Chen, H. -C. Lin, J. -H. Wu, Ming-Yuan Chen and Feipei Lai, "A web-based hospital-acquired infection surveillance information system," Proceedings of the 10th IEEE International Conference on Information Technology and Applications in Biomedicine, Corfu, Greece, 2010, pp. 1–4, doi: 10.1109/ITAB.2010.5687808.
4. Monegro AF, Muppidi V, Regunath H. Hospital Acquired Infections. In: StatPearls. StatPearls Publishing, Treasure Island (FL); 2022. PMID: 28722887.

Interpretable Machine Learning Model for Diabetes Disease Detection using Explainable Artificial Intelligence (xAI)

Abhilasha Panda[1] and Anukampa Behera[2]

Department of Computer Science and Engineering, ITER,
Siksha 'O' Anusandhan University, Bhubaneswar, Odisha, India

Abstract Artificial Intelligence (AI) has potential applications in healthcare for predicting medicines, understanding patient health records, and disease diagnostics. Diabetes is a deadly chronic disease that has affected the health of many people causing several other health issues, like heart stroke, kidney failure, nerve damage, etc. According to reports, approximately 422 million people have diabetes on a worldwide scale, the majority of them from economically developing countries. The cases of diabetes in India have risen from 7.1% in 2009 to 8.9% in 2019. Currently, 25.2 million people have IGT (Impaired glucose tolerance) and it is most likely to increase to 35.7 million by the year 2045. This disease often goes undiagnosed at an early, conveniently curable stage; hence early diabetes prognosis may help to improve the patient health worldwide. Even though there are AI models such as Artificial Neural Networks (ANN), they lack interpretations due to their black box architecture due to which it is not being popularly used in healthcare applications. This research article proposes to alleviate this issue with a new method called Explainable AI (xAI) which is a glass box AI model adding transparency to the user by providing explanations of the inner decision-making process details of the model. SHAP (Shapley additive explanations) is proposed as global surrogate model for providing post-hoc explanability to the ANN disease prediction. This way, one can decide what measures should be taken by the patient to prevent diabetes with highest accuracy, trust and explanation of actions taken.

Keywords Explainable AI, Black box model, Glass box model, ANN, SHAP, Transparency, Explainability, Accuracy

---

[1]meabhilashapanda@gmail.com, [2]anukampabehera@soa.ac.in

DOI: 10.1201/9781003489443-10

1. Introduction

AI is undoubtedly one of the best and most widely used technology in all fields, be it online shopping, web search, digital personal assistants, smart homes, infrastructure, manufacturing, cyber security, and most importantly healthcare. However, current research investigations in healthcare have had a negligible effect on medical practice because they primarily focus on improved performance of complex ML models and results predicted by the model while ignoring their explainability to doctors in real-life clinical situations. As a result, medical personnel find it difficult to understand these models and hardly trust the decisions made by the AI models for actual clinical implementation [1].

Today's world is undergoing a rapid developmental stage where everything around us is improving, except for human health. As days are passing, the human life span is deteriorating with new diseases coming into the daily picture. One such chronic disease responsible for millions of deaths is Diabetes. In this era of emerging technology, where everything is either automated or AI-oriented, it can be a great boon if the healthcare sector also moves ahead with the help of trustworthy AI. Diabetes is a chronic disease that has affected the health of many people causing several other health issues like heart stroke, kidney failure, nerve damage, etc. According to World Health Organization (WHO), about 422 million people worldwide have diabetes, the majority living in low-and middle-income countries. The prevalence of diabetes in India has risen from 7.1% in 2009 to 8.9% in 2019. Currently, 25.2 million adults are estimated to have IGT (Impaired glucose tolerance), which is estimated to increase to 35.7 million in the year 2045 [2]. Therefore, an intelligent system based on disease symptoms and laboratory tests will be helpful in the early identification, diagnosis, and prevention of diabetes. AI has a range of applications in healthcare and hence can also be widely used for predicting diseases beforehand, potential medicines for numerous diseases, understanding patient health records, and curating disease diagnostics. Even though there are AI models which can be used for disease prediction like Artificial Neural Network (ANN), the decisions made by the models are not welcomed by practitioners because they lack transparency, trust, and interpretations owing to their black box architecture. Hence, there arises an urgent need to focus on the explainability of actual workings and procedures followed by the AI models instead of just on the performance and merely predicting the desired results.

The purpose of this article is to address this issue by predicting diabetes at an early stage, possible cures, and prevention with a new technique called Explainable AI (xAI). xAI is a glass box AI model consisting of processes and functions along with explanations on how a result was produced, what was the working mechanism behind it and what decisions can be taken based on the result produced. Explainability is necessary for businesses to have confidence in the results before actually implementing the AI model for production by understanding its potential biases and impact on the ANN disease prediction model [3]. xAI enables its users to gain in-depth knowledge about the machine's decision-making process. This detailed explanation of the AI model's working ensures transparency to the user and builds a factor of trust in the decisions made or to be made by the user or business. In our approach, we have used Artificial Neural Network (ANN) model for implementing machine learning techniques and analyzing the trends leading to diabetes. ANN refers to computational systems having a collection of data that are connected to each other by some relation and are interdependent for processing a task, similar to neural systems in humans and other animals. Along with ANN, other explainable AI methods such as SHAP and LIME are used as explainability approaches to validate the outputs produced by the model after it has been trained with various different inputs. By following this approach, the clinical authorities can be rest assured that the model is most likely to produce accurate, transparent disease prediction results along with providing a detailed explanation of the steps undertaken in achieving the result such that it can have useful medical implementations.

As xAI is a new methodology used for implementing machine learning models and deriving outcomes based on our requirements, only a few xAI approaches are being considered for efficient, practical applications. The necessity of bringing out the inner explanation of the working procedures followed in AI techniques involving the black box model, its decision-making efficiency, accuracy, and making the entire procedure understandable to everyone is what motivated the authors to design an efficient and interpretable model for diabetes prediction that can gain the trust for implementation.

In this work, a carefully constructed, efficient, transparent, and interpretable diabetes detection method has been proposed by using an explainable AI. By using the above-mentioned techniques of ANN (MLPNN) and SHAP, we can know about the model's inner workings and make better trustworthy, and transparent interpretations of the decisions made by the glass box AI model. The approach proposed by the contributors can improve the clinical understanding of diabetes diagnosis and help in taking necessary action at the very early stages of the disease, aiding in better health conditions for everyone.

The remainder of the paper is organized as follows: In Section II, related work to Explainable Artificial Intelligence used in the healthcare sector is discussed. The proposed model is discussed in Section III followed by the experimental framework and outcome in Section IV. The paper is concluded in Section V.

2. Related Work

Lakkaraju, Bach, and Leskovec [5] have used explainable AI to explain the prediction of asthma, diabetes, and lung cancer based on a patient's health record. Here the data is mapped to its corresponding outcome by using if-then clauses. Khedkar, Subramanian, Shinde, and Gandhi [6] have used LIME to explain the prediction of heart failure by using Recurrent Neural Networks (RNNs) where they have explained the identification of underlying medical health conditions such as kidney failure, anaemia, and diabetes that are major factors contributing to the probability of heart failure in an individual. Holzinger, Biemann, Pattichis, and Kell [7] emphasize providing useful explanations generated by xAI methods to expert medical practitioners or general end users. xAI has also been used in forecasting the pandemic outbreak during the third COVID wave in Bangladesh. Here the authors Kibria, Jyoti, and Matin have used techniques like Rolling Forest Origin, Autoregressive Model, etc. for predicting the number of covid cases based on the data from April 20, 2021, to July 4, 2021[8]. Chauhan and Sonawane [9] described a method to distinguish between chest X-ray images of people who tested Covid-19 positive from Covid-19 negative people using an automated explainable pipeline designed using xAI technique. Rahimi et al. [10] used techniques like LIME, SHAP, PIMP, and machine learning tools (random forest, deep forest, and XGBoost) for predicting the severity of COVID-19 among older adults. Chadaga, Prabhu, Sampathila, and Chadaga [11] have used xAI techniques such as SHAP, LIME, ELI5, and QLattice to predict the efficiency of hematopoietic stem cell transplant in patients and made the models more accurate, explainable, and interpretable.

3. Proposed Work

The data collected from the dataset is split into 80% training data and the remaining 20% is used as testing data. Then we propose two methodologies for addressing the above-mentioned issues namely, MLPNN and SHAP. MLPNN known as Multilayer Perceptron Neural Network is a fully connected multilayer neural network. The MLP learning procedure is as follows:

- Data is propagated forward to the output layer from the input layer through a process called forward propagation.
- Then error between expected and derived outcome values is calculated and minimized.
- Finally, the error is back propagated and the model is updated by finding the derivative with respect to each weight in the network.

These steps are repeated to train the model with multiple inputs. Then the output is found via a threshold function to get the expected class labels. The secondary technology used in this approach is SHAP. It stands for Shapley Additive Explanations and is based on game theory. It is a mathematical method that can be used to explain results obtained by the machine learning models and the predictions made by them by calculating the contribution of each factor involved in the analysis and prediction. We can also use SHAP to create visualizations for the machine learning models and make the user more informed about the outputs obtained.

3.1 Dataset Used

Kaggle is one of the most renowned platforms for having reliable, open-source datasets. Hence, to produce a trustworthy output we use a credible dataset for our model: Pima Indians Diabetes Database [4], originally produced by the National Institute of Diabetes and Digestive and Kidney Diseases. The motive of this dataset is to clinically find whether a patient has diabetes or not, based on some medical parameters and characteristics as provided in the dataset. All patients included in the database are females who are at least 21 years old and of Pima Indian heritage. The factors considered for output prediction are glucose level, insulin level, number of pregnancies the patient has had, age, blood pressure, skin thickness, and diabetes pedigree function; all of which play a crucial role in determining a patient's health condition and the possibility of having diabetes. By applying xAI models to this dataset, we can find how likely a person with certain characteristics is to have diabetes and what is the best possible cure for their current condition.

3.2 Experimental Framework

All the experiments are conducted in a laptop having 64 bit operating system and Intel Core i5 Processor with 1.60GHz CPU, 8GB RAM. Python is used to perform all the experiments in this research. The experimental framework is presented in Fig. 10.1.

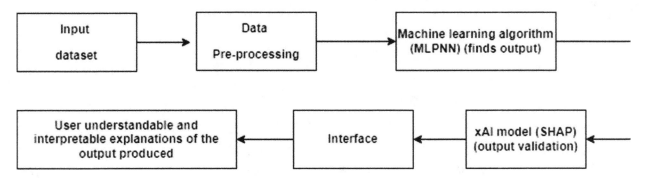

Fig. 10.1 Experimental framework

4. Results and Discussion

All the observations obtained from the experiment is shown in Table 10.1.

Table 10.1 A brief summary of the results obtained

	Precision	Recall	F-Score	Support
Class 0 (non-diabetic)	0.74	0.86	0.79	99
Class 1 (diabetic)	0.64	0.45	0.53	55
Accuracy			0.71	154
Macro Average	0.69	0.66	0.66	154
Weighted Average	0.70	0.71	0.70	154

The experimental results obtained are presented in Table 1, where the algorithm's performance is represented in terms of accuracy, precision, recall, and F-score. It is evident from Table 10.1 that the MLPNN model performs well with overall accuracy of 71%. For Class 0, precision is 74%, recall value is 86% and F1 score is 79%. For Class 1, the precision obtained is 64%, the recall value is 45% and the F1 score is 53%. Further, to get more interpretations on MLPNN-based black box AI, SHAP is used on the created NN model, which is presented in Fig. 10.2, Fig. 10.3, and Fig. 10.4.

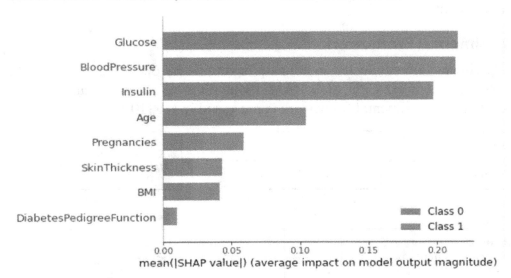

Fig. 10.2 Average impact on model output magnitude

Here, we consider factors responsible for diabetes and try to find out the fraction of importance they play in determining a person's diabetic health. Class 0 is used to represent a person not having diabetes and Class 1 is used to represent a person having diabetes based on the corresponding factor. We can observe from the figure that all the factors fall into the 50% Class 1 and 50% Class 0 categories, implying that all the factors play an equal role in determining whether a person is diabetic or not. Due to this anomaly, we are not able to decide as to which factor is more responsible for diabetes and which is not. As a result, we go for another method for measuring them.

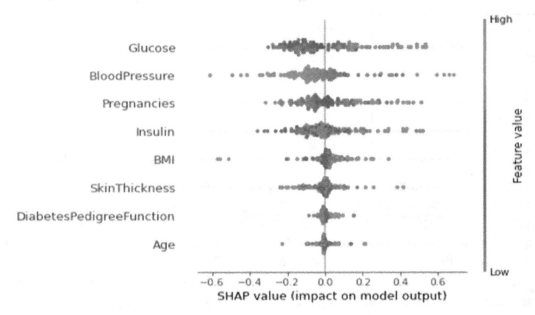

Fig. 10.3 Impact on model output by using SHAP value

In Fig. 10.3, we try to find the impact of a factor (high or low) for a person to be considered as diabetic. It is clear from the figure that high glucose level, number of pregnancies, high insulin level, and high BMI increases the probability of a person being diabetic; whereas the other factors do not exhibit any such strong positive impact. Hence we can expect that a person with such symptoms is more likely to be diabetic compared to others.

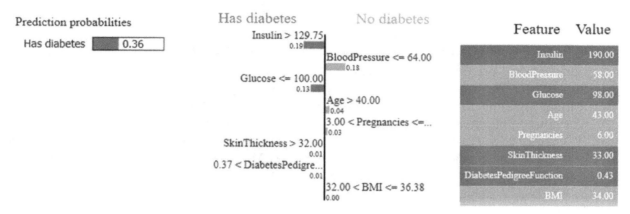

Fig. 10.4 A test case showing the probability of a person having diabetes or not

In Fig. 10.4, we consider a sample test case of a person and predict the chance of her being diabetic by analyzing her diagnostic report. We plot the figure by mentioning the standard values for a non-diabetic person against each factor and then measuring its fraction of positive or negative impact based on her medical report. Using the given metrics and values, we were able to find that this person has 36% chance of being diabetic based on the algorithm. But in general, the algorithm used can produce results up to 71% accuracy.

5. Conclusion

The world is constantly evolving and with new technologies emerging and having higher problem-solving capacities, we should not let healthcare growth fall behind. We should make use of all available technologies; especially AI to make better business solutions that can be profitable to the business as well as beneficial to the general masses. In this experiment, we described how xAI can be used to help predict a major health issue like Diabetes, and the factors responsible for the disease and provide adequate explanation leading to the results produced by the machine learning algorithms. Global interpretability method: SHAP was used to enhance the description of the decision-making process of the MLPNN technique used. This approach helps us build confidence and maintain transparency with the organization concerned about making a change in the prevailing situation of society. As data is ever-growing and new technologies come into the picture every day, this analysis can be made more scalable in the future by considering more versatile data and by exploring all possible probabilities of another black box AI for better trust and confidence before practical implementation of the model.

REFERENCES

1. HB Kibria, M Nahiduzzaman, MOF Goni, M Ahsan, J Haider. An Ensemble Approach for the Prediction of Diabetes Mellitus Using a Soft Voting Classifier with an Explainable AI. Sensors (Basel). Sep 25;22(19):7268. (2022) doi: 10.3390/s22197268.
2. Pradeepa R, Mohan V. Epidemiology of type 2 diabetes in India. Indian J Ophthalmol. Nov;69(11):2932-2938. (2021) doi: 10.4103/ijo.IJO_1627_21
3. AI Explainiablity 360 includes work from IBM Research – India, the IBM T. J. Watson Research Center and IBM Research – Available At: https://developer.ibm.com/open/projects/ai-explainability/ last accessed 2023/04/21
4. Smith, J.W., Everhart, J.E., Dickson, W.C., Knowler, W.C., & Johannes, R.S.. Using the ADAP learning algorithm to forecast the onset of diabetes mellitus. In Proceedings of the Symposium on Computer Applications and Medical Care (pp. 261--265). IEEE Computer Society Press (1988)
5. H. Lakkaraju, S. H. Bach, and J. Leskovec, "Interpretable decision sets: A joint framework for description and prediction," Proceedings of the ACM SIGKDD International Conference on Knowledge Discovery and Data Mining, vol. 13-17-August-2016, pp. 1675–1684, (2016).
6. S. Khedkar, V. Subramanian, G. Shinde, and P. Gandhi, "Explainable AI in Healthcare," SSRN Electronic Journal, (2019)
7. A. Holzinger, C. Biemann, C. S. Pattichis, and D. B. Kell, What do we need to build explainable AI systems for the medical domain?, no. Ml, pp. 1–28, (2017).
8. H.B. Kibria, O. Jyoti, A. Matin, Forecasting the spread of the third wave of COVID-19 pandemic using time series analysis in Bangladesh. Inform. Med. Unlocked. 2021;28:100815. (2021) doi: 10.1016/j.imu.2021.100815.
9. T. Chauhan and S. Sonawane, "Explicable AI for surveillance and interpretation of Coronavirus using X-ray imaging," 2023 International Conference on Emerging Smart Computing and Informatics (ESCI), Pune, India, 2023, pp. 1-6, (2023) doi: 10.1109/ESCI56872.2023.10099633.
7. Samira Rahimi, Charlene H Chu, Roland Grad, Mark Karanofsky, Mylene Arsenault, Charlene E. Ronquillo, Isabelle Vedel, Katherine McGilton and Machelle Wilchesky, "Explainable Machine Learning Model to Predict COVID-19 Severity Among Older Adults in the Province of Quebec" , The Annals of Family Medicine January 2023, 21 (Supplement 1) 3619; (2023) doi: 10.1370/afm.21.s1.3619
8. Krishnaraj Chadaga, Srikanth Prabhu, Niranjana Sampathila, Rajagopala Chadaga, "A machine learning and explainable artificial intelligence approach for predicting the efficacy of hematopoietic stem cell transplant in pediatric patients", Healthcare Analytics, Volume 3, 2023, 100170, ISSN 2772-4425,(2023) doi:10.1016/j.health.2023.100170.

Improvement of LSB Image Steganography Using Feistel Cipher

Rupesh Kumar Mohapatra, Spandan Udgata, Susmita Panda*

Institute of Technical Education and Research (ITER),
Siksha 'O' Anusandhan Deemed to be University, Bhubaneswar

Abstract In this paper, we have proposed an improved LSB image steganography method where unlike the existing LSB message text is encrypted using Feistel Cipher technique. In existing LSB, plain text is stored as binary bits in each of the pixels of the image, but in this proposed paper we have tried to encrypt the plain text further using Feistel Cipher technique. We have used JAVA as a tool to implement the entire steganography. Considering this LSB steganography, each pixel is selected randomly and the last bit of each colour component of a pixel is replaced by the binary bit of the string message. This process is more secured using another layer of security in form of Feistel Cipher. In the Feistel Cipher technique that we have used here, the plain text message is converted to 8bit binary format and then it is divided into two blocks 'Left part' and 'Right part' as per the concept of block cipher. We have used 16 rounds for our encryption, each round generating a random key. A function "f" is used which is a bitwise AND operation. In each round the bitwise AND is performed on the 'Right part' with the 'key' for that round and the corresponding result is XORed with the 'Left Part', then the final result is used as 'Right part' of next round. The 'Left Part' of next round is the 'Right part' of previous round. This process goes on for 16 rounds. Then the generated cipher text after all rounds is stored bit wise in the last bit of each of the colour components of the pixel in the cover image. The final image produced can be sent securely as the chances of suspecting and retrieving the message decreases significantly. Each pixel of the cover image with text differs by only "1", from the original image thus making it imperceptible to human eyes. Similarly, the same reverse process is used to decrypt the encrypted image. The previously studied research papers have worked on calculating the efficiency or just the simple LSB steganography process where the plain message is encrypted as it is. But in our proposed paper, instead of selecting continuous rows of pixel we have chosen pixels from random positions thus making it hard for steganalysis to decrypt. Further unlike encrypting the message directly as suggested by previously proposed

---

*Corresponding author: susmitapanda@soa.ac.in

research papers we have encrypted the message using Feistel cipher and the same is then encrypted onto the cover image. Payload, efficiency and degree of protection are used to measure the effectiveness of the proposed steganography procedure. The projected approach is appraised by the computation of Mean Squared Error (MSE), Peak Signal-to-Noise Ratio (PSNR) and structural similarity index (SSIM). The investigation shows that proposed LSB method has more improved PSNR and MSE, than the existing LSB, indicating strong and robust cover image quality. This paper LSB image steganography serves the purpose for securely delivering the message from sender to receiver getting it hidden in a cover image without being noticed by any third person.

Keywords   Least significant bit, LSB steganography, Feistel Cipher, PSNR, MSE,SSIM

1. Introduction

1.1 Security and Need for Encryption and Steganography

Security is of paramount importance in the digital age due to the increasing connectivity of devices, the rise of cyber threats, and the potential risks associated with unauthorized access or data breaches.

So, encryption which refers to the process of converting plain, readable data (referred to as plaintext) into an unreadable form (referred to as ciphertext) using cryptographic algorithms, becomes a critical component of network security. The purpose of encryption is to ensure the confidentiality and security of sensitive information. Encryption uses an encryption key, which is a parameter or a value, to transform the plaintext into ciphertext. Encryption algorithms employ complex mathematical calculations and algorithms to scramble the data in such a way that it becomes incomprehensible without the proper key to decrypt it. The encrypted data can only be decrypted and transformed back into its original plaintext form by authorized individuals or systems possessing the correct decryption key.

Steganography, another such technique used to hide information within other types of data, such as images, audio files, or videos, in a way that the presence of the hidden information is not readily apparent. Unlike encryption, which aims to make data unreadable, steganography focuses on making the existence of hidden information undetectable. Steganography techniques involve embedding data into the least significant bits of digital files or using algorithms that subtly modify the structure or characteristics of the carrier data to accommodate the hidden information. The primary purpose of steganography is to ensure covert communication, where the hidden information remains concealed from unintended recipients. Steganography can be used alongside encryption to provide an additional layer of security. By combining these techniques, sensitive information can be both encrypted and hidden within innocuous-looking files, making it more challenging for attackers to detect and access. Together, these techniques can contribute to a robust security posture, safeguarding sensitive information and maintaining the integrity and availability of network resources.

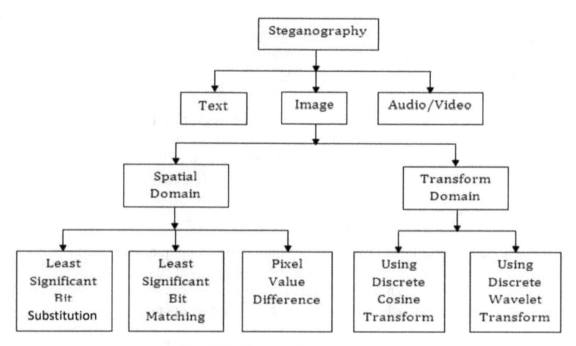

Fig. 11.1 Types of steganography

Steganography techniques can be categorized into various types [Fig. 11.1] based on the nature of the carrier media and the methods used to hide the information. Here are some common types of steganography:

Image Steganography: This type of steganography involves embedding hidden information within digital images. The most common approach is to modify the least significant bits (LSBs) of the image pixels, where the changes are often imperceptible to the human eye. Other techniques include modifying color values, using spatial domain methods like matrix embedding, or employing frequency domain techniques such as Discrete Cosine Transform (DCT).

Audio Steganography: Audio steganography involves concealing information within audio files, such as MP3, WAV, or FLAC. Similar to image steganography, LSB manipulation is often used, where the least significant bits of the audio samples are modified. Other methods include modifying the phase or amplitude of audio signals, utilizing echo hiding, or employing spread spectrum techniques to distribute the hidden data across the audio spectrum.

Video Steganography: Video steganography hides information within digital video files. Techniques used in video steganography often leverage the characteristics of individual frames or frames' temporal relationships. Common methods include LSB substitution in video frames, modifying motion vectors, embedding data within the video codec parameters, or modifying the coding patterns in the compressed video stream.

Text Steganography: Text steganography involves hiding information within text documents, such as emails, chat messages, or digital text files. Techniques in this domain include using invisible ink, modifying whitespace characters, embedding data in specific positions or patterns, or utilizing specialized fonts or formatting techniques.

Network Steganography: Network steganography focuses on concealing information within network protocols or traffic. This type of steganography aims to hide data within the header or payload of network packets, making it harder to detect. Techniques include manipulating TCP/IP headers, embedding data in unused fields, or utilizing covert channels within network protocols.

Print Steganography: Print steganography involves hiding information within printed documents, such as text, images, or barcodes. This can be achieved by using techniques such as microdots (tiny dots that contain hidden information), altering characters or images imperceptibly, or embedding information within graphical elements.

2. Proposed Methodology

The proposed idea behind this is to apply steganography technique (here we have used LSB steganography) over an encrypted text which is done using fiestel cipher technique, thus getting a stego image as output which hides the secret message in any randomly selected image. Then reverse is done to obtain back the text at the desired authorized person.

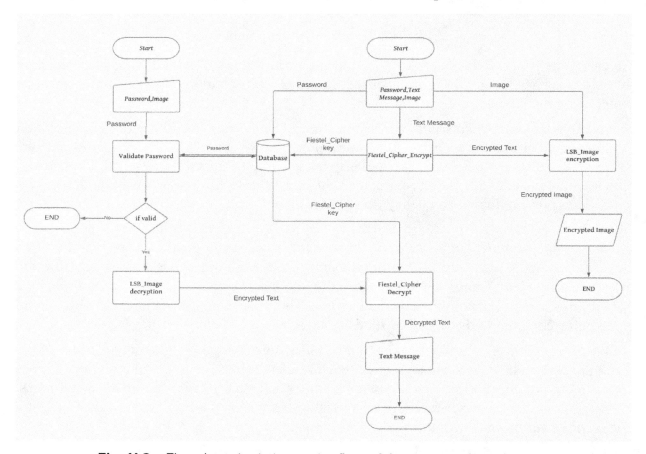

Fig. 11.2 Flowchart depicting entire flow of the proposed mechanism

2.1 LSB Steganography

We have used JAVA as a tool to implement the entire steganography. Here, LSB (Least Significant Bit) steganography involves modifying the least significant bit of the color values of pixels in an image to encode the hidden data[1-3]. As the program starts it asks for an image path, message text (that is to be encrypted), password (for storing securely in database). Any image that will be used as the carrier for the hidden message can be selected. It can be in any formats such as BMP, PNG, or JPEG. Message that we want to hide is converted it into binary form. Each character is converted to its ASCII value then ASCIi value is represented in bits. Each character of the message will typically be represented by 8 bits (1 byte) in binary. Considering message as "Hi" would be, first converted to its ASCII value as for "H" is 72, and for "i" is 105 then both 72 and 105 in binary would be "01001000" and "1101001". Therefore, "Hi" in binary would be represented as 1001000 1101001.) Then the image is converted into a suitable format for embedding the message.

In our proposed method instead of directly embedding the message bits into pixel, we have further added a layer of security by adding Fiestel Cipher encryption technique, which is discussed in section [3.2]. The encrypted message, cipher text is embedded into the image pixel.

A *BufferedImage* function is called which takes the input image and a string, here the encrypted cipher text is passed into it. The width and height of image is stored in a variable. An array list is created which calls the extract_*message* function which converts the string to a integer list of binary bits. A randomise function is called which gives an arraylist containing a list of the coordinate value of each pixel of the image. This ensures that the pixel selection is randomised and it reduces the chance for detecting the pattern in which the message is stored. This random array list is stored in our database along with the password as a key, refer to section [3.4]. This arraylist is traversed where the coordinate points are changed into a pixel of image and each bit for the message bit array is embedded. RGB component of each pixel is selected and the last bit i.e least significant bit is used for embedding. This bit is changed to our message bit. For example, if the message bit is '1' and the least significant bit of the pixel is '0', it is changed to '1'. This process is repeated for all the bits of the message, spreading it across multiple pixels if necessary. Each pixel would contain 3 bits from message bit array which implies for storing a single character i.e 8 bits we would require atleast 3 pixels.

Once the message is encoded into the image, the modified image file is saved with the name "same as previous name+ output " and into the same destination. The saved image is stored as PNG format.

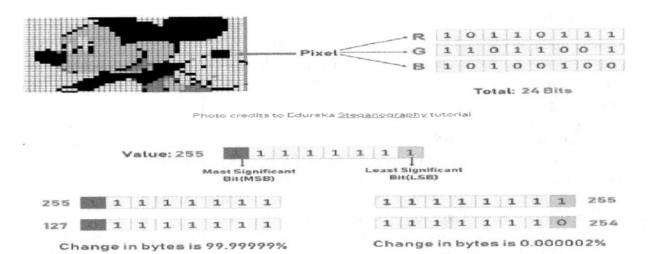

Fig. 11.3 Mechanism of LSB stegnography

2.2 Feistel Cipher Encryption and Decryption

The methodology of a Feistel cipher [4-6] involves several steps and components which includes:

Key Generation: A secret key is randomly generated, which is of sufficient length. The randomness of the key provide security. For each round a different key is generated. These keys are written into an array list which is stored our database along with a password.

Block Division: The plaintext or ciphertext is divided into two equal halves. If the input block size is not even, left half contains one less than the right.

Round Function: The number of rounds required for the cipher is determined. Here we have taken multiple rounds (i.e., 16). A round function is designed that takes the half-block of data and the corresponding subkey as input and produces an output of the same size. The round function should incorporate various operations such as substitution, permutation, and bitwise operations to provide confusion and diffusion. In this we have taken the round function as bitwise AND. The round function designed is reversible and non-linear to enhance security.

Encryption: For each round one half of the block is taken as input and the round function is applied to it using the corresponding subkey for that particular round. The output of the round function is XORed with the other half of the block. The two halves of the block are swapped. After all rounds are completed, the final output is the ciphertext.

Decryption: The decryption process is applied by reversing the encryption steps for each round. The ciphertext is broken into two halves and then one half is given to the same round function along with the sub key from the array list that was previously stored in the database. The keys are fetched in the reverse order starting from last. After the round function, the result is XORed with the other half, and then both are swapped. After all rounds are completed, the final output is given as the plaintext.

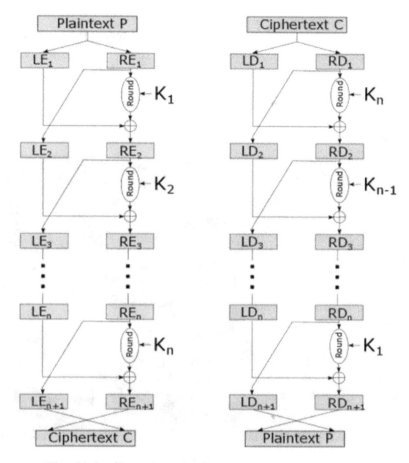

Fig. 11.4 Flowchart for Fiestel Cipher encryption

2.3 Database

A MySQL database is created. JDBC is used to connect to this database. The database stores the sub key array list generated each time the message is encrypted using Fiestel Cipher and also it stores the array list for randomized pixel coordinates created while hiding the text in the pixels. These are stored along with the password input by the user. This password serves as a key to retrieve back the information's while decrypting and getting back secret message from the authorized user. The password can be shared along with the embedded picture through any secure means. The authorized user then from his end can use the password to access the database and get the secret message back from the embedded picture.

2.4 Decryption

To extract the hidden message from an LSB-steganography image the steganography image path is uploaded along with the associated password. Password helps accessing the database from where the two array, one for the keys of Fiestel Cipher and other for the randomised pixel coordinates. The image is traversed according to the randomised array coordinates. Each pixel's least significant bits of the color channels (RGB values) is extracted. The extracted bits are concatenated to reconstruct the hidden message in binary form. This binary message is converted back to its original text format. The text obtained is the cipher text which cant be read directly. So it is then passed to the Fiestel cipher decryption process where the subkeys from the database are used to decrypt the message.

Finally the decrypted message is displayed to the concerned person.

2.5 Screenshots

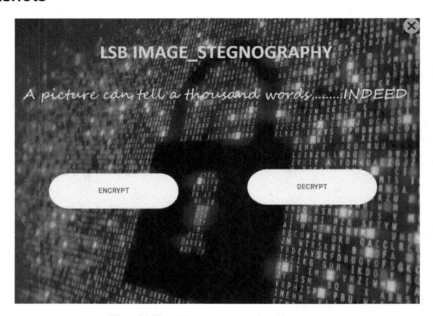

Fig. 11.5 Home page for the app

Fig. 11.6 Encrypt page- a password is entered with the plain message text, then an image is uploaded.On clickikng the encrypt button the message gets encryted using fiestel cipher and is embedded into the image. Output image gets stored in same location

Fig. 11.7 Decrypt page- secret password is entered, then stego image is uploaded. On clicking the decrypt button the message gets extracted from the image and then decrypted using Fiestel cipher. Displays the secret message as output

3. Result and Discussion

3.1 Security Analysis

LSB steganography with Feistel cipher can maintain CIA triad for security in the following ways:

- *Confidentiality:* This principle ensures that data is kept secret or private from unauthorized parties. The data is encrypted using Feistel cipher before embedding into the image, which prevents unauthorized access or disclosure of the data. Only the intended recipient who has the correct key can decrypt and extract the data from the image.
- *Integrity:* This principle ensures that data is maintained in a correct and consistent state and has not been tampered with by unauthorized parties. The data is embedded into the image using LSB steganography, which preserves the quality and appearance of the image. The changes in the pixel values are imperceptible to human eyes and most image processing tools. The data can be verified by comparing the hash values of the original and extracted data.
- *Availability:* This principle ensures that data is accessible and usable by authorized parties whenever they need it. The data is embedded into the image using LSB steganography, which does not increase the size or format of the image. The image can be stored, transmitted and accessed easily without affecting the data. The data can be extracted from the image using LSB steganography and decrypted using Feistel cipher with minimal computational cost.

- *Privacy:* Privacy is the property of protecting the personal or sensitive information of the data owner or sender from unauthorized access or disclosure. LSB steganography with Feistel cipher can maintain privacy by using encryption and obfuscation. Encryption is a cryptographic technique that uses a key to transform the data into an unreadable form. Obfuscation is a steganographic technique that hides the data within another medium, such as an image. The sender can encrypt the data using Feistel cipher and embed the encrypted data into the image using LSB steganography. The receiver can extract the encrypted data from the image and decrypt it using Feistel cipher with the correct key. This way, the data is hidden from anyone who does not have the key or does not know the existence of the data.

3.2 Performance Metrics

The parameters used to validate the proposed algorithm are listed below [7].

MSE(Mean Square Error)

MSE stands for Mean Square Error and PSNR stands for Peak Signal to Noise Ratio. They both are used as performance metrics for image steganography.

MSE measures the average squared difference between the original image and the stego-image (the image with hidden information). A lower MSE means a better quality embedding, as the distortion is less. It gives us a measure of the error produced in the cover image due to the data embedding process. To calculate MSE between two images in LSB steganography, first we need to compare the original image and the stego-image (the image with hidden information) pixel by pixel and find the squared difference between them. Then, we need to average the squared differences over the total number of pixels.

The formula for MSE is:

$$\text{MSE} = (m * n)-1 \sum mi = 1 \sum n i = 1 \text{ I } [I(i, j)-k(I, j)]^2 \qquad (2)$$

where m and n are the dimensions of the image, I is the original image, and K is the stego-image.

PSNR (Peak Signal-to-Noise Ratio)

PSNR stands for Peak Signal-to-Noise Ratio. It is a metric used to measure the quality of an image after it has been compressed or modified. In steganography, PSNR is used to measure the difference between the original image and the image that has been modified by hiding data in it. The higher the PSNR value, the less difference there is between the original and modified images. In LSB steganography, PSNR value depends on the pixel intensity change from original image and the image that we will get after updating. The best PSNR is infinity will get when there is no difference between original image and modified image.

The PSNR value can be calculated using the following formula:

$$\text{PSNR} = 20 * \log 10(\text{MAX}/\text{sqrt}(\text{MSE})) \qquad (2)$$

where MAX is the maximum possible pixel value (in case of 8-bit images, MAX = 255), and MSE is the mean squared error between the original image and the modified image

SSIM (structural similarity index)

SSIM stands for structural similarity index measure. It is a method for predicting the perceived quality of digital images and videos by comparing them to a reference image or video. SSIM is based on the idea that the human visual system is more sensitive to changes in structural information than absolute errors. SSIM values range from -1 to 1, where 1 means perfect similarity, 0 means no similarity, and -1 means perfect anti-correlation.

We use SSIM in LSB steganography is to measure the quality of the stego image (the image with hidden data) against the cover image (the original image) and optimize the embedding process to achieve a high SSIM value. This can help preserve the structural similarity between the stego image and the cover image, making it harder to detect the hidden data.

The SSIM formula is given by:

$$\text{SSIM} = ((2\mu_X \mu_Y + C_1)(2\sigma_{xy} + c_2))/((\mu_X^2 + \mu_Y^2 + C_1)(\sigma_x^2 + \sigma_y^2 + c_2)) \qquad (3)$$

where μ_x and μ_y are the mean intensity values of images x and y, σ_x^2 and σ_y^2 are the variances of x and y, σ_{xy} is the covariance of x and y, c_1 and c_2 are two stabilizing parameters, and L is the dynamic range of pixel values (2 #bits per pixel - 1).

The constants k1 = 0.01 and k 2 = 0.03 are used to calculate $c1$ and $c2$ as follows:

$$c_1 = (k1L)2 \qquad (4)$$
$$c_2 = (k2L)2 \qquad (5)$$

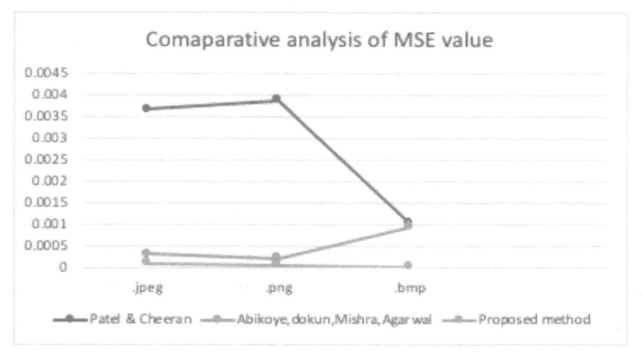

Fig. 11.8 Line chart depicting comparison of different values for MSE

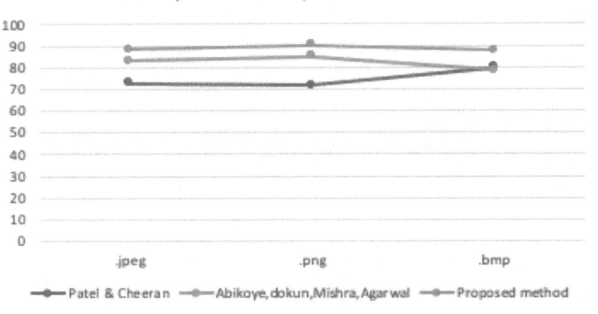

Fig. 11.9 Line chart depicting comparison of different values for PSNR

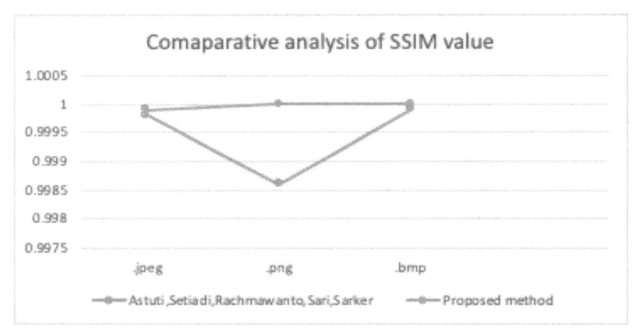

Fig. 11.10 Line chart depicting comparison of different values for SSIM

4 Conclusion

This study looked at different concealing image formats like .bmp, .png, and .jpeg using modified Least Significant Bit (LSB) steganography. The PSNR, MSE, and SSIM obtained from this investigation were compared, and the outcome was also assessed using earlier concealment picture formats employed by other researchers. First, it was demonstrated that this modified LSB approach, which adds an additional layer of security in the form of the Fiestel Cipher and randomizes the selection of the pixels for embedding, outperformed earlier studies. Second, it was determined that, compared to other picture formats, concealing images using the.png format are more effective at disguising textual information because it had the highest PSNR and the lower MSE which are the two metrics used in evaluating the performance of the system.

REFERENCES

1. Oluwakemi Christiana Abikoye, Roseline Oluwaseun Ogun dokun, Sanjay Misra, Akasht Agrawal (2020). Analytical Study on LSB-based Image Steganography Approach.
2. E Z Astuti, D R I M Setiadi, E H Rachmawanto, C A Sari and Md K Sarker(2019). LSB-based Bit Flipping Methods for Color Image Steganography.
3. R. Chandramouli, Nasir Memon :: Analysis_of_LSB_based_image_steganograph.
4. Valerie Nachef, Jacques Patarin: Feistel Ciphers
5. https://www.geeksforgeeks.org/lsb-based-image-steganography-using-matlab/
6. https://en.wikipedia.org/wiki/Feistel_cipher
7. Patel, F. R., & Cheeran, A. N. (2015). Performance Evaluation of Steganography and AES encryption based on different formats of the Image. Performance Evaluation.

Efficient Data Management for Blood Banks: Integrating Data Warehousing, Database Management, and Location Tracking APIs

Sakshi, Anitesh Raj*, Swati sucharita Roy, Bharat J R Sahu

ITER, Siksha 'O' Anusandhan deemed to be University, Bhubaneswar, Odisha, India

Abstract Blood banks provide life-saving blood and blood products to patients in need. However, they require efficient data management and analysis techniques to optimize inventory management, organize donor information, and blood inventory data. This work proposes the design and implementation of a data-driven Blood Bank Management System using data warehousing (a technique for storing large amounts of structured data) to store and organize information on blood inventory. The system is implemented using the Python programming language and a PL/SQL database. The system also updates the donor locations automatically using APIs that access the location data from the donors' mobile devices. The system has several features such as a distributed search and filter system for donors and blood banks, a donation reminder, and statistics on successful donations. It also improves the accuracy and reliability of blood bank data and helps to improve the quality of healthcare facilities by using a centralized database that can easily keep track of blood products, expiry dates, quantities, and locations. The automatic update of donor locations is a key feature of the system that enables the system to match donors with nearby blood banks and notify them of urgent requests for blood donations. This feature not only saves time and resources for both donors and blood banks, but also increases the chances of finding compatible donors for rare blood types and emergency situations. This study is limited by the fact that the original database management system for blood banks was only tested for a single Blood Donation Organization. Further research and implementation of data warehousing and other business techniques can effectively help in exploring the effectiveness of the program in different Blood Bank settings, identifying problems that may arise, and finding solutions. In this article we have proposed a novel approach to the blood bank management system that integrates data warehousing, database management, and APIs for location tracking. It demonstrates how data-driven solutions can improve the performance and sustainability of blood banks and contribute to better health outcomes for patients who need blood transfusions.

---

*Corresponding author: Aniteshraj10@gmail.com

DOI: 10.1201/9781003489443-12

Keywords Healthcare management, PL/SQL, Data warehousing, Blood bank, Management system

1. Introduction

Blood transfusion is a vital medical procedure that can save lives and improve health outcomes for patients with various conditions. However, blood transfusion also poses significant challenges and risks, such as blood shortages, contamination, wastage, and incompatibility. Therefore, efficient data management for blood banks is essential to ensure the optimal use of blood products and to minimize the potential harm to donors and recipients. Data management for blood banks involves collecting, storing, processing, analysing, and sharing information about blood donors, donations, inventories, distribution, and utilization. However, existing data management systems for blood banks are often fragmented, outdated, or inadequate to meet the growing and complex demands of blood transfusion services. This paper argues that efficient data management for blood banks requires integrating data warehousing, database management, and location tracking APIs to ensure the quality, safety, and availability of blood products for transfusion.

Data warehousing is a technique that consolidates data from multiple sources into a centralized repository for analysis and reporting. Database management is a process that organizes data into structured and accessible formats for storage and retrieval. Location tracking APIs are application programming interfaces that enable the tracking of the location and movement of blood products using GPS or RFID technologies. By integrating these three components, blood banks can achieve the following benefits:

- Enhance the accuracy and timeliness of data collection and reporting.
- Improve the efficiency and effectiveness of blood inventory and distribution.
- Reduce the risk of errors, fraud, or misuse of blood products.
- Increase the transparency and accountability of blood transfusion services.

The proposed article discusses the current state of data management for blood banks, the challenges and opportunities of integrating data warehousing, database management, and location tracking APIs, and the implications and recommendations for blood transfusion policy and practice.

2. State of the Art

Data management systems for blood banks are systems that can handle the information of blood bags, donors, recipients, and transactions in a blood bank. They can also assist the blood test results, the quality, availability and the distribution of blood. They can reduce the paper work and save time for blood bank operations, as well as promote a healthy community of donors and recipients.

Several studies have proposed and implemented different data management systems for blood banks using various techniques and technologies. Some of the examples are:

- A web portal by e-RaktKosh [1] provides a centralized blood bank management system for India. The portal claims that it can enforce drug and cosmetic act, national blood policy standards and guidelines for blood bank operations. It can also provide features such as Aadhar linkage, decision support, dashboard, statutory reports
- Alimul Rajee et al. [2] presents a web-based blood bank management system that stores, processes, retrieves, and analyses data about blood bank administration. The paper claims that their system can supervise blood inventory management and other blood bank-related activities using online storage, updating, and retrieval facility.
- And several other papers that describes a database management system for blood bank applications using Oracle software. The paper claims that their system can provide data entry, sorting, and searching capability, report production, and statistical analysis functions for blood bank records.

These are some of the existing data management systems for blood banks in the literature. However, there are still some gaps and limitations in these systems, such as:

- Lack of integration of data warehousing, database management, and location tracking APIs for blood bank operations.
- Lack of performance evaluation using real-world data from a blood bank in India.
- Lack of security and privacy protection for the data of donors and recipients.
- Lack of scalability and adaptability to different contexts and scenarios.

Therefore, there is a need for a novel approach that addresses these gaps and limitations and provides an efficient data management system for blood banks.

3. Problem Statement and Motivation

There are several challenges faced by blood banks around the world, such as forecasting the demand and supply of blood, finding suitable locations for blood donation campaigns, securing the blood supply chain and protecting the data privacy of donors and recipients, managing the inventory and distribution of blood units, and ensuring the safety and compatibility of blood transfusions. Therefore, there is a timely requirement for an effective and secure management system for blood banks.

The existing data management systems for blood banks have some gaps and limitations that hinder their efficiency and quality. Some of these gaps and limitations are:

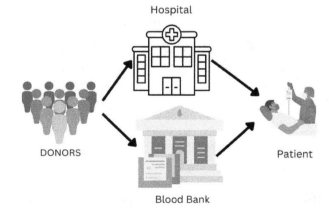

Fig. 12.1 Diagram illustrating the lack of centralization in current blood bank management systems

- Lack of integration of data warehousing, database management, and location tracking APIs for blood bank operations. Data warehousing can help to store and analyse large amounts of data from different sources and provide decision support. For example, data warehousing can help to forecast the demand and supply of blood based on historical data, seasonal trends, demographic factors, etc.

- Database management can help to organize and manipulate data in a structured and consistent way. For example, database management can help to store and retrieve the information of blood bags, donors, recipients, and transactions in a blood bank. Location tracking APIs can help to track and monitor the location of blood bags, donors, recipients, and vehicles. For example, location tracking APIs can help to find the nearest available blood bag or donor in an emergency situation. Integrating these technologies can improve the performance, accuracy, and reliability of blood bank operations [3].

- Lack of performance evaluation using real-world data from a blood bank in India. India faces a shortage when it comes to the amount of blood donated. The gap in demand and supply is widened due to mismanagement and inefficient databases. Evaluating the performance of a data management system using real-world data from a blood bank in India can help to validate its effectiveness and applicability in a challenging context. For example, evaluating the performance of a data management system using real-world data from a blood bank in India can help to measure its impact on reducing the wastage of blood, increasing the availability of blood, enhancing the quality of blood, etc [4].

- Lack of security and privacy protection for the data of donors and recipients. The data of donors and recipients is sensitive and confidential, as it contains personal information such as name, address, contact number, blood group, and medical history. This data should be protected from unauthorized access, modification, or disclosure. However, the existing data management systems for blood banks do not provide adequate security and privacy measures to safeguard this data. For example, the existing data management systems for blood banks do not use encryption or authentication techniques to protect the data from hackers or intruders. They also do not comply with the legal or ethical standards for data privacy such as consent or confidentiality.

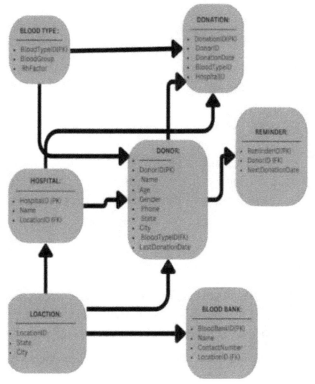

Fig. 12.2 Flowchart for proposed system

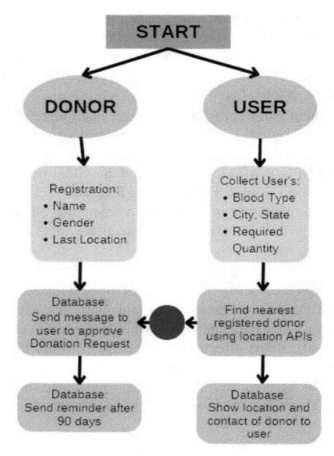

Fig. 12.3 ER Diagram for possible database

- Lack of scalability and adaptability to different contexts and scenarios. The data management systems for blood banks should be able to scale up or down according to the changing needs and demands of the blood bank operations. They should also be able to adapt to different contexts and scenarios such as emergencies, disasters, pandemics, etc. However, the existing data management systems for blood banks are not flexible enough to accommodate these variations. For example, the existing data management systems for blood banks are not able to handle sudden spikes or drops in the demand or supply of blood due to natural calamities or outbreaks. They are also not able to adjust to different regulations or requirements in different regions or countries.

4. Solution

The study findings highlight the existing challenges in data management within blood banks, such as the lack of standardization, manual processes, and inadequate data security measures. In response to these challenges, a comprehensive solution is proposed.

The proposed solution focuses on developing an advanced blood bank system that incorporates:

- Standardized data management practices, streamlines processes, enhances data security measures, and facilitates efficient donor-recipient matching.

- Through the adoption of this proposed solution, coordination and collaboration between the various blood banks involved could be greatly enhanced.
- Many of the manual processes currently in place may be automated thereby increasing overall efficiency.
- Most importantly, the privacy as well as confidentiality of each individual donor's information would be meticulously safeguarded.

Conclusion

The proposed approach, if implemented, could dramatically streamline and optimize the lifesaving mechanisms involved in the intricate transfer of this vital red fluid between suffering individuals. The implementation of standardized data management practices will enable better data exchange and sharing among blood banks, leading to improved coordination and timely matching of donors with recipients. Automation of manual processes will reduce errors and save valuable time and resources, allowing blood banks to focus on their core mission. Additionally, robust data security measures will protect sensitive donor information, ensuring compliance with privacy regulations and maintaining public trust.

REFERENCES

1. Government Blood Management System: https://www.eraktkosh.in/
2. Alimul Rajee, Junaid Ahmed, Hridoy Chandra Das, Md Fuad Hasan, Title: Blood management system, 2022[2]
3. Sri A. S. S. M. Pravallika, M. Tharun, O. Satish Kumar, K. Sridevi, K. Balaji, Title: A Systematic Review on Blood Bank Information Systems, 2023
4. P A Morel, Title: Database management systems for blood bank applications, 1998[3]

PicoPager: An Opensource Raspberry-Pi Pico W Based IP Paging System

Manish Kumar Tiruwar[1] and Bharat Jyoti Ranjan Sahu[2]

Institute of Technical Education and Research, Siksha 'O' Anusandhan,
ITER College Rd, Jagmohan Nagar, Bhubaneswar, Odisha, India

Abstract In restricted environments such as schools, hospitals, and medical facilities, the use of handheld electronics is often limited, necessitating alternative communication methods. Pagers have traditionally been employed in these critical spaces; however, they face constraints in scalability, reach, and flexibility. To overcome these limitations, we propose an innovative solution in the form of a TCP/IP paging system that leverages Internet Protocol (IP) networks. Our project focuses on the implementation of an open-source TCP/IP-based simplex paging system using the Raspberry Pi Pico W microcontroller. It is important to note that the scope of our paging system is limited to areas where legally and feasibly transmitting within the 2.4GHz frequency range is possible. This solution aims to enhance communication capabilities in restricted environments by utilizing TCP/IP networks, offering improved scalability, reach, and flexibility compared to traditional paging systems.

Keywords Picopager, IP pager, IP transreciever, Simplex communication, Raspberry Pi Pico W

1. Introduction

In certain environments such as schools, hospitals, and medical facilities, the use of handheld electronics like phones, tablets, and walkie-talkies may be restricted due to various reasons, including privacy concerns, interference with medical equipment, or maintaining a focused learning environment. As a result, alternative communication methods are required to facilitate effcient and reliable communication within these critical spaces. Pagers have long been utilized

---

[1]tiruwar.nibaran@gmail.com, [2]bjrsahu@gmail.com

DOI: 10.1201/9781003489443-13

in such environments; however, traditional paging systems often have limitations in terms of scalability, reach, and flexibility.

To address these limitations and provide an innovative solution, we propose a novel TCP/IP paging system based on the Raspberry Pi, a low-cost, credit-card-sized single-board computer. Our system utilizes an ad hoc network, a decentralized network infrastructure that allows devices to establish direct communication without the need for centralized infrastructure or pre-existing network infrastructure.

The core idea behind our TCP/IP paging system is to leverage the capabilities of the Raspberry Pi Pico W to create a scalable and flexible communication platform. By utilizing the TCP/IP protocol suite, which forms the foundation of the Internet, we can take advantage of its robustness, reliability, and widespread adoption. The use of an ad hoc network allows our system to operate in environments where there may be limitations on existing network infrastructure or restrictions on the use of wireless devices.

The Raspberry Pi Pico W serves as the central hub of our system, acting as a paging server that coordinates the transmission of messages to individual pagers or pager groups. It communicates with the pagers over the ad hoc network, establishing direct connections and enabling effcient and real-time communication. The pagers, designed to be compact and portable, receive and display messages sent by the server, ensuring reliable and instant communication between the server and the intended recipients.

By utilizing the Raspberry Pi Pico W and TCP/IP technology, our TCP/IP paging system offers several advantages over traditional paging systems. Firstly, it provides improved scalability, allowing for the inclusion of a larger number of pagers and accommodating the growing needs of the environment. Secondly, the system offers greater reach, as the ad hoc network enables direct communication between devices without the need for existing network infrastructure. Finally, the flexibility of our system allows for customization and adaptation to the specific requirements of different environments, ensuring optimal performance and seamless integration.

Fig. 13.1 A top view of our implementation

Fig. 13.2 A top view of 3d printed filed in PLA

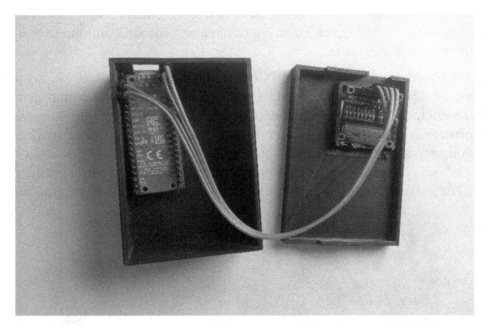

Fig. 13.3 A top view of snap fit Raspberry Pi Pico W and SSD1306 oled

Pico Pager 61 is our implementation of an Open-Source IP Pager. While PicoPager is based on Raspberry Pi Pico W [1], which is a cheap dual core 16-bit microcontroller with an onboard CYW43-series wifi chip [2]. PicoPager includes a SSD1306 [3] oled for display. For wire management3 2 we have included a 3d printed case [4] (STL Files).

The case can accommodate a 18650 battery to make the Pico Pager truly wireless. The software implementation relies heavily on Arduino's Wifi Library[5]. To utilize the onboard wifi chip (CHW43-series), our wifi implementation borrows from Esp8266 Arduino Core [6][7][8].

Especially for TCP stack implementation as in Wificlient[7] and Wifiserver[[8] classes. To utilize oled (SSD1306) a custom micropython script was written.

Rather than Buying expensive paging system, my paging system provides the following benefits: (a)allows user to define the Sub-network, (b)create Custom Encryption for message passing, (c) and monitor all messages over the Webpage interface, (d) gives users the right to repair and resell.

Apart from primary objective; For wire management: I have designed a 3D Printed Case 2 (parametric, snap fit). The firmware will be released under GPLv3. The hardware will be released under OHL CERN v2-S. All STL files have been released under Attribution-ShareAlike 3.0 (CC BY-SA 3.0). By releasing this as an Open Source project, we hope that the online Tinkerer Community will continue to develop the project, and I expect collaboration in future.

A Github documentation with software release will be provided [9].

2. Literature Survey

Pico Pager is an IP based simplex communication device which works over the TCP/IP protocol. IP paging system utilize their existing IP network, and broadcast messages to various nodes connected over the network. The transmitter (base node) transmits data over the IP network. The transmitter collects data from the paging controller. My aim is to develop a paging system like Motorola ADVISOR II [10], but in an IP paging Layout. The ADVISOR II pager provided alphanumeric messaging that was ideal for demanding business environments.This product has been discontinued. Motorola made two Advisor models [11] that could communicate on different combinations of UHF, VHF, and 900-megahertz bands (the frequency was user selectable). They offered at-the-time blazing fast transmission rates of 1600, 3200, or 6400 baud. The communications protocol was a one-way system called Flex; it was created by Motorola and used primarily for its pagers (a later version, called ReFlex, was two-way).

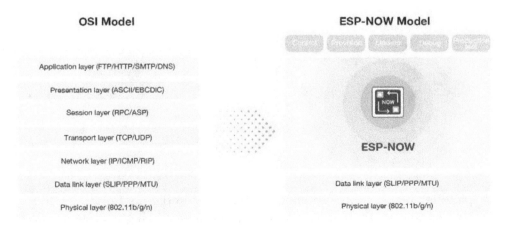

Fig. 13.4 A 3-layer ESP-NOW protocol vs full OSI stack

While there is ongoing chip shortage in the world, my aim is to provide an affordable, reliable paging system. I was inspired to use Expressif's ESPNOW, [12]protocol; which is only a

3 layer protocol 4. But since The protocol is not compatible with IP paging, I opted for a full OSI implementation.

3. Methodology

3.1 System Architecture Design

Raspberry Pi Pico W has an on-board 2.4GHz wireless interface using an Infineon CYW43439. The WiFi driver is limited as of now, but fully functional for sending and receiving data. Since Combined STA/AP mode is currently not supported, we are using the setup as a one - to -many configuration. (hence the Paging System.) Due to Onboard Flash limitations the no. of connected receivers is limited to 16 . PicoPager is an IP Paging System, where one host (Raspberry Pi Pico W) is broadcasting a String, and receivers display the message on (SSD1306) OLed according to the preprogrammed function. (which concatenates the broadcasted string).

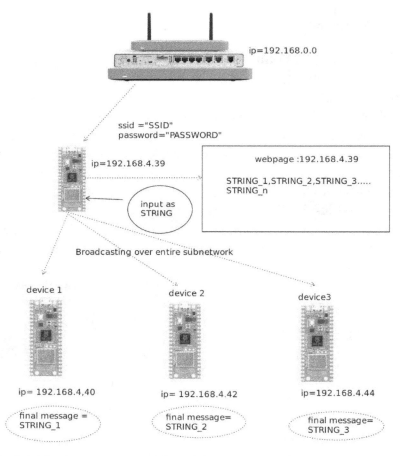

Fig. 13.5 A pictorial block representation of how my implementation. At the core of our implementation is the paging controller. It connects to a given network with password and SSID given. It then collects string from user via terminal window. And broadcasts a encypted string on its IP. This broadcasted string can be viewed in the browser on any device in the network. The nodes have the SSID, Password, IP of Paging controller, Node id. According to the Node Id, each node concatenates the string, decrypts the string and displays on SSD1306.

3.2 Raspberry Pi Pico Programming

Since we wanted to optimise the usage of system resources such as memory, processing power, and input/output(I/O) capabilities. We opted to program in C utililising the SDK 1.5.1[13] provided by Raspberry Pi.

The software implementation relies heavily on Arduino's Wifi Library[5]. To utilize the onboard wifi chip (CHW43 -series), our wifi implementation borrows from Esp8266 Arduino Core[6][7][8]. Especially for TCP stack implementation as in Wificlient[7] and Wifiserver[[8] classes. To utilize oled (SSD1306) a custom script was written.

A Github documentation with software release will be provided [9].

3.3 Implementation

After hardware connection is estabilished, install firmware PicoPager.uf2 on each of your devices. Then by default a Secret.py file is present. Change the parameters in the file. Include SSID, Password, Mode, and String to be broadcasted. After changing parameters reboot the Pico W.

For detailed documentation please refer to github page [9].

Fig. 13.6 A top view of our implementation where random number is sent to each of the nodes. Raspberry Pi 4B is used to create a local network. And the IP paging controller is used to send random number to each of the nodes

4. Conclusion

In summary, the implementation of a TCP/IP paging system using Raspberry Pi Pico has proven to be a versatile and effective solution for communication in restricted environments. By harnessing the capabilities of the Raspberry Pi Pico microcontroller and leveraging TCP/IP networking protocols, we have developed a custom firmware that facilitates reliable message transmission and seamless interaction with pagers. The ad hoc network configuration ensures flexibility and scalability, while our optimization techniques ensure effcient utilization of system resources. Rigorous testing and integration have confirmed the system's functionality, performance, and compatibility. The customizable nature of this solution demonstrates its potential to greatly improve communication in environments like schools, hospitals, and industrial facilities, where traditional handheld devices face limitations. Overall, the successful implementation of this TCP/IP paging system underscores its effectiveness as a reliable and scalable communication solution in restricted spaces.

Results. Picopager has achieved all criteria we had set: (a)allows user to define the Sub-network, (b)create Custom Encryption for message passing, (c) and monitor all messages over the Webpage interface, (d) gives users the right to repair and resell.

Acknowledgments. I would like to thank my supervisor Dr. Bharat Jyoti Ranjan Sahu, for providing guidance which was critical in completing the project.

I would thank the Arduino Community for providing software and library support for PicoPager. Most of my hardware and software debugging issues were solved.

REFERENCES

1. Raspberry Pi Ltd.: Raspberry Pi Pico W Datasheet. (2023). Raspberry Pi Ltd.. Pdf Datasheet at https://datasheets.raspberrypi.com/picow/pico-w-datasheet. pdf
2. Cypress Semiconductor Corp: Infineon CYW43439 DataSheet. (2021). Cypress Semiconductor Corp. Pdf Datasheet at https://www.mouser.com/datasheet/2/196/Infineon CYW43439 DataSheet v03 00 EN-3074791.pdf
3. Solomon Systech: SSD1306 Datasheet. (2008). Solomon Systech. Pdf Datasheet at https://cdn-shop.adafruit.com/datasheets/SSD1306.pdf
4. Tiruwar, M.K.: PICOPAGER. Stl files at https://www.tinkercad.com/things/2c5UP6EOW2Y (2023)
5. Arduino.cc: arduino-libraries/WiFi. Library files at https://github.com/arduino-libraries/WiFi
6. ESP8266Forum: Esp8266 Arduino Core. Library files at https://github.com/esp8266/Arduino
7. ESP8266Forum: Esp8266 Wificlient. Library files at https://github.com/arduino-libraries/WiFi/blob/master/src/WiFiClient.h
8. ESP8266Forum: Esp8266 WiFiserver. Library files at https://github.com/arduino-libraries/WiFi/blob/master/src/WiFiServer.h
9. Tiruwar, M.K.: PICOPAGER. Documentation and releases athttps://github.com/manishkumartiruwar/PICOPAGER
10. Motorola: ADVISOR II™ Flex Alphanumeric Pager. Documentation and releases athttps://www.motorolasolutions.com/content/dam/msi/docs/business/products/pagers/advisor ii/ documents/static files/advisor ii flex spec sheet.pdf

11. IEEE: THE CONSUMER ELECTRONICS HALL OF FAME: MOTOROLA ADVISOR PAGER. Documentation at https://spectrum.ieee.org/the-consumer-electronics-hall-of-fame-motorola-advisor-pager
12. Espressif: ESP-NOW. Description at https://github.com/arduino-libraries/WiFi/blob/master/src/WiFiServer.h
13 Pi, R.: SDK 1.5.0. Release at https://github.com/raspberrypi/pico-sdk

Multivariate Machine Learning Approaches for Dynamic Prediction of Air Quality and Estimating Heatwave Occurrence

Abhinandan Roul[1], Shubhaprasad Padhy[2], Sambit Kumar Sahoo[3], Ayush Pattanayak[4], Manoranjan Parhi[5], Abhilash Pati[6]

Department of Computer Science and Engineering, Siksha 'O' Anusandhan (Deemed to be) University, Bhubaneswar, Odisha, India

Abstract Pollutants in the air lead to degradation in the air quality which leads to global warming and unpredictable heatwave occurrence. Air pollution has been a leading cause of respiratory diseases and premature deaths. To forecast AQI and temperature, a multivariate approach based on AR-Net and Temporal Fusion Transformer (TFT) is proposed. A combination of atmospheric and meteorological variables as input features to train the model, including temperature, humidity, wind speed, and pollutant concentrations (PM2.5, PM10, SO2, NO2). An open-source dataset of Air quality in India is used to train the models and evaluate the experiments. With a MAPE of 7% for AQI and MAPE of 4% for heatwave prediction, it is concluded that the results demonstrate appreciable accuracy in predicting AQI and the occurrence of heatwaves.

Keywords Air quality, Time series forecasting, Deep learning, Climate change, Heatwaves

1. Introduction

Air quality is a critical factor that affects public health and the environment. Accurate prediction of air quality is essential in providing timely warnings and supporting decision-making. In addition, the occurrence of heat waves can exacerbate the impact of poor air quality on human health [1]. Traditional methods of air quality prediction rely on statistical models that use data from air quality monitoring stations. However, these models are limited by the availability and

---

[1]abhinandanroul@outlook.com, [2]pshubhaprasad@outlook.com, [3]sahoosambitkumar1234@gmail.com, [4]ayush766604@gmail.com, [5]manoranjanparhi@soa.ac.in, [6]abhilashpati@soa.ac.in

DOI: 10.1201/9781003489443-14

quality of data, as well as the complexity of the underlying processes that affect air quality and heatwave occurrence.

The use of multivariate machine learning approaches for dynamic prediction of air quality and estimating heatwave occurrence has significant implications for public health, urban planning, and environmental policy [2]. Accurate prediction of air quality and temperature can support decision-making related to health advisories, transportation planning, and energy consumption, while the estimation of heatwave occurrence can support emergency response planning and heat mitigation strategies.

In recent years, there has been a growing interest in the use of machine learning techniques to improve air quality prediction and estimate heatwave occurrence. This paper presents a study on the application of multivariate machine learning approaches, specifically Neural Prophet and Temporal Fusion Transformer, to predict air quality dynamics and estimate heatwave occurrence in a metropolitan area.

The process of predicting air quality index (AQI) and temperature involves several steps, including data collection, pre-processing, and feature extraction. First, data such as air quality measurements, meteorological data, and demographic data are collected from multiple sources, such as air quality monitoring stations and weather stations. The data is then pre-processed to remove outliers and missing values and to normalize the data for consistency [3]. Feature extraction involves selecting relevant variables that affect air quality and temperature, such as temperature, humidity, wind speed, and pollutant concentrations. These variables are used to train a machine learning model, such as Neural Prophet or Temporal Fusion Transformer, to predict AQI and temperature over time. The model is evaluated using metrics such as mean absolute error and root mean square error to assess its accuracy and effectiveness. Overall, the multivariate machine learning approach to AQI and temperature prediction involves a complex process of data collection, pre-processing, and feature extraction, followed by model training and evaluation to produce accurate and reliable predictions.

The study utilizes a combination of air quality, meteorological, and demographic data to develop a model that can accurately predict air quality levels and estimate the occurrence of heat waves. The results of the study demonstrate the potential of multivariate machine learning approaches in improving air quality prediction and estimating heatwave occurrence, with the ability to capture complex patterns and relationships in data.

This paper contributes to the field of air quality prediction and heatwave estimation by introducing a novel approach that utilizes multivariate machine learning techniques. The results of the study have implications for urban planning, public health, and environmental policy, as they can provide insights into the factors that affect air quality and heatwave occurrence and support decision-making related to these issues [4].

1.1 Research Gap and Research Questions

A variety of research has been carried out using the multivariate machine learning approaches for dynamic prediction of air quality and heat wave occurrence that had significant implications for public health, urban planning, and environmental policy decision-making related to health advisories. It should be noted that these studies are meant for transportation planning, and energy consumption, while the estimation of heatwave occurrence can support emergency response planning and heat mitigation strategies.

The following **research questions (RQs)** were taken into account for this investigation:

RQ1: What is the main goal behind air quality index prediction and how it is related to heat wave mitigation?

RQ2: What are the different factors that affect air quality and heatwave occurrence and support decision-making related to these issues?

RQ3: What are the benefits of measuring the air quality index in time series?

RQ4: What are the preventive measures against the change in AQI and heat waves?

RQ5: What are the major drawbacks faced during the prediction of particular time series?

1.2 Motivations and Objectives

The motivation behind using multivariate machine learning approaches for dynamic prediction of air quality and estimating heatwave occurrence is to improve the accuracy and timeliness of air quality predictions and heatwave occurrence estimates. This is important because poor air quality and heatwaves can have significant negative impacts on human health, the environment, and the economy. Multivariate machine learning approaches can use multiple variables and data sources to build models that can better capture the complex relationships between air quality, weather patterns, and other environmental factors. By incorporating real-time data, these models can provide more accurate and timely predictions, enabling authorities to take action to mitigate the impact of poor air quality and heatwaves.

The objective of using these approaches is to develop models that can accurately predict air quality and estimate the occurrence of heatwaves, to improve public health outcomes and reduce the economic and environmental costs associated with poor air quality and heatwaves. These models can also help policymakers to develop more effective strategies for managing air quality and mitigating the impacts of heatwaves.

1.3 Contributions

The significant contribution of the work can be noted as the following.

- Identification of future air quality and temperature prediction with ease.
- Comparison of various deep learning techniques to select the best model for each use case, i.e., AQI and heatwave prediction.
- Collection of new data from Telangana State Pollution control board, for fine-tuning to India region.
- Provides an easily implementable methodology for prediction, which is cost-effective.

1.4 Paper Layout

The remainder of the paper is organized as the following. Section 2 explains the recent research in this area. Section 3 explains the methodology and modeling techniques. This section is followed by the results and discussion section, where we compare the performances of various time series models. Section 5 mentions the conclusion, future scope, and areas of improvement of the paper.

2. Related Works

With the help of combined daily meteorological observation and captured daily fire pixel data from the Moderate Resolution Imaging Spectroradiometer (MODIS), Feng et al.'s work [5] focused to train their models using BPNN ensembles. In order to estimate daily fire pixel counts, the study used the climatological biomass combustion residues data from the Fire Inventory from NCAR (FINN). These estimates were then used to power the WRF-Chem regional air quality model Significant improvements in the precision of daily PM2.5 concentration estimates in Southern China were made by integrating the BPNN-ensemble-forecasted everyday biomass fire pollutant particulates. A decrease in the average inaccuracy of modeled surface PM2.5 values from -9.1% to -1.2% mirrored this improvement.

Using machine learning (ML) technologies, such as Support Vector Machines (SVM), random forests, and artificial neural networks, Khan et al.'s [6] study were aimed at creating a heatwave prediction model that is resilient to climate change. The study investigated the association between various ocean-atmospheric features and heatwave days (HWDs), stressing the necessity for a rolling approach in creating a robust climate forecasting model. With an a%NRMSE of 36, an R2 of 0.87, an md score of 0.76, and an rSD of 0.88 throughout the validation period, SVM outperformed the other ML algorithms in terms of predicting HWDs. These findings highlight the potential of the SVM model as a trustworthy tool for heatwave forecasting in the context of climate change.

According to a study by Asadollah et al. [7], a novel hybrid approach called decision tree (ABR-DT) and Ada-Boost Regression may be used to estimate annual heatwave days (HWDs) in Iran using synoptic predictors. The most effective structure was created by reducing the many predictors and their properties using the principal component analysis. Using just the particular humidity and wind components as predictors, the grid-point-specific performance evaluation demonstrated the superiority of ABR-DT, which displayed a correlation coefficient (CC) of 0.860 and a mean absolute error (MAE) of 6.929 as its metrics. The geographical performance indicators throughout Iran's eight distinct climate areas also demonstrated ABR-DT's superior performance, increased by 185 and 19%, respectively, to the CC and MAE of its two alternatives.

This study describes how model uncertainty is handled in accordance with the phases for developing ANN models, conducted by Cabaneros et al. [8]. The experiment which was based on 128 studies published between 2000 and 2022, shows that input uncertainty received more attention than the structural, parameter, and output uncertainties. Neuro-fuzzy networks have been used the most, then ensemble techniques. The application of techniques that may quantify uncertainty, such as bootstrapping, Monte Carlo simulation, and Bayesian analysis, was also constrained. It recommends creating and using methodologies that can manage and quantify the unpredictability related to the creation of ANN models.

This work suggests an air quality prediction model based on the improved VLSTM with multichannel input and multi-route output (IVLSTM-MCMR), according to research by Zhu et al. [9]. The IVLSTM and MCMR modules are part of the suggested model. The suggested IVLSTM module is created by strengthening the VLSTM inner structure in order to minimize the number of factors that contribute to the convergence's acceleration. A multichannel data input model (MC) with better linear similarity dynamic time wrapping is added in the MCMR module to choose the appropriate data for the IVLSTM input. A multi-route output model (MR), which outputs the results of several target stations with various attributes by various routes, is created to incorporate the findings from MC.

According to a study by Zeng et al. [10], deep learning (DL) algorithms for time series data forecasting, such as the long short-term memory (LSTM) neural networks and recurrent neural network (RNN) have garnered a lot of attention recently and have been used to forecast air quality indexes (AQIs). In this paper, a novel forecasting model that combines the Nested Long Short-Term Memory (NLSTM) neural network and Extended Stationary Wavelet Transform (ESWT) is presented for forecasting PM2.5 air quality. In terms of several error metrics, including, MAE, MAPE, RMSE absolute error, and R2 the findings demonstrate that the suggested technique surpasses state-of-the-art forecasting methods and recently published studies.

Sarkar et al. [11] researched a study that mentioned, various error-prone approaches, including R-Squared (R2), Root Mean Square Error (RMSE), and Mean Absolute Error (MAE) methods, are included in order to forecast the AQI value for Particulate Matter (PM2.5) m at a certain location in Delhi. This method combines the Gated Recurrent Unit (GRU) and Long-Short-Term Memory (LSTM) deep learning models to predict the AQI of the environment. Several machine learning (ML) and deep learning (DL) models, including LSTM, K-nearest neighbor (KNN), linear regression (LR), GRU, and support vector machine (SVM), are also trained on the same dataset to compare their performances with the proposed hybrid (LSTM-GRU) model. With an MAE value of 36.11 and an R2 value of 0.84, it is discovered that the suggested hybrid model performs superiorly.

A climate model was suggested in a study by Dumas et al. [12] as a different method of predicting the occurrence of intense, long-lasting heatwaves. This new method will be helpful for several important scientific objectives, such as the analysis of climate model statistics, the development of a quantitative proxy for resampling uncommon occurrences in climate models, the investigation of the effects of climate change, and eventually, forecasting. used large-class under-sampling, transfer learning, and 1,000 years' worth of climate model data to train a convolutional neural network. The trained network performs much better than the untrained network in predicting the presence of prolonged, intense heatwaves using the observed snapshots of the surface temperature and the 500hPa geopotential height fields.

Li et al. [13] proposed CEEMDAN-mvMDE-BVMD-RSO-KELM, a new hybrid AQI prediction model. First, CEEMDAN decomposes the AQI series into many intrinsic mode function (IMF) components, and mvMDE computes complexity of each component. Then, BVMD, a VMD improved by the BES method, is presented to tackle the problem of determining the decomposition level K and penalty factor of VMD, and BVMD is utilized to execute secondary decomposition of high complexity components. Then, the RSO-KELM method is developed, which optimizes the penalty coefficient and kernel parameter of KELM. RSO-KELM predicts the values of all IMF components. Reconstructing the forecast outcomes of all IMF components yields the final prediction results. The RMSE was found to be 2.921, MAE of 2.23 and MAPE of 0.04.

Yan et al. [14] proposed a multi-time, multi-site forecasting model for AQI with deep learning models with a focus on spatiotemporal clustering. The study was done on the Beijing area with data from China national environment monitoring center. The long short-term memory (LSTM) model was shown to be the best model for multiple-hour forecasting, while the convolutional neural network (CNN), LSTM, and BPNN all performed well in next-hour forecasting for the overall prediction.

Ren et al. [15] proposed a methodology using Takagi-Sugeno fuzzy model. It comprises of a hierarchical clustering based method, which is an improvement of BIRCH (Balanced Iterative Reducing and Clustering using Hierarchies). By including a refinement step for managing clusters with various forms, this technique enhances the BIRCH approach. The number of clusters is automatically determined using a cluster validity measure. The estimate of the model order selection and model parameters is then stated as a sparse optimization problem that can be addressed via global optimization. The model can learn the spatial-temporal correlations and localized trend patterns of multivariate time series data on air quality.

Das et al. [16] undertook a study in which the team used LSTM, RNNs and MLPs to predict PM10 and SO2 air pollutants. The modification of error term of traditional models were done, and missing data was handled. Dataset consisted of Basaksehir district, Istanbul. On evaluation it was found that LSTM performed better than RNN and MLP models.

Janarthanan et al. [17] proposed an SVM and LSTM combined model for AQI prediction. They used the Grey Level Co-occurrence Matrix (GLCM) to calculate the mean square error, mean, and standard deviation. Modified Fruit Fly optimization (MOFA) was used to improdve the extracted features. The data for the experiment was from Chennai region. The MSE for PM2.5 was found to be 0.179 and R2 was 0.821.

Table 14.1 Literature survey summary of papers

Work	Author	Methodology	Findings	Advantages	Limitations
5	Feng et al.	• MODIS-recorded fire occurrences • BPNN ensembles-predicted fire occurrences	• Applied to the normalized mean bias between the simulated (PM2.5) concentrations is the NCAR Fire Inventory (FINN).	• The BPNN ensembles effectively predicted the fire pixel's daily fluctuation	• The accuracy of the BPNN model was just 70%.

Work	Author	Methodology	Findings	Advantages	Limitations
6	Khan et al	• Elimination method was used to identify the input variables by using recursive feature which was based on SVM. • Developed using window technology over a 5-year time step.	With an %N-RMSE of 36, R2 of 0.87, the SVM's greater ability to forecast HWDs was exhibited.	The forward rolling model based on SVM performed better in predicting heatwaves.	To make better decisions on the likelihood of heatwaves, forecast uncertainty may be evaluated.
7	Asadollah et al.	Decision tree and Ada-Boost Regression- A hybrid technique for heatwave forecast	ABR-DT performed better, proven from Grid point based evaluation. Correlation is 0.860, and MAE = 6.929, which used wind and specific humidity as input features.	Hybrid models performed better.	Inaccurate results due to alteration of predictor variables
8	Cabaneros et al	• Bootstrap method • Bayesian method • Fuzzy method • Sensitivity simulation • Genetic algorithm • Monte Carlo analysis • Ensemble technique	The emergence of ANN models has made it possible for researchers to produce precise AP forecasts without the theoretical knowledge necessary for conventional physics-based models.	The creation and use of techniques that can address and quantify the uncertainty that surrounds the creation of ANN models.	It is necessary to conduct more study on how to better report the accuracy and uncertainty of ANN model findings.
9	Fang et al	A deep learning network module with fewer parameters and greater use of previous data is initially presented as an IVLSTM structure.	The suggested IVLSTM-MCMR performs well in terms of precision and efficiency because it is able to react fast to variations in time series.	The more precise air quality forecast models FFA and STE both perform better than conventional methods.	The ability to understand periodicity and enhance forecast accuracy is more challenging.
10	Chen et al.	A framework for AQI forecast using deep learning, with wavelet transform and zero mean normalization.	Reduction in absolute error and R2 index, indicates that model fits the data accurately.	Exhibits better performance than others. ESWT NLSTM, framework is better in real life.	• Inconsistent data decomposition. • Unable to fit properly.

Work	Author	Methodology	Findings	Advantages	Limitations
11	Sarkar et al	• Data collection • Data pre-processing which includes Data Inputation, Data Aggregation, Data Normalization, and Feature Selection	• For forecast the AQI of additional cities suggested method can be expanded. • The MAE score of 36.11 and the R2 value of 0.84 demonstrate the hybrid model's superior performance.	To compare model performances with the suggested hybrid (LSTM-GRU) model to provide more precise information, models are developed on an identical dataset.	Exploring and using model hyper-parameter optimisation is possible.
12	Dumas et al.	Large-class undersampling, transfer learning, and 1,000 years' worth of climate model outputs used in CNN	TPR = 56.3 (12.9), FPR = 2.1 (without Transfer learning)	Predicted for three distinct degrees of rigour, beginning as soon as 15 days before the event (30 days before the event's conclusion)	Less positive occurrences are included in the datasets, which makes training's learning task more challenging.

3. Proposed Model

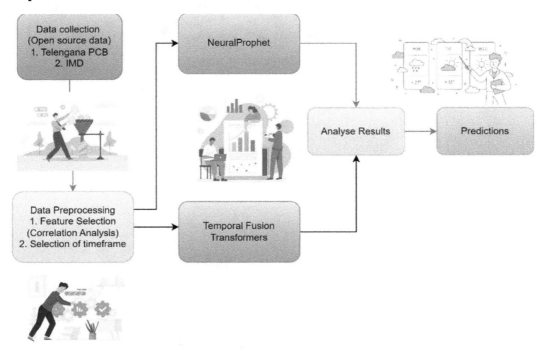

Fig. 14.1 Process flow for methodology

The proposed model and the process flow diagram for the methodology is depicted in Fig. 14.1. It discusses about the following steps.

- Data Collection: From provided sources as mentioned in previous slide.
- Pre-processing: Feature selection done using Correlation analysis.
- Min-Max Normalization is done for AQI, and temperature.
- Modelling: Trained using Neural Prophet, Temporal Fusion Transformer.
- Hyperparameter tuning: For determining best model

3.1 Data Pre-processing

Data is collected from the Telangana pollution control board, IMD, and Kaggle dataset (open source). The collected data is then compiled into a preferred CSV file for easy manipulation. This is followed by a selection of time duration for analysis. A correlation analysis is done to find out which features have a higher impact on AQI prediction.

	City	PM2.5	PM10	NO	NO2	NOx	NH3	CO	SO2	O3	Benzene	Toluene	Xylene	AQI	Air_quality
City	1.000000	0.062446	-0.021395	0.047182	-0.078589	0.024948	-0.182565	-0.255886	-0.101327	0.001530	0.002070	-0.062730	-0.047806	-0.121778	-0.053770
PM2.5	0.062446	1.000000	0.520767	0.435245	0.358287	0.375938	0.164615	0.093695	0.120451	0.166064	0.030539	0.132609	0.087813	0.633654	0.408614
PM10	-0.021395	0.520767	1.000000	0.433864	0.352396	0.416523	0.200548	0.023932	0.165468	0.208398	0.027522	0.126400	0.030257	0.447686	0.208080
NO	0.047182	0.435245	0.433864	1.000000	0.465848	0.747409	0.155804	0.221445	0.174681	0.020788	0.045202	0.151930	0.084393	0.438039	0.259407
NO2	-0.078589	0.358287	0.352396	0.465848	1.000000	0.561938	0.151731	0.361750	0.388413	0.289435	0.042179	0.297913	0.188965	0.531952	0.259199
NOx	0.024948	0.375938	0.416523	0.747409	0.561938	1.000000	0.129433	0.231878	0.205912	0.089095	0.049004	0.184764	0.085447	0.436282	0.212592
NH3	-0.182565	0.164615	0.200548	0.155804	0.151731	0.129433	1.000000	-0.014252	-0.048679	0.076295	0.004093	0.020040	-0.030328	0.097916	0.077250
CO	-0.255886	0.093695	0.023932	0.221445	0.361750	0.231878	-0.014252	1.000000	0.476784	0.043311	0.066022	0.285559	0.181215	0.650475	0.215848
SO2	-0.101327	0.120451	0.165468	0.174681	0.388413	0.205912	-0.048679	0.476784	1.000000	0.160489	0.038955	0.273314	0.249808	0.454182	0.170843
O3	0.001530	0.166064	0.208398	0.020788	0.289435	0.089095	0.076295	0.043311	0.160489	1.000000	0.023982	0.135498	0.084323	0.199100	0.030637
Benzene	0.002070	0.030539	0.027522	0.045202	0.042179	0.049004	0.004093	0.066022	0.038955	0.023982	1.000000	0.693310	0.094190	0.052027	0.037358
Toluene	-0.062730	0.132609	0.126400	0.151930	0.297913	0.184764	0.020040	0.285559	0.273314	0.135498	0.693310	1.000000	0.289731	0.288749	0.160761
Xylene	-0.047806	0.087813	0.030257	0.084393	0.188965	0.085447	-0.030328	0.181215	0.249808	0.084323	0.094190	0.289731	1.000000	0.186991	0.092216
AQI	-0.121778	0.633654	0.447686	0.438039	0.531952	0.436282	0.097916	0.650475	0.454182	0.199100	0.052027	0.288749	0.186991	1.000000	0.467065
Air_quality	-0.053770	0.408614	0.208080	0.259407	0.259199	0.212592	0.077250	0.215848	0.170843	0.030637	0.037358	0.160761	0.092216	0.467065	1.000000

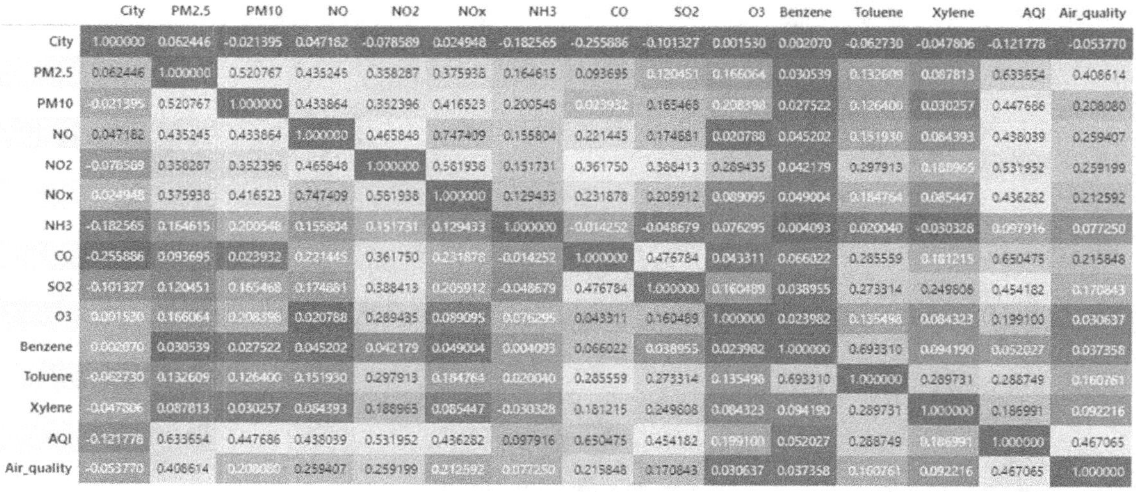

Fig. 14.2 Correlation analysis

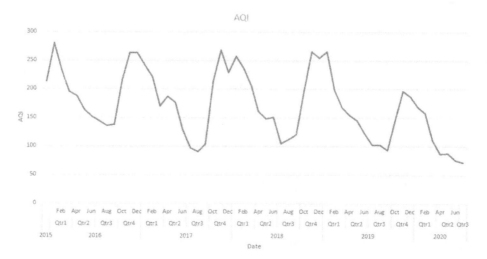

Fig. 14.3 AQI shows a repeating pattern

From Fig. 14.2, we can notice that PM2.5, PM10, NOx, and CO show a positive correlation coefficient with AQI. The positive correlation observed suggests that as the levels of PM2.5, PM10, NOx, CO increase, the AQI also increases and vice-versa. It's important to note that the positive correlation observed provides an indication of the statistical relationship between these variables.

In Fig. 14.3, the AQI levels for a region across the year is visualized, which indicates the seasonal trend.

3.2 Neural Prophet

We propose a methodology to use time series forecasting with all air quality parameters such as SO2, NO2, PM10, PM2.5, NOx, CO, Benzene, Toluene, and Xylene to model a multivariate approach.

First, the modeling was done with a univariate approach using, Meta's Neural Prophet framework. The model components include seasonality, trend, events, regressors, autoregression, covariates, and global modeling. It is the intersection of traditional methods and recent deep-learning methods. We fine-tuned the model with a trend regularization of 2, with a yearly seasonality.

The system learns to identify change points, or dates when a distinct deviation in trend happens. These dots are evenly initialized along the time axis. The cost of each of these linear regressions is then added to the model's total loss. Gradient descent reduces loss and hence improves regression. Specifically, the optimal parameter of linear regression is the slope, often known as the growth value.

Here, the use of yearly seasonality has been set to True, which enables the model to learn patterns that repeat every 365 days. From EDA, it was clear that the AQI follows a yearly seasonality. Here, the model assumes that the target (AQI) is a periodic and continuous function, which can be expressed as a Fourier series.

$$S_P(t) = \Sigma_{j=1}^{k} \left(a_j \cdot cos\left(\frac{2\pi jt}{p}\right) + b_j \cdot sin\left(\frac{2\pi jt}{p}\right) \right) \qquad (1)$$

Where k indicates the number of Fourier terms for the seasonality with periodicity (p).

In cases with multiple seasonality, n values differ, for each periodicity. The practice of regressing a variable's future value versus its past values is known as auto-regression (AR). A important component of many forecasting applications is auto-regression. The number of prior values contained is typically referred to as the AR (p) model's order p. Hence, a coefficient θ_i fits each historical value. Each coefficient θ_i determines the size and direction of the impact of a certain historical value on the projection.

$$y_t = c + \Sigma_{i=1}^{i=p} \theta_i \cdot y_{t-i} + e_t \qquad (2)$$

3.3 Temporal Fusion Transformers (TFTs)

The TFT, a novel attention-based structure, integrates multi-horizon forecasting with excellent accuracy and comprehensible temporal dynamics. To learn temporal correlations at different scales, TFT uses interpretable self-attention layers for long-term reliance and recurrent layers for local processing. TFT uses specialized components to select critical features and a series of gating layers to suppress redundant components, enabling greater performance in a range of conditions. The multivariate modeling was done using TFT architecture with using PyTorch. The time-varying known values were set as the features and AQI was set as unknown values.

3.4 AQI and Heatwave Prediction

We defined a time series forecasting problem with a maximum prediction length of 150 days (5 months), using the Pytorch Forecasting library. The data is assumed to be stored in a pandas DataFrame with a column "time_idx" representing the time index, and a column "AQI" representing the target variable to be predicted. There is also a categorical variable "City" which identifies the city to which each time series belongs.

Then a training dataset was created using the TimeSeriesDataSet class from the Pytorch Forecasting library. The training dataset is defined as a subset of the original data, with a training cutoff that is 150 days before the end of the time index. The encoder length is set to 720, which is twice the maximum prediction length, to allow for a long enough history to capture seasonal patterns. The dataset is grouped by city, and the target variable is normalized using the soft plus function and normalized by the group. The time index, as well as several real-valued features related to air quality, are included as time-varying known reals.

A validation dataset is also created using the same TimeSeriesDataSet class, with predict = True to indicate that it should be used for predicting the last 150 days of each time series. The validation dataset is created from the training dataset to ensure that the categorical encoding and normalization parameters are consistent.

Finally, the code creates data loaders for the model using the 'to_dataloader' method of the TimeSeriesDataSet class. The batch size is set to 128 for the training data loader and 1280 (10 times the training batch size) for the validation data loader. The PyTorch Data Loader class is used to load data in batches for efficient training of the model.

Next, the code sets up a PyTorch Lightning Trainer object to train and evaluate the model. The trainer is configured to run for a maximum of 30 epochs and log results to a TensorBoardLogger. Early stopping is used to stop training if the validation loss does not improve by at least 1e-4 for 10 consecutive epochs. The learning rate is also monitored and logged during training.

The TFT model is defined using the TemporalFusionTransformer class. This method initializes the model with the same categorical encoding and normalization parameters as the training dataset. Hyperparameters such as the learning rate, hidden size, attention head size, dropout rate, and output size (which determine the number of quantiles to predict) are used to train the model. The loss function is set to the QuantileLoss, which minimizes the quantile loss between the predicted and actual quantiles. The reduce_on_plateau_patience parameter reduces the learning rate by a factor of 10 if the validation loss does not improve after 4 epochs. The heatwave prediction is done, using TFT architecture, implemented using PyTorch forecasting. The collected data is preprocessed and trained for 30 epochs with 3 attention heads, a learning rate of 0.03.

4. Results and Discussions

Evaluation metrics are critical in time series models because they quantify performance and assess forecast accuracy. They allow for model comparison, model selection, and insights into the model's capacity to recognize trends and produce precise forecasts, thereby supporting informed decision-making.

4.1 Mean Absolute Percentage Error (MAPE)

MAPE (Mean Absolute Percentage Error) is a frequently used statistic to assess the precision of predictions provided by regression models. It calculates the percentage difference between a variable's expected value and its actual value.

$$M = \frac{1}{n}\sum_{t=1}^{n}\left|\frac{A_t - F_t}{A_t}\right| \tag{3}$$

Where, n **is the** iteration of the summation function, A_t **is the** Actual Value, F_t **is the** Predicted value

4.2 Root Mean Squared Error (RMSE)

Root Mean Squared Error (RMSE) is a statistic used to assess the accuracy of regression model predictions. It is calculated by taking the square root of the average of the squared discrepancies between the expected and actual values.

$$RMSE = \sqrt{\sum_{i=1}^{n}\frac{(\hat{y}_i - y_i)^2}{n}} \tag{4}$$

Where, $\hat{y}_i$ refers to the predicted values, y_i refers to the observed values, and n is the number of observations.

These two metrics are very useful for evaluation regression workloads, we efficiently tried to use this. Different experiments were conducted using NeuralProphet, TFTs, and NBeats models to determine the efficacy of each on AQI Prediction and heatwave prediction. The primary datasets used for this work are sourced from the Telangana state pollution control board.

4.3 Prediction of Each Feature

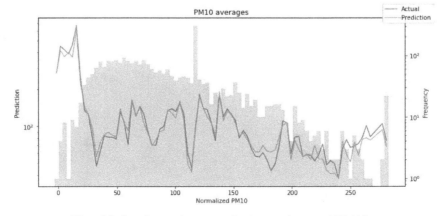

Fig. 14.4 Actual vs prediction values of PM10

128 Prospects of Science, Technology and Applications

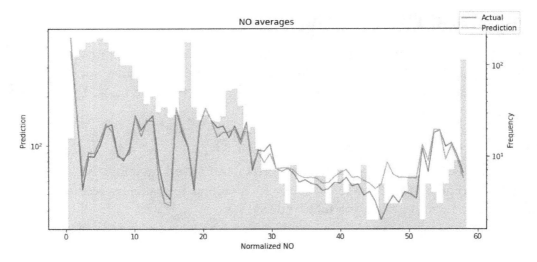

Fig. 14.5 Actual vs prediction values of NO

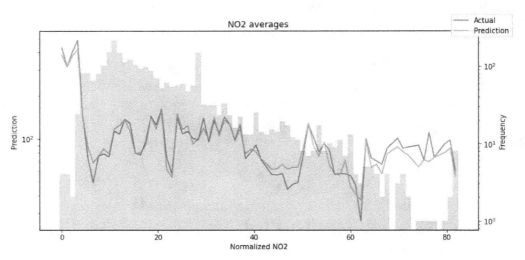

Fig. 14.6 Actual vs prediction values of NO2

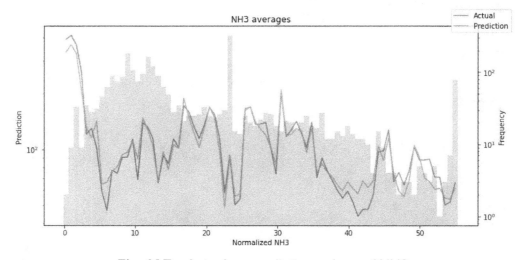

Fig. 14.7 Actual vs prediction values of NH3

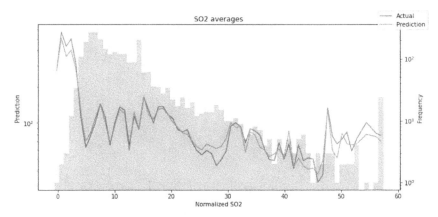

Fig. 14.8 Actual vs prediction values of SO2

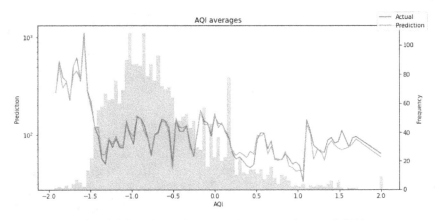

Fig. 14.9 Actual vs prediction values of AQI

The above i.e., Fig. 14.4 to Fig. 14.9 indicate the actual vs the predicted value for the PM10, NO, NO2, NH3, SO2 and AQI Components respectively. The blue line indicates the actual values of the parameter and the orange line indicates the predicted values. The figures show that it is possible to accurately predict the features using TFT and it provides a feasible technique to predict the AQI.

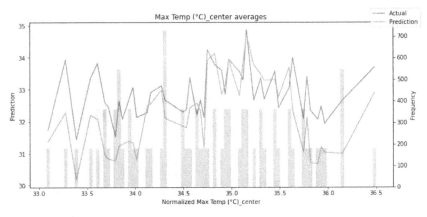

Fig. 14.10 Actual vs prediction values of Max Temp

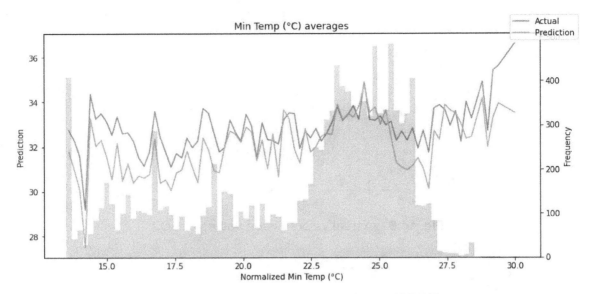

Fig. 14.11 Actual vs prediction values of Min Temp

Figure 14.10 and 14.11 depict the actual vs prediction values of Max Temp and Min Temperature.

On experimentation with different models and hyperparameters, it is found that multivariate forecasting using the Neural Prophet model with additive future regressors performed best for AQI prediction. SO2 and PM10 were used as the future regressors. The mean absolute percentage error (MAPE) is found to be 7%. For Heatwave determination, the best result is obtained by using TFT Model with a MAPE of 4%. The table below indicates the performance comparison of various models.

Table 14.2 Comparison of results from various models

Model	AQI	Heatwave (Max Temp)
SARIMA	MAPE = 19.38% RMSE = 22.38	MAPE = 23.33%, RMSE = 8.67%
LSTM	MAPE = 18.99%, RMSE = 19.72	MAPE = 5.1%, **RMSE = 2.32**
Neural-Prophet (Univariate)	MAPE = 0.17, (17%), RMSE = 0.3	MAPE = 0.05 (5%), RMSE = 0.09
NeuralProphet (Multivariate)	Future Regressors as SO_2, PM_{10} MAPE = 0.07 (7%) RMSE = 0.05	Future Regressors as max. Humidity, min. temp and precipitation MAPE = 0.1263 (12.63%) RMSE = 0.084
Temporal Fusion Transformer (TFT) (Multivariate)	MAPE = 0.14, (14%) RMSE = 24.6	MAPE = 0.04, (4%) RMSE = 1.66

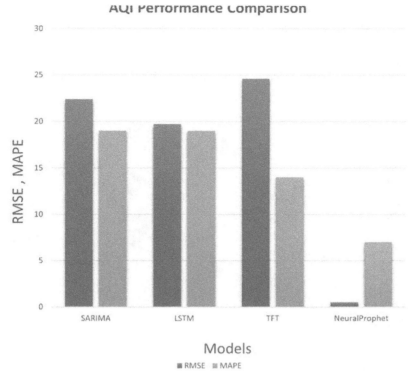

Fig. 14.12 AQI Error comparison

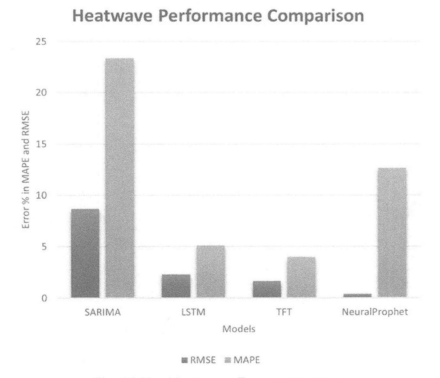

Fig. 14.13 Heatwave Error comparison

The Fig. 14.12 and 14.13 depicts the MAPE and RMSE error for AQI and heatwave predictions respectively. Here, the lower is the error the better is the performance of the model.

132 Prospects of Science, Technology and Applications

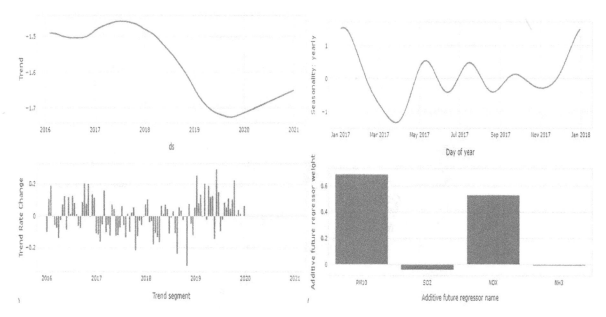

Fig. 14.14 Neural prophet model metrics for Multivariate AQI

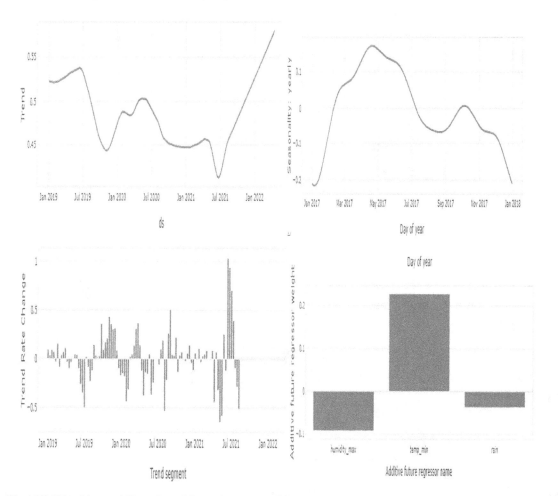

Fig. 14.15 Neural Prophet Model metrics for Temperature prediction (for Heatwave)

The visualization of the trend of data with seasonality is depicted in Fig. 14, and Fig. 15 for AQI and heatwave prediction. It is seen that AQI, PM10, and NOx values hold the highest weight as additive future regressors. A high weight for the variables indicates that the additional time series data provided is highly correlated with the time series being forecasted. This suggests that the exogenous variable included in the data may have a significant impact on the time series being forecasted and can help improve the accuracy of the forecast. Similarly for heatwave prediction, it is found that the maximum humidity and minimum temperature of the day provided a significant role in prediction.

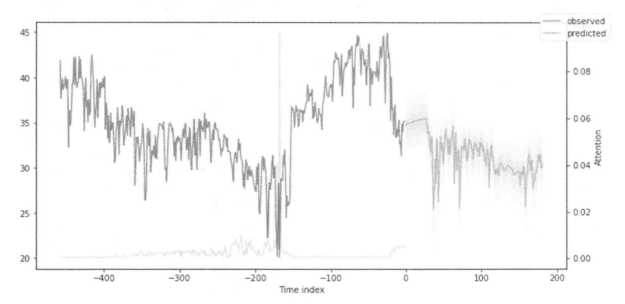

Fig. 14.16   Max Temp Prediction using TFT (for future timeline)

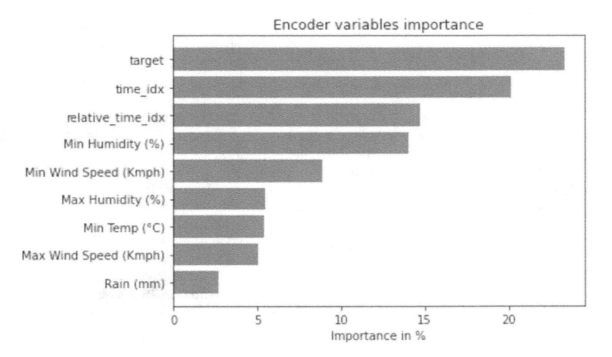

Fig. 14.17   Encoder variable importance as per TFT model

The fig 16 indicates the maximum temperature prediction. The blue line indicates the past data and the orange line depicts the future predictions. The fig 17 indicates the encoder variables importance as per the TFT model.

5. Conclusion and Future Scope

The study on AQI and Heatwave prediction is an accurate and reliable way to forecast and predict weather events. The research conducted on open-source data from IMD as well as Telangana PSB focused on the use of the latest DL architectures for univariate and multivariate forecasting techniques. The correlation analysis determined that PM10, PM2.5, SO2, and NOx were correlated with the AQI, and minimum temperature and humidity correlated with the temperature. This provided us an insight into the selection of variables as additive regressors for future forecasts. The use of NeuralProphet, Temporal Fusion Transformer(TFT) with extensive hyperparameter tuning provided us with the best results,i.e., MAPE of 7% on AQI (univariate) and MAPE of 4% on TFT (multivariate) for heatwave (Max Temp) prediction.

There are minor issues that need to be clarified for further studies such as a high weight for the Additive Future Regressor (AFR) variable does not necessarily imply causation. The high correlation between the AFR and the time series being forecasted may be simply a coincidence. Therefore, it is important to carefully interpret the results of the model and consider the underlying data-generating process when using the AFR in the neural prophet model.

REFERENCES

1. Kiyan, A., Gheibi, M., Akrami, M., Moezzi, R., Behzadian, and K.: A Comprehensive Platform for Air Pollution Control System Operation in Smart Cities of Developing Countries: A Case Study of Tehran. Environmental Industry Letters. Vol. 1 No. 1 (2023): Environmental Industry Letters (EIL), https://doi.org/10.15157/EIL.2023.1.1.10–27, (2023).
2. Balogun, A.-L., Tella, A., Baloo, L., Adebisi, N.: A review of the inter-correlation of climate change, air pollution and urban sustainability using novel machine learning algorithms and spatial information science, http://dx.doi.org/10.1016/j.uclim.2021.100989, (2021).
3. Kalajdjieski, J., Zdravevski, E., Corizzo, R., Lameski, P., Kalajdziski, S., Pires, I.M., Garcia, N.M., Trajkovik, V.: Air Pollution Prediction with Multi-Modal Data and Deep Neural Networks, http://dx.doi.org/10.3390/rs12244142, (2020).
4. Ravindra, K. et al.: Generalized additive models: Building evidence of air pollution, climate change and human health, http://dx.doi.org/10.1016/j.envint.2019.104987, (2019).
5. Feng, X., Fu, T.-M., Cao, H., Tian, H., Fan, Q., Chen, X.: Neural network predictions of pollutant emissions from open burning of crop residues: Application to air quality forecasts in southern China, http://dx.doi.org/10.1016/j.atmosenv.2019.02.002, (2019).
6. Khan, N., Shahid, S., Ismail, T.B. et al. Prediction of heat waves over Pakistan using support vector machine algorithm in the context of climate change. Stoch Environ Res Risk Assess 35, 1335–1353 (2021). https://doi.org/10.1007/s00477-020-01963-1

7. Asadollah, S.B.H.S., Khan, N., Sharafati, A., Shahid, S., Chung, E.-S., Wang, X.-J.: Prediction of heat waves using meteorological variables in diverse regions of Iran with advanced machine learning models, http://dx.doi.org/10.1007/s00477-021-02103-z, (2021).
8. Cabaneros, S.M., Hughes, B.: Methods used for handling and quantifying model uncertainty of artificial neural network models for air pollution forecasting, http://dx.doi.org/10.1016/j.envsoft.2022.105529, (2022).
9. Fang, W., Zhu, R., Lin, and J.C.-W.: An air quality prediction model based on improved Vanilla LSTM with multichannel input and multiroute output, http://dx.doi.org/10.1016/j.eswa.2022.118422, (2023).
10. Zeng, Y., Chen, J., Jin, N., Jin, X., Du, Y.: Air quality forecasting with hybrid LSTM and extended stationary wavelet transform, http://dx.doi.org/10.1016/j.buildenv.2022.108822, (2022).
11. Sarkar, N., Gupta, R., Keserwani, P.K., Govil, M.C.: Air Quality Index prediction using an effective hybrid deep learning model, http://dx.doi.org/10.1016/j.envpol.2022.120404, (2022).
12. Jacques-Dumas, V., Ragone, F., Borgnat, P., Abry, P., Bouchet, F.: Deep Learning-Based Extreme Heatwave Forecast, http://dx.doi.org/10.3389/fclim.2022.789641, (2022).
13. Li, G., Tang, Y., Yang, H.: A new hybrid prediction model of air quality index based on secondary decomposition and improved kernel extreme learning machine, http://dx.doi.org/10.1016/j.chemosphere.2022.135348, (2022).
14. Yan, R., Liao, J., Yang, J., Sun, W., Nong, M., Li, F.: Multi-hour and multi-site air quality index forecasting in Beijing using CNN, LSTM, CNN-LSTM, and spatiotemporal clustering, http://dx.doi.org/10.1016/j.eswa.2020.114513, (2021).
15. Ren, Z., Ji, X.: On prediction of air pollutants with Takagi-Sugeno models based on a hierarchical clustering identification method, http://dx.doi.org/10.1016/j.apr.2023.101731, (2023).
16. Das, B., Dursun, Ö.O., Toraman, S.: Prediction of air pollutants for air quality using deep learning methods in a metropolitan city, http://dx.doi.org/10.1016/j.uclim.2022.101291, (2022).
17. Janarthanan, R., Partheeban, P., Somasundaram, K., Navin Elamparithi, P.: A deep learning approach for prediction of air quality index in a metropolitan city, http://dx.doi.org/10.1016/j.scs.2021.102720, (2021).

Software Bug Classification Using Machine Learning Approach

Sandeep Soumya Sekhar Mishra[1]

Department of Computer Science and Engineering, ITER,
SOA Deemed to be University, Bhubaneswar, Odisha, India

Swadhin Kumar Barisal[2]

Center for Data Science, ITER, SOA Deemed to be University,
Bhubaneswar, Odisha, India

Abstract Software bug classification (SBC) is a crucial aspect of ensuring software reliability, performance and trustworthiness. Different traditional methods for bug classification are based on analyzing the data manually and it is time-consuming and subjective. The work presents an ML-based system of software bug categorization. The proposed system utilizes supervised learning paradigms like Decision Tree Classifiers, Random Forest Classifiers (RF) and Artificial Neural Networks (ANN) to train and test the classification models on a labelled dataset. Various features, such as day, test workers, number of faults and fault class are used to make three datasets. The obtained datasets are treated as the input for the machine-learning models. The datasets are separated to learning and evaluation sets for learning of models and valuation. The effectiveness of the approach is evaluated by considering multi-class classification scenarios. Valuation results present the accuracy of ML methods by using evaluation metrics like Root Mean Square Error (RMSE), Precision, Recall and F1 Score. Significant assessment parameters considered are RMSE and F1 Score. The optimum results are obtained for the Decision Tree, which are RMSE of 0.043, Precision is 0.63, Recall of 0.65 and F1 Score of 0.64. It indicates that the proposed system is giving promising results in the case of Decision Trees and Random forests.

Keywords Software bug classification, Machine learning algorithms, Precision, Recall, F1 score, RMSE

---

[1]sandeepmishra432@gmail.com, [2]swadhinbarisal@soa.ac.in

DOI: 10.1201/9781003489443-15

1. Introduction

The presence of bugs or faults in software is unavoidable in software development. A bug is a fault or defect within the software that leads the software to behave in ways that are neither expected by the users nor intended by the developers. It is generally created during the development process of any software. The main causes of the presence of software bugs are undefined requirements, reporting errors, programming or coding errors, poor communication, software complexity, insufficient testing and environmental issues. Bugs can produce various issues including program crashes, incorrect calculations, unexpected behavior, memory leaks, user interface problems, compatibility issues, security vulnerabilities, data corruption or loss, system instability etc. These issues can interrupt the proper functioning of the software applications and result in financial losses for organizations.

Timely identification and resolution of software bugs are essential for ensuring qualitative and dependable software. Bug classification involves categorizing bugs, based on their characteristics, attributes and severity. By categorizing the bugs, bug triage becomes more streamlined and thereby allowing the researchers to focus on the critical issues at first.

Traditional bug classification methods are time-consuming, subjective, and prone to human errors. The researchers are dependent on analyzing bug reports manually.

With the increasing complexity and scale of modern software applications, nowadays Machine Learning approaches have gained significant attention in SBC, due to their ability to analyze large and complex datasets. Machine Learning algorithms assimilate the systematic and interconnected arrangements that are contained in the dataset more effectively.

Several works have been conducted earlier on software bug classification using various techniques and approaches. Deep Learning-Dependent Bug Classification methods like Convolutional Neural Networks (CNN), Ensemble Learning Techniques like Random Forest and AdaBoost, Multimodal Learning, a combination of machine learning techniques with network analysis, machine learning algorithms with natural language processing techniques etc. are some examples.

The purpose of the approach is to investigate the proficiency of different ML techniques for SBC. By applying various algorithms to different datasets, the faulty modules are identified and removed. As a result, the accuracy and efficiency of the software is increased. The findings of this experiment can add value to the improvement of SBC methods, benefiting software developers and improving overall system performance.

This paper is organized to demonstrate the structure and layout of the proposed work. It consists of several key sections. The introduction provides background information, problem statements, objectives and research questions about the work. The literature review presents existing research and related works done in this field. The basic concept section describes a fundamental idea of a particular subject. The Proposed Approach section emphasizes on proposed design concepts and methods employed. The experimental details section provides applied requirements, environments as well as results. A comparison of proposed work with recent work done in this field is described in contrast to the recent state-of-the-art section. The conclusion section summarizes the work and gives direction for future scope. Before going to the proposed approach, some previous works done in this field are described below in the literature review section.

2. Literature Review

A literature study or review is a systematic investigation of published articles/paper studies on a particular subject matter or area of interest or research question. The — motto of this review is to furnish a comprehensive perception of the recent mode of research on a specific subject, identify areas of incomplete or insufficient knowledge in the previous studies, and establish a foundation for further investigation. Several recent studies contributed to the field of software bug classification using machine learning techniques, are described in this section. These studies have shown various approaches, algorithms, and techniques to improve the accuracy and efficiency of bug classification processes. Here is an overview of some recent works.

F Lopes, J Agnelo, C. A Teixeira, N Laranjeiro and J Bernardino [1] proposed Orthogonal Defect Classification (ODC), for the classification of software bugs. They evaluated ML algorithms like k-nearest Neighbors, Support Vector Machines, Naïve Bayes, Nearest Centroid, Recurrent Neural Networks, and RF. They used unstructured textual bug reports for classification. The experimental results revealed that using solely reports, it was difficult to classify certain ODC attributes automatically. They suggested that, if larger datasets are used, the accuracy of correctly classified instances in a classification task can be enhanced.

Saritha, Srilatha, Navya, Sri Lavanya and N. MD. Jubair (2022) [2] proposed a new planned model to promote the standardization of the XGBoost model by automatically boosting its parameters. They suggested that the intended models were intelligent in the prediction of software system defects and identified the categorization of bug severity. In future work, they said that the metrics of software systems within the learning process can be a method to enhance the correctness of the forecast model to make more accurate and precise predictions.

Ahmed Hafiza Anisa, Bawany Narmeen Zakaria and Shamsi Jawwad Ahmed in [3] applied Context-aware Predictive Bug (caPBug) classification framework by Natural Language Processing (NLP) and supervised ML methods. The dataset was obtained from Mozilla and Eclipse repositories. They used ML methods i.e., Naive Bayes, Decision Tree, RF and Logistic Regression to classify as well as order the bug according to the details. By using the RF classifier, the CaPBug framework achieved a correctness of 88.78% in categorizing. Similarly, for the CaPBug framework, 90.43% of accuracy was attained in forecasting the importance of fault reports. They planned to further develop the CaPBug framework by taking additional input points in the sample. Deep learning techniques may be implemented to improve the robustness as well as outcomes.

3. Basic Concepts

SBC using machine learning techniques involves the application of ML algorithms to automatically classify software bugs grounded on their characteristics & patterns. The objective is bug management and resolution by efficiently identifying and prioritizing software bugs.

In this section, the basic concepts of the proposed approach of software bug classification using ML techniques are described. Decision Tree, RF, and Multi-Layer Perceptron (MLP) are used for model training, model evaluation and bug classification.

The decision tree (DT) is a popular supervised ML technique. It is implemented for regression as well as classification problems. DT are widely used in various domains due to their transparency, comprehensibility, and capability to interpret categorical and numeric data. It is a tree-structured classifier consisting of roots, nodes, edges, and leaves. In the tree, the root node denotes the whole dataset and internal nodes are the features of the considered dataset. The branches represent the outcomes after the decision leading to further nodes or leaves. The leaf node represents the final result past classification. The algorithm for the decision tree is called the Classification and Regression Tree Algorithm (CART). In this method, decisions are utilized for the attributes and rules are established according to the attributes in the dataset. It asks questions and decisions are taken after the yes or no outcomes. However, the limitation is that a small change within any node may affect the performance of other nodes []. Decision Trees overfit the training dataset, which results in unsatisfactory prediction of hidden data. The complexity of the tree is reduced by removing the unnecessary branches and nodes and it is called Pruning. This enables handling overfitting and increases the model's capability to predict new data. They can be further enhanced implementing ensemble techniques such as RF or Extreme Gradient Boosting to improve performance.

RF is another popular ensemble technique which associates several decision trees to predict results. It is broadly applied for classification as well as regression problems. It offers improved accuracy and robustness compared to individual decision trees. It can handle large datasets. It is a collection of multiple decision trees, which trains the generated trees on the selected subsets of the training dataset. Individual decision trees are trained independently and contribute to the final prediction. It selects subsets of train sets, known as bootstrap samples, to train each decision tree. This process is known as *Bagging*. For every node of the decision tree, RF arbitrarily chooses a subset of the dataset to consider for splitting. Random feature selection further enhances the diversity among the decision tree and reduces the correlation between them. The Random Forest combines the predictions of individual trees through voting, where the majority class is chosen as the final output. In the case of regression problems, the outputs of each tree are averaged for obtaining the conclusive output. Random Forest can handle missing data, and outliers by using surrogate splits and averaging over multiple trees. It is reliable for noisy or incomplete data, making it suitable for real-world datasets. Random Forest can easily distribute the computation across multiple processors for the faster training (Parallelization) as each decision tree can be trained independently. RF is applied for more accurate prediction, generalization against overfitting, and the capability to understand complex data. RF is broadly applied in various domains, including finance, healthcare, and image recognition, where accurate and reliable predictions are required.

Multilayer Perceptron (MLP) is a widely used neural network architecture that is applied to various activities like classification, regression, pattern recognition etc. It belongs to the category of ANN which has multiple layers of connected nodes, neurons. Neurons take the inputs, calculate a sum of inputs multiplied by weights and biases, and use an activation function to evaluate the results. Each neuron in the input layer represents a feature or an attribute of the input data. MLP can have one or more hidden layers that are responsible for learning complex representations of the input data. Biases are added to each neuron to introduce a shift or offset in the output. An activation function is implemented with the weighted sum of inputs in each neuron to introduce nonlinearities. The output layer of an MLP calculates the concluding output depending on the transformed input of the unseen layers. The total number of neurons for the output layer relies on the specific task, particularly binary classification, multiclass classification, or regression. MLP performs forward propagation to compute the output by passing the input through the network layer by layer, applying the weights and biases, and applying the activation function at each neuron. Backpropagation is used to train the MLP by iteratively adjusting the weights and biases depending on the difference between the projected outcome and the actual target result. In this step, the gradient loss function using the weights and biases is calculated and used as the optimization for the model performance.

In model training, a machine learning model is trained using labelled bug data. ML algorithms such as decision trees and random forests are implemented. Similarly deep learning models like neural network is also employed. The model learns patterns and relationships from the training data to make predictions about unseen bugs.

In model evaluation, the trained model is first predicted. It is then assessed by applying evaluation parameters like root mean square error (RMSE), precision, recall, and F1 score. This assessment helps in determining the accuracy of the bug classification and identifying any potential issues or areas for improvement.

In bug classification, the trained and evaluated model is used to classify new unseen bugs. Bug reports, based on learned patterns and features are fed into the trained model. The model assigns the bugs to the appropriate bug categories.

Bug classification using ML techniques is an iterative process. The model performance can be enhanced by refining the attribute selection algorithm, exploring different algorithms, or incorporating domain-specific knowledge. Regular updates and fine-tuning of the model based on new bug data and feedback are essential for enhancing its accuracy and effectiveness. It is called Iterative Improvement.

Overall, software bug classification using ML techniques leverages the capability of ML algorithms to automate the process of categorizing as well as prioritizing bugs. It helps in efficient bug management, enabling developers to assign resources effectively and resolve software issues promptly.

4. Proposed Model

The proposed work aims to develop a software bug classification system using machine learning techniques. The focus is on leveraging supervised learning algorithms to accurately classify and organize bugs as per their characteristics and impact. The work involves dataset collection, feature extraction, model training, testing and evaluation. Different ML algorithms, for instance, DT, RF and MLP are explored as well as compared. The accuracies of the categorization systems are assessed using different assessment parameters.

5. Detailed Description

For the proposed approach to the model, at first, the bug datasets are collected. Various ML methods, such as DT, RF, and MLP are fit and evaluated to obtain faulty datasets. The trained model is utilized to classify new bug data into appropriate categories or classes based on their features. Evaluation metrics like RMSE, precision, recall and F1 score are implemented to calculate and contrast the outputs for the models.

Proposed model utilizes machine learning techniques to analyze and classify bugs considering key factors like days, number of test workers, number of faults, and fault classes. The model starts by collecting data on the number of days during which the bug classification system operates. This information helps in understanding the temporal aspect of bug occurrences and patterns. Next, the model considers the number of test workers involved in the bug classification process. This factor accounts for the human resources dedicated to identifying, analyzing, and classifying bugs. Models also consider the faults encountered within the system. By analyzing the frequency and distribution of faults, the model can identify common patterns and trends. Furthermore, the model incorporates information about different fault classes. Fault classes represent distinct categories or types of bugs. By classifying bugs into specific fault classes, the model enables more targeted bug resolution strategies and assists in prioritizing bug fixes.

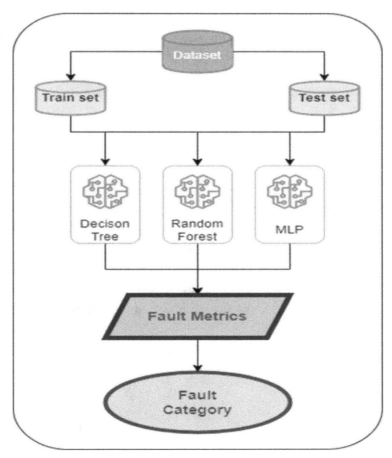

Fig. 15.1 Architecture of suggested method

The suggested methods are described in Fig. 1. Obtained inputs are partitioned into learn and valuation sets in 80:20 ratio. The train sets are used for training our models and are fitted with ML models like DT, RF and MLP. The obtained test set is then predicted with the ML models to categorize the faults. The valuation of the models are calculated by applying the assessment parameters like RMSE, Precision, Recall and F1_score.

6. Experimental Details

Some key components and steps for the proposed work on "Software bug classification using machine learning approach" are typically included in this section.

6.1 Experimental Setup

The experiment is carried out using a CPU having Intel(R) Core (TM) i5-8250U @ 1.60GHz 1.80 GHz processor, 8GB Random-Access-Memory and 250GB SSD. The computer program is Win10 Pro (64-bit) version 22H2. The codes used are in Python language, version 3.9. The Integrated Development Environment (IDE) used is SPYDER.

The experimental setup begins with the selection and preparation of a suitable dataset. Three datasets i.e., DS1, DS2 and DS3 are created. They consist of days and relevant attributes including the number of test workers, the number of faults as well as the respective fault classes. The DS1 dataset has 58 rows and 4 columns. DS2 has 110 rows and 4 columns. Similarly, DS3 contains 109 rows and 4 columns. The fault classes column is the output. ML algorithms like DT, RF and MLP are considered for bug prediction. The selected machine learning models are trained using the train sets. The remaining data i.e., the test set is then used for evaluation purposes. The predictabilities of the models are calculated by implementing assessment parameters, RMSE, recall, precision and F1 score. The average of the metrics of all considered datasets is calculated as the final result.

6.2 Results Discussion

Root Mean Square Error (RMSE), Precision, Recall and F1 Score are considered as evaluating parameters. Among these, the RMSE and F1 Score are considered key evaluation metrics.

The first used evaluation metric is RMSE, which is calculated by taking the root over of mean square error. Mean Square Error (MSE) be used for evaluating the average of squared variances among the calculated and given values. Variation between the estimated and real value of SFP output is evaluated. The algorithm works fine if the obtained RMSE value is less. Eq. (1) is the expression to find RMSE.

$$\text{RMSE} = \sqrt{\frac{1}{N} \Sigma_{i=1}^{N} (d_i - p_i)} \qquad (i)$$

Where:

Σ = "sum",

d_i = given output for i_{th} data,

p_i = predicted output for i_{th} data,

n = sample size

The next evaluation metric is Precision. It is measured by dividing the total count of accurately estimated positive occurrences (true positives) by the sum of count of occurrences projected as positive (both true positives and false positives). It focuses on how accurately the model calculates the positive predictions. The precision value ranges from zero to one. A greater result of the precision demonstrates the improved output of the model. Obtained precision values of the experiment are within the limit and are shown in Table I. Eq. (2) is used to calculate the precision value.

$$\text{Precision} = \frac{True\ Positives}{(True\ positives + False\ Positives)} \qquad (2)$$

Where:

True Positives: is the sum of accurately assessed positive occurrences. False Positives: is the sum of occurrences assessed positive and are originally negative.

The next evaluation parameter applied is Recall. It is a performance metric used in classification tasks. It evaluates the ratio of accurately assessed positive occurrences beyond the total true positive occurrences. Recall result varies between zero and one. A greater result within the limit shows the improved output of the model. The values for Recall are also within the limit and are shown in Table I. The recall value is calculated using (3).

$$\text{Recall} = \frac{True\ Positives}{(True\ positives + False\ Negatives)} \qquad (3)$$

Where:

True Positives: is the sum of accurately assessed positive occurrences. False Negatives: is the sum of occurrences assessed as negative but are originally positive.

The last metric used is F1 Score. The evaluation metric, F1 Score is used for classification problems to combine precision and recall in a single value. It considers both false positives (precision) and false negatives (recall) and provides a trade-off between the two metrics. The F1 score is especially convenient, once the goal stands to find a balance between precision and recall in spam detection tasks. It allows for a comprehensive evaluation of a model's performance by examining together the capability to prevent false positives as well as the capability to find true positives. The F1 Score occurs between zero and one. A greater result within the range shows the improved output of the model. In this study, the F1 Score value is appreciable and is shown in Table 15.1. Equation 4 shows the formula used to compute F1 Score.

$$F1\ Score = 2 * \frac{(Precision * Recall)}{(Precision + Recall)} \qquad (4)$$

Where:

Precision signifies the ratio of accurately assessed positive occurrences out of all the collective occurrences assessed as positive. The Recall represents the ratio of accurately assessed positive occurrences out of all real positive occurrences.

Table 15.1, 15.2 and 15.3 shows the results of evaluation metrics for the DT, RF and MLP respectively.

Table 15.1 Evaluation metrics for decision tree model

Dataset	RMSE	Precision	Recall	F1-Score
DS1	0.00	0.60	0.60	0.60
DS2	0.13	0.53	0.60	0.56
DS3	0.00	0.75	0.75	0.75
Average	0.043	0.63	0.65	0.64

Table 15.2 Evaluation metrics for random forest model

Dataset	RMSE	Precision	Recall	F1-Score
DS1	0.13	0.40	0.40	0.40
DS2	0.13	0.60	0.60	0.60
DS3	0.00	0.75	0.75	0.75
Average	0.09	0.58	0.58	0.58

Table 15.3 Evaluation metrics for multi-layer perceptron

Dataset	RMSE	Precision	Recall	F1-Score
DS1	0.18	0.20	0.20	0.20
DS2	0.18	0.52	0.53	0.52
DS3	0.38	0.43	0.35	0.37
Average	0.25	0.38	0.36	0.36

The above experimental results reveal that, for the decision tree, the average value of RMSE is 0.043, Precision of 0.63, Recall of 0.65, and F1-Score of 0.64. Similarly, for the random forest, the average of RMSE, Precision, Recall, and F1-Score are 0.09, 0.58, 0.58, and 0.58, respectively. In the case of the multi-layer perceptron (MLP), the average value of RMSE is of 0.25, Precision of 0.38, Recall of 0.36, and F1- Score of 0.36. For model evaluation, we consider RMSE and F1-Score.

The below Figs 15.2, 15.3, 15.4 and 15.5 show the graphical representation of the above averaged calculated evaluation metrics.

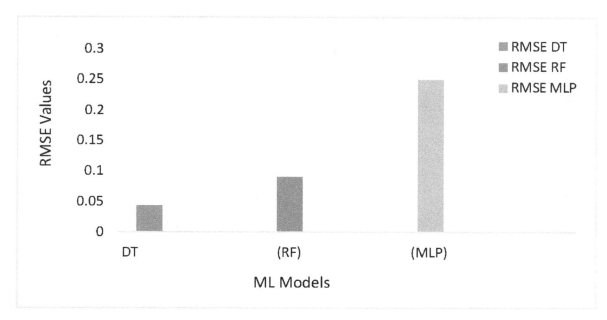

Fig. 15.2 Graphical representation of RMSE of different ML techniques

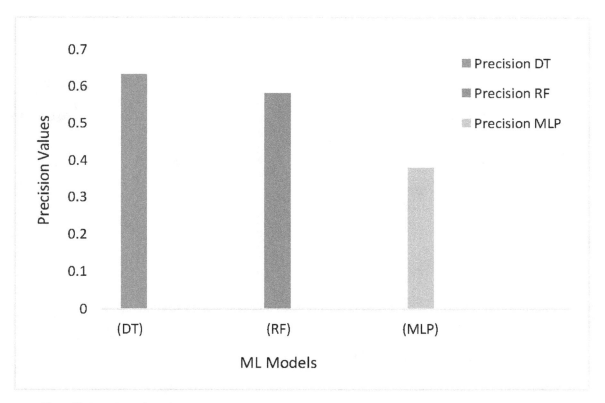

Fig. 15.3 Graphical representation of precision values of different ML techniques

146 Prospects of Science, Technology and Applications

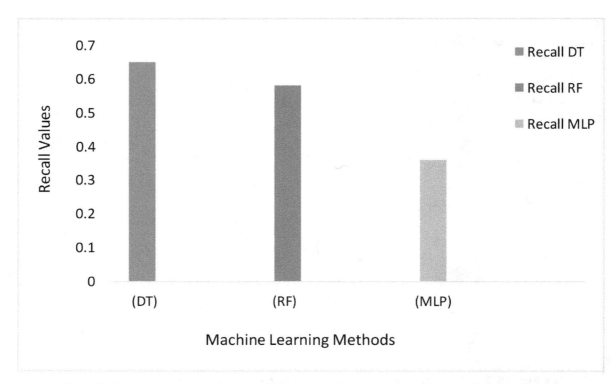

Fig. 15.4 Graphical representation of recall score of different ML techniques

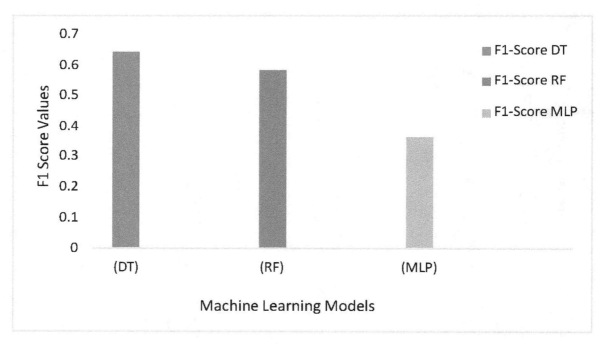

Fig. 15.5 Graphical representation of F1 score for different ML techniques

7. Comparison with State-of-the-arts

This section highlights different developments made within the realm of software bug classification leveraging ML algorithms. It determines the efficiency of these approaches in accurately analyzing and classifying bugs in software development processes.

Yu et al. recommended a bug categorization approach based on deep learning techniques. They have used a deep neural network architecture to learn features from bug data. It achieved improved classification accuracy in comparison to the conventional ML methods. However, their approach required a large amount of labelled data for training.

Li et al. offered a bug categorization approach applying a combination of traditional machine learning algorithms and feature selection techniques. They used training models such as SVM, Random Forest, and AdaBoost and achieved good classification accuracy, but the effect of attribute choice on classification assessment needs further exploration.

In this paper, ML algorithms in particular decision trees, random forests and multi-layer perceptron are used for software bug classification. The evaluation parameters are RMSE, Precision, Recall and F1 Score. This approach shows appreciable results in all the above evaluation metrics.

8. Conclusion

Software bug classification using machine learning techniques has emerged as a promising approach to mechanizing the identification and categorization of bugs.

In this paper, bug classifiers in particular decision trees, random forests and multi-layer perceptron (ANN) are proposed for analyzing and classifying software faults. The datasets contain features like days, number of test workers, number of faults and fault classes. RMSE, Precision, Recall and F1 Score are implemented to assess the predictability of proposed methods. Among all three algorithms, the decision tree shows better values for RMSE i.e., of 0.043, Precision is of 0.63, Recall is of 0.65 and F1 score is of 0.64. Whereas the multi-layer perceptron (ANN) shows values of RMSE as 0.18, Precision as 0.52, Recall as 0.53 and F1 score as 0.52, which is not satisfactory as compared to decision tree and random forest. Application of different Machine Learning algorithms has shown significant improvements in accuracy and efficiency compared to traditional manual bug classification. It is planned to examine the suggested model on a wide range of datasets, which can be revolutionary in the field of the software engineering industry for early detection and classification of software bugs.

REFERENCES

1. Lopes, F., Agnelo, J., Teixeira, C. A., Laranjeiro, N., & Bernardino, J. Automating orthogonal defect classification using machine learning algorithms, *Future generation computer systems 102*, 932-947 (2020).

2. Doppalapudi, S., Badduri, S., Bommu, N., Kaparouthu, S. L., & Basha, N. M. J. A model for automated bug classification using machine learning(2022).
3. Ahmed, H. A., Bawany, N. Z., and Shamsi, J. A. Capbug-a framework for automatic bug categorization and prioritization using nlp and machine learning algorithms. *IEEE Access 9*, 50496-50512 (2021).
4. Hammouri, A., Hammad, M., Alnabhan, M. and Alsarayrah, F.: Software bug prediction using machine learning approach, International journal of advanced computer science and applications 9(2), 2158-107 (2018).
5. Y. Tohman, K. Tokunaga, S. Nagase, and M. Y.: Structural approach to the estimation of the number of residual software faults based on the hyper-geometric distribution model, IEEE Trans. on Software Engineering 15(3), 345–355 (1989).
6. Gupta, Dharmendra Lal and Saxena, K.: Software bug prediction using object-oriented metrics, Sadhana 42, 655-669 (2017).

Hindi Image Captioning Using Deep Learning

Soumyarashmi Panigrahi[1], Jaydev Sutar[2], Dibya Ranjan Das Adhikary[3]

Department of Computer Science and Engineering, Siksha 'O' Anusandhan Deemed to be University, Bhubaneswar, Odisha, India

---

Abstract In the era of Artificial Intelligence (AI), the real-world need for Image Captioning (IC) lies in making visual content accessible, understandable, and useful across diverse applications, from aiding the visually impaired to enhancing e-commerce and enriching social media engagement. An image needs to have a description, which is the idea behind image captioning. There has been a lot of study already done in this area, but it has primarily concentrated on creating image descriptions in English language because the majority of the datasets for existing image captions are in English language, and language must not create a constraint on the IC generator model. So we have tried to generate the caption in Hindi language. Hence, we used Flickr8k with Hindi caption as our dataset. We have incorporated a diverse strategy to achieve the goal by combining Recurrent Neural Networks (RNN) and Convolutional Neural Networks (CNN). To achieve and maintain the knowledge for a longer period of time, we have also utilized Long Short-Term Memory (LSTM), a sort of RNN. Additionally, many applications of the same have also been covered. The results obtained have significant role to the understanding and management of IC in the practical field.

Keywords Convolutional neural networks (CNN), Recurrent neural networks (RNN), Long short-term memory (LSTM)

---

1. Introduction

Image Captioning (IC) is a process in which we have to generate a text, which represents and explains the picture given as input. Basically, it tells us more about the picture which makes the task of an observer easier to analyze what is happening and what exactly is the picture trying to represent. During the time of big data, pictures were among the most commonly found

---
[1]soumya.rashmi.iter@gmail.com, [2]jaydevsutar@gmail.com, [3]dibyadasadhikary@gmail.com

DOI: 10.1201/9781003489443-16

data types on the internet, and the requirement to add notes and categorize them has grown. Hence, we undertook an interesting multi modal topic where we have used the approach of combining both image and text processing to develop a useful Deep Learning (DL) application i.e., IC. Most studies on captioning images have concentrated on captions that consist of only one sentence. The format has a restricted ability to describe as only one syllable is capable of depicting a small portion of an image. It generates a model that breaks down both photos and captions into their component parts using a Long Short-Term Memory (LSTM) model and a Natural Language Processing (NLP) technique to identify linguistic regions in images. The motivation behind undertaking this work stems from a genuine interest in exploring a captivating area of research that has witnessed remarkable advancements. The prospect of delving into an evolving field of computing, one that extends beyond the confines of the standard curriculum, held great appeal. Moreover, IC serves as a solution to the challenge of connecting visual and textual information, empowering machines to comprehend images and produce meaningful descriptions. By formulating unique captions for images, we aim to bridge the gap between visual and textual domains in a manner that has not been previously covered.

The manuscript is organized as follows; Firstly, a brief description of the topic along with the main motivation is summarized. Next the research papers of various authors from where we got the ideas are being explained. Then the various methods and algorithms which are used have been mentioned. Then the results which we have got while implementing the model is being explained by showing the screenshots of graphs and preparing tables. Finally, we concluded our work by summarizing and the future scope.

2. Literature Survey

Several research studies have explored on IC in English and Hindi languages using different DL techniques. These studies have investigated the application of DL algorithms to accurately describe the captions for the meaning of diverse images. Few of the researchers' work has been elaborated below.

The model by S. Das et al. [1] was based on military image captioning. They primarily used CNN-RNN as their preferred framework and to reduce the gradient descent issue and encode the images, they used the Inception model. They used Flickr8k with a few modifications of some military images as their dataset. The model by A. Hani et al. [2] was built using the Encoding Decoding Model for captioning images. They employed both template-based and retrieval-based captioning. Their dataset was Flickr8k. In retrieval-based captioning, the created captions are placed in one scope while training images are placed in another. Ankit Rathi et al. [3] have generated Hindi captions for the images. They used Flickr8k with Hindi captions as their dataset. S. K. Shukla et al. [4] used a CNN to scan an image and identify the objects. To generate captions that are both contextually and grammatically appropriate for recognized objects in an image, attention mechanisms are employed in tandem with RNN and LSTM. For those who need to cross roadways safely but are sight challenged, this model is quite helpful. Their collection includes over 82,000 pictures, each annotated with at least five different captions. Aishwarya Maroju et al. [5] used RESNET-LSTM model to generate captions instead of CNN-RNN and VGG model by using Flickr8k as their dataset. Suhyun Cho et al. [6] have suggested method calls for modifying the object's dictionary and moving away from using learning data to give the dictionary connected to the domain object priority. These results in the generation of a variety of image captions by in-depth descriptions of the objects needed for each specific area. They have used COCO as their dataset.

As several models have already developed describing the image captions in English language and only few researches are there who described the generated images in Hindi language. Hence, we have used the Flickr8K dataset with Hindi captions as our dataset. We found out that the structure of RNNs puts limitations on their ability to store long-term memories, which is restricted to only a certain extent; it is also very hard to train the traditional RNN. Therefore, we have used the CNN-LSTM method to tackle this problem, as LSTM retains and stores information for a longer period of time. This, as a result, improves the accuracy of our model.

3. Proposed Approach

CNN[7,8] receives the input image. In order to extract the image's features, CNN analyses the input image via a number of convolutional layers. The output of the CNN layer is a feature vector that contains the features of the image. The feature vector is then feed to the LSTM[9,10] cells. Based on the prior to words in the caption and the feature vector input, each cell uses the feature vector to input and generate a word for the caption. The caption for the input image is produced by the LSTM as a series of words.

CNN is frequently utilized as the model's encoder component. The CNN's goal is to take the input image and extract visual elements from it. The CNN produces a feature representation of the input image that meaningfully encodes the visual content. Important aspects of the image, like objects, forms, and textures, are captured by these visual elements. The decoder component of the model receives the retrieved visual characteristics from the CNN.

The decoder part of the model is an LSTM that creates textual descriptions based on the visual elements that are retrieved from the input image. LSTM is suitable for producing coherent and contextually relevant captions because it excels at handling sequential input and capturing long-term dependencies. The CNN's visual properties are used by the LSTM to produce a series of words or tokens that make up the caption. The LSTM uses the current input token, the prior hidden state, and the prior cell state at each time step to predict the subsequent word in the caption.

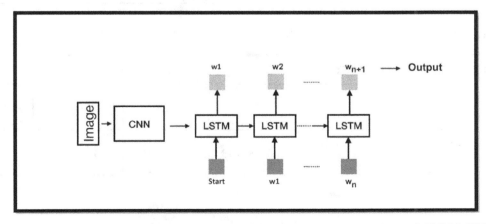

Fig. 16.1 A model representation of how the IC generator consisting of CNN-LSTM works

4. Experimentation and Model Evaluation

To analyze the overall performance of the technique on the Flickr8k with Hindi caption dataset, an extensive experimentation is conducted using Google COLAB platform. We have used Tesla T4 as our GPU which is provided by Google Colab. We have run the model on the system having Intel Core i5 10th gen 8 cores as our CPU which is having 8 Gigabytes of RAM and 256 gigabytes of NVME SSD. We have used Google Colab Version Python 3.10 as the environment to run our model and used Google Drive to store the dataset and trained model. Some of the terms used for the model experimentation and evaluation are as follows:

- *Epoch:* An epoch is an iteration where the model sees and learns from all the training examples. During an epoch, the training data are processed by the model in mini-batches, which are smaller chunks of the total dataset. The model modifies its parameters and weights based on the computed loss and the selected optimization strategy for each mini-batch, which has a set number of training samples.
- *Bilingual Evaluation Understudy (BLEU)*[11,12]: An indicator called BLEU is used to assess how well machine translations or text-generation activities like image captioning are performed. It evaluates how closely a generated output, like a caption, matches one or more reference outputs, like captions that were written by humans. A higher score shows greater similarity between the generated output and the human references. The BLEU score goes from 0 to 1, with a higher value indicating better performance.

BLEU Score = Brevity Penalty(BP) * Geometric Average Precision Score(GAPS)   (1)

Table 16.1 Bleu score in various epoch values

EPOCH	BLEU-1	BLEU-2	BLEU-3	BLEU-4
20	0.026316	0.000000	0.000000	0.000000
25	0.531463	0.334349	0.205846	0.123432
30	0.540158	0.335217	0.205972	0.123511

Table 16.2 Comparison of a. Rathi et al.[3] Model with our proposed model

MODEL	DATASET	BLEU-1	BLEU-2	BLEU-3	BLEU-4
A. Rathi et al.[3] Model	Flickr8k with Hindi captions	0.4136	0.278	0.201	0.09
Our Proposed Model	Flickr8k with Hindi captions	0.531463	0.334349	0.205846	0.123432

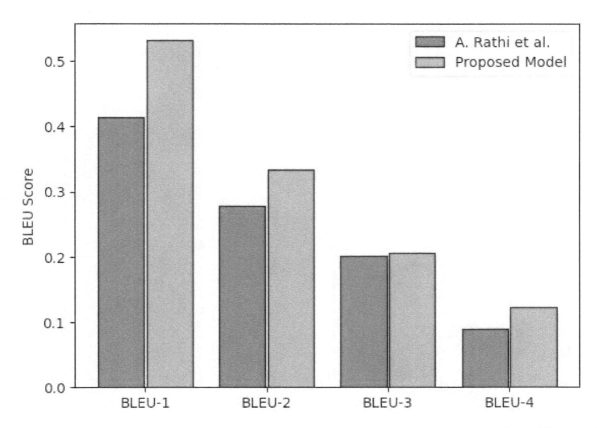

Fig. 16.2 Comparison of A. Rathi et al.[3] model Vs. our proposed model at 25 epochs BLEU-1(unigram), BLEU-2(bigram), BLEU-3(trigram) and BLEU-4(4-gram)

In comparing A. Rathi et al.'s model with the proposed model after 25 epochs of training, a notable difference emerges in their respective performance metrics which is depicted in Fig. 16.2. A. Rathi et al.'s model achieves BLEU-1, BLEU-2, BLEU-3, and BLEU-4 scores of 0.4136, 0.278, 0.201, and 0.09, respectively. In contrast, the proposed model exhibits superior performance with BLEU-1, BLEU-2, BLEU-3, and BLEU-4 scores of 0.531463, 0.334349, 0.205846, and 0.123432. Our proposed model in BLEU-1 has 28.49%, BLEU-2 has 20.26 %, BLEU-3 has 2.41 %, and BLEU-4 score has 37.14 % improvement in BLEU Score. These results suggest that the proposed model, even at the 25-epoch mark, outperforms A. Rathi et al.'s model across all evaluated BLEU metrics, demonstrating its potential to generate more accurate and contextually relevant text translations.

Furthermore, the marked improvement in BLEU scores for the proposed model underscores its ability to capture finer nuances and context within the translated text, especially as the n-gram size increases. This enhanced performance can be attributed to potential advancements in model architecture, data preprocessing, or training techniques employed in the proposed model. As a result, the proposed model holds promise for applications in machine translation tasks where higher BLEU scores indicate a closer alignment with human-generated translations, ultimately contributing to more effective and precise language translation capabilities. In Fig. 16.3, some of the generated Hindi caption generated for the input image using the proposed approach is provided.

Fig. 16.3 Image caption generated from various input images

5. Conclusion

To develop an IC generator for our work, we have used the CNN-LSTM model, reviewed the limitations of the conventional technique, and concentrated on the approach we are using currently. Experimental results indicate that this work has a greater scope and room for development, despite the fact that IC has numerous applications in various sectors. There are very few studies to create the captions in Hindi, even if picture captioning techniques based on DL have developed tremendously in recent years. In order to clearly describe the 8000 photographs with diverse points of view, we used the Flickr8k dataset, which provides 5 Hindi captions for each related image. The fact that this model is data-dependent means that it cannot anticipate terms outside of its vocabulary. It's crucial to keep in mind that errors might also happen on behalf of the computer because no machine is error-free all the time.

In the future with improvement of hardware specification to train the model on larger datasets like MSCOCO can increase the accuracy as the system's performance is crucial. For creating more accurate and useful models, additional large datasets like PASCAL 1K and AIC might can be used, for the model to perform much better.

REFERENCES

1. S. Das, L. Jain and A. Das, "Deep Learning for Military Image Captioning", 21st International Conference on Information Fusion (FUSION), 2018.
2. A. Hani, N. Tagougui and M. Kherallah, "Image Caption Generation Using A Deep Architecture", International Arab Conference on Information Technology (ACIT), pp. 246–251, doi: 10.1109/ACIT47987.2019.8990998, 2019.
3. A. Rathi, "Deep learning apporach for image captioning in Hindi language", International Conference on Computer, Electrical & Communication Engineering (ICCECE), pp. 1-8, doi: 10.1109/ICCE 48148.2020.9223087, Kolkata, India, 2020.
4. Sujeet Kumar Shukla, Saurabh Dubey, Aniket Kumar Pandey, Vineet Mishra, Mayank Awasthi, Vinay Bhardwaj, "Image Caption Generator Using Neural Networks", International Journal of Scientific Research in Computer Science, Engineering and Information Technology (IJSRCSEIT), ISSN : 2456-3307, Volume 7 Issue 3, pp. 01–07, May-June 2021.
5. Aishwarya Maroju, Sneha Sri Doma, Lahari Chandarlapati,"Image Caption Generating Deep Learning Model", International Journal of Engineering Research & Technology, Vol. 10 Issue 09, 2021.
6. Suhyun Cho, Hayoung Ho, "Generalised Image captioning for multilingual support", Vol. 13, Issue 4, 2023.
7. Bhatt, Dulari, et al. "CNN variants for computer vision: History, architecture, application, challenges and future scope." Electronics 10.20 (2021): 2470.
8. Al-Jamal, Anbara Z., Maryam J. Bani-Amer, and Shadi Aljawarneh. "Image captioning techniques: a review." 2022 International Conference on Engineering & MIS (ICEMIS). IEEE, 2022.
9. Indumathi, N., et al. "Apply Deep Learning-based CNN and LSTM for Visual Image Caption Generator." 2023 3rd International Conference on Advance Computing and Innovative Technologies in Engineering (ICACITE). IEEE, 2023.
10. Deorukhkar, Kalpana Prasanna, and Satish Ket. "Image Captioning using Hybrid LSTM-RNN with Deep Features." Sensing and Imaging 23.1 (2022): 31.
11. Al-Kabi, Mohammed N., et al. "Evaluating English to Arabic machine translation using BLEU." International Journal of Advanced Computer Science and Applications 4.1 (2013).
12. Chun, Pang-Jo, Tatsuro Yamane, and Yu Maemura. "A deep learning-based image captioning method to automatically generate comprehensive explanations of bridge damage." Computer-Aided Civil and Infrastructure Engineering 37.11 (2022): 1387-1401.

An Integrated Model for Smart Healthcare Solutions With 5G Network Slicing

Swati Sucharita Roy[1], Bharat Jyoti Ranjan Sahu[2], Shatarupa Dash[3]

Department of Computer Science and Engineering, FET-ITER, Siksha 'O' Anusandhan (Deemed to be) University, Bhubaneswar, INDIA

Abstract The IoT (Internet of Things) is a framework of interconnected devices or things with autonomous configuration and performance capabilities. The connected devices in the IoT infrastructure act intelligently, intelligent processing is performed, and communication is more informative. It is operating in a variety of industries, including agriculture, manufacturing, retail, transportation, and healthcare. Without human involvement, IoT does remote monitoring, predictive maintenance, facility management, industrial efficiency, connected items, and many more tasks. In this article, we focus on the impact of the IoT on the healthcare system with the integration of 5G network slicing. IoT gathers and accumulates essential data autonomously, performs analysis on those stored data for future decision-making, which influences the system's overall performance. The existing traditional healthcare system has many flaws, such as insufficient healthcare centers with specialized and qualified staff, medical and other traditional equipment used to generate patient reports, and the high cost of competent people in all areas of the health centers. We have analyzed the existing techniques followed by various healthcare centers, proposed a 5G network slicing based more efficient smart healthcare infrastructure, and discussed ML's role in its advancement. This proposed system can provide a new direction and enhance the facilities provided by various healthcare units across the globe.

Keywords IoT, ML, 5G, Network slicing, Smart Healthcare

1. Introduction

The IoT is an integrated collection of autonomous devices that perform independently and have distinct identities. IoT devices can collect and exchange data on their own; they are linked by

---

[1]swatisucharitaray@gmail.com, [2]bharatjyotisahu@soa.ac.in, [3]shatarupadash@soa.ac.in

An Integrated Model for Smart Healthcare Solutions With 5G Network Slicing 157

various communication technologies such as Wi-Fi (Wireless Fidelity), ZigBee, and Bluetooth and can be controlled remotely. IoT creates an environment in which heterogeneous devices are connected and operate autonomously. Kevin Ashton introduced the term IoT at the beginning of 1999 for supply management. IoT influences various fields, including healthcare, smart cities, electricity management, traffic congestion control, smart waste management, disaster prediction and management, industry, and transport. Figure 17.1 represents many IoT-influenced application areas, such as smart homes, smart agriculture, smart manufacturing, smart energy/grid, smart city, and smart retail, to improve system efficiency and productivity. IoT makes life more comfortable, more accessible, or more convenient. IoT with ML for prediction and 5G for efficient wireless communication saves time and effort while providing faster and more reliable services. Different application areas in IoT have different requirements in operations and communication. Figure 17.2 categorizes the various requirements for various IoT applications.

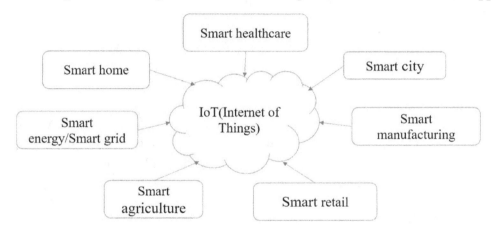

Fig. 17.1 Application areas of IoT

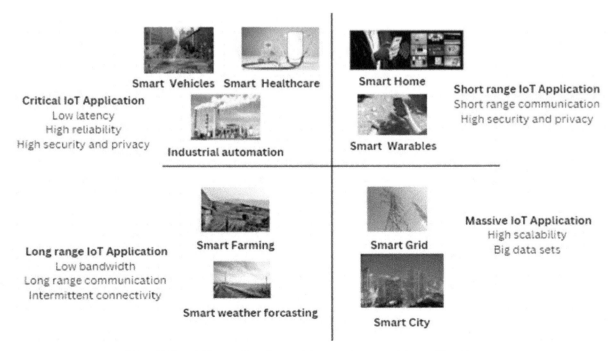

Fig. 17.2 IoT application with requirement specifications

1.1 Role of IoT in the healthcare system

IoT is an interconnected network where many heterogeneous devices or objects are connected through the computer network and can collect, store, share, and analyze data for required decision-making and can access information anytime, anywhere. IoT adopts various technologies like NFC (Near Field Communication), Wi-Fi, ZigBee, Z-wave, Bluetooth, and WSN (Wireless Sensor Network) to connect physical devices to the Internet and build the system based on the user's requirements. IoT has vast application areas such as traffic control, irrigation systems, environment monitoring systems, smart cities, IIoT (Industrial Internet of Things), and smart grid, and one of them is the smart healthcare system. Proper diagnosis and timely treatment can save a life. Nowadays, healthcare concerns are spreading uncontrollably; due to delayed response, improper diagnosis, delayed treatment, and a lack of competent healthcare facilities, many patients suffer significantly, and many die, negatively impacting our society.

Different methods and technology are being used to improve the healthcare system daily. However, there are still certain challenges and complications in the existing traditional healthcare system. The IoT has numerous applications in the healthcare system to resolve healthcare-related issues. It automates manual tasks, speeds up performance, improves the facilities, and becomes more convenient for patients, medical practitioners, healthcare administrators, and society. Healthcare with IoT becomes more convenient for any category of people and provides uninterrupted services. IoT is the integration of various computing and communication devices, digital and mechanical objects having the data generation and transferring capability automatically, i.e., without human involvement. IoT uniquely identifies the devices or objects, and devices seamlessly interact and communicate.

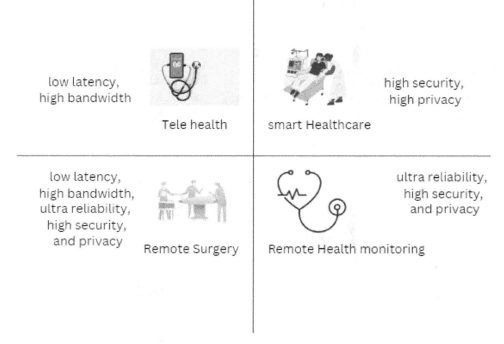

Fig. 17.3 Various Dimensions of the Healthcare System with Their Requirements

1.2 Integration of 5G and Machine learning with IoT

ML extracts hidden information from the data collected through various interconnected devices or sources in IoT. IoT with ML works intelligently to increase decision-making efficiency in various fields like healthcare, business, smart cities, electricity management, green energy, traffic congestion control, waste management, disaster prediction and management, agriculture, industry, and transport. Various ML methods like SVM (Support Vector Machine) and LSTM (Long Short-term Memory) strengthen IoT to enlighten hidden patterns from a massive amount of data for optimum prediction [1]. IoT makes healthcare systems smarter and performs intelligent processing. Integrating ML with smart healthcare becomes more efficient and convenient compared to legacy healthcare systems. ML and 5G combined with IoT enable the healthcare system to maintain healthcare information autonomously, perform real-time monitoring of patient activities, and perform disease diagnosis automatically.

Although IoT with AI(Artificial Intelligence)/ML is more efficient for autonomous and reliable healthcare systems, network communication may affect the system's overall performance. 5G may provide faster and more reliable communication. However, it may increase complexity and overall cost due to heterogeneous communication services. To manage the performance, we need to understand different applications of smart healthcare and their requirements. In this work, we have used network slicing to allocate resources based on the application requirements. Separate slices (S_1, S_2...S_n) are used for separate requirements (r_1, r_2...r_n). Slice isolation may support avoiding interference and make available the required resources in mission-critical applications like remote surgery and remote patient monitoring. Smart Healthcare has different applications like remote health monitoring, remote surgery, smart hospital, and Telly health consultation. Reliability, high bandwidth, ultra-low latency, security, and privacy are major concerns in the healthcare system. In this article, we have two-fold the work: firstly, we discuss the requirements of smart healthcare system, and secondly, we focus on efficient communication, remote data access and processing using 5G network slicing. Figure 17.3 shows different resource requirements (bandwidth, latency, reliability, privacy, and security) in various healthcare applications(AP_1, AP_2,...AP_m).

2. Literature Review

IoT enables healthcare systems for remote monitoring of patients, drug delivery, and reduced overall cost. IoT connects patients with healthcare professionals. IoT can alleviate the burdens on healthcare systems to preserve quality and standards. ML can assist IoT for early disease prediction more accurately in healthcare sectors. In [1], Amani Aldahiri et al. discussed various machine learning algorithms for classification and prediction in healthcare sectors. ML analyzes healthcare datasets to do disease prediction and decision-making by integrating the benefits of IoT for generating medical data and IoT to reduce strain to preserve quality treatment. BAN(Body Area Network) or BSN(Body Sensor Network) integrates WSN (Wireless Sensor Network) with MEMS(Micro Electro Mechanical Systems) for patient's health conditions monitoring like cardiac arrest and Alzheimer's disease [1]. In [2], authors intend to overview the influence of IoT in healthcare and explain different layers in IoT-based healthcare systems, such as the perception layer, network layer, process and storage layer, and application layer. Furthermore, focuses on the security and privacy of the system. In [3], the authors systematically analyzed various features and benefits of HIoT(healthcare Internet of Things). In [4], authors focus on various smart healthcare applications and propose a model to reduce privacy and security issues in healthcare.

Healthcare monitoring, health assistance, disease diagnosis, drug recommendation, sensing healthcare data, and maintaining healthcare DB (database) are different IoT based healthcare applications [5]. Various AI/ML algorithms including Naïve Bayes algorithm, decision tree, random forest, SVM (Support Vector machine), KNN (k nearest neighbors), k-means clustering, linear regression, and logistic regression can be used to analyse data acquired from the IoT environment for disease prediction and patient healthcare condition prediction. In [6], authors explore how the combination of IoT and ML plays an important role in healthcare and how blockchain for secure healthcare system. IoT environment gathers and provides the data for analysis, and ML uses various methods to predict critical diseases, including heart attack, cancer, and diabetes. In [7], authors adopted IoT for smart healthcare systems to gather data through various sensing technologies and then communicate the collected data to the server for analysis and prediction.

People in remote areas face many challenges with healthcare issues; they do not have the facilities for disease diagnosis and timely treatment. Many researchers are trying to overcome healthcare issues and make available healthcare facilities near patients at low cost. In [8], authors introduced a modern low-cost healthcare system to provide healthcare-related facilities based on IoT and ML, which sense the patient's health-related data and process it for prediction and decision-making about serious health issues and treatment. In [9], authors intend to highlight the benefits of IoT in monitoring the behavior of patients with critical conditions like neurological illness, mental disorder, and mental trauma, then use deep belief neural network for analysis on those collected data from the IoT based infrastructure.

IoT enables advanced healthcare environment for collecting, communicating, and analyzing healthcare data. In [10], usages, challenges, and issues related to the healthcare system with the combination of IoT and ML are overviewed. In [11], a model has proposed to predict the patient's body condition and disease by analyzing the data generated from the IoT-based healthcare environment. The system collects the patient's body temperature, heart rate, and oxygen label through sensors, sends it to the Raspberry Pi, and then exchanges it to the cloud server using ML techniques to predict diseases and health conditions. Data generated from different sources and healthcare systems is considered for processing. IoT stores the generated smart healthcare data into the cloud server.

Communication plays an important role in healthcare sectors for on-demand, secure, real-time data communication. 5G plays vital role for more efficient communication compared to existing communication technologies like 3G, 4G. To enhance the healthcare communication, network slicing approach makes virtual slices on the same physical network based on application requirements like remote surgery and remote patient monitoring. Network slicing enables flexible, scalable, and on-demand resolution in massive IoT applications. In [12], the authors systematically analyzed the features and role of 5G network slicing in IoT realization. Furthermore, it discusses its requirements considering different network-slicing-based IoT applications such as healthcare, smart transportation, IIoT(Industrial Automation /Industrial IoT), smart home, smart city, smart grid, military applications, UAVs(Unmanned Arial Vehicles) and drones, AR, VR, and gaming. In [13], the authors proposed a model for slice access selection in a 5G network based on application requirements and proposed network slicing architecture to share the same infrastructure with different needs. In the article [14], the authors modelled a logical architecture for 5G network slicing and focused on slicing management, resource allocation between different 5G network slices, and mobility management within different access networks. In [15], the primary focus is function isolations within slice components (intra-slice isolation) and end-to-end delay. The author proposed model for optimal slice allocation for more reliable, efficient, and secure application services.

3. Proposed Features of Smart Healthcare

Smart Healthcare has features like transparency, ubiquity, intelligence, adaptability, responsiveness, sensitivity, and personalized [1]. These features must be considered to establish a more efficient smart healthcare system. Our main goal is to focus on timely diagnosis of diseases with more accuracy, best and most affordable treatment for the patient, and provide more care and facilities to the kids and senior citizens as well as poor patients, which can be achieved by optimal prediction and decision making. The primary purpose of integrating ML with IoT is prior disease diagnosis and prevention with more accuracy. IoT supports collecting healthcare data with more reliability and store in the cloud server for analysis. The healthcare system can be improved through optimum prediction and decision-making by applying various ML models like SVM and LSTM on those medical data.

IoT integrates Healthcare data like electronic medical data, image and lab results, genomic data, environment data, mobile healthcare data, family health history, drug data, and public healthcare data gathered from various sources, performing analysis on those data for decision making, which will help in the future advancement of the healthcare system.

This proposed model focuses on the impact of IoT in the healthcare system compared to the traditional methods used to date. The proposed model has three sections such as,

   I. Collection of healthcare data
  II. Communication, storage, and access to healthcare data
 III. Analysis, prediction, and decision-making based on collected healthcare data.

I. Collection of healthcare data

In the first section, various IoT devices used, like BP monitor for BP (blood pressure) monitoring and ECG sensors, are implemented for primary health check-ups of the patient, lab test reports, and medical data collection. Various wearable or sensor devices are used to generate health data, and then these gathered data are collected and communicated for storage and future analysis and prediction.

II. Communication, storage, and access to healthcare data

In the second section, the collected data are transferred or communicated through various communication gateways like Wi-Fi, ZigBee, Bluetooth, NFS (Near Field Communication), Z-wave, LTE (long-term evaluation), 5G and then these collected data stored in the IoT cloud server for analysis, prediction and decision making. Healthcare data are very sensitive and should be secured from unauthorized people. It should be accessed only by the respective medical practitioner, patient, or family members. In this smart healthcare system, data storing, signal enhancement, feature extraction, and classification are performed autonomously and maintained in the database for analysis. Various security mechanisms are applied to make it secure against unauthorized access.

IoT and wireless communication technologies are evolving rapidly, changing how we live and work. IoT demands uninterrupted, reliable, and consistent connectivity, which is being catered by the 5G network. 5G networks have extremely low latency, high bandwidth, availability, full coverage and high reliability to meet the performance needs of real-time, reliable, and secure communications.

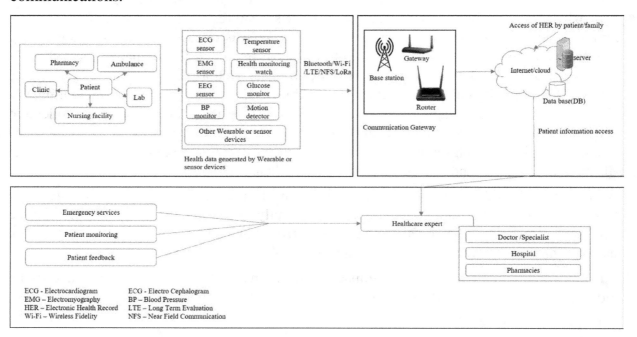

Fig. 17.4 Overall configuration of IoT-based healthcare system

With increased massive wireless data traffic, mission-critical applications in smart healthcare demand efficient resource allocation, which can be supported by 5G network slicing. Network slicing creates virtual networks based on different requirements like low latency, reliability, speed, and Quality of Service (QoS). Network slicing is used for faster, more reliable, and more efficient communication services and tries to maintain uninterrupted and consistent connectivity. 5G network slicing divides the physical network into multiple logical networks (i.e., network slices) to specialize each slice with specific characteristics and network capabilities [16]. SDN (Software-defined networking), NFV (Network Functions Virtualization), and cloud computing enable network slicing to enhance communication [12].

III. Analysis, prediction, and decision-making based on collected healthcare data

In the third section of the model, the main focus is on analysis and prediction. Analysis and prediction carried out on stored healthcare data for future decision making. Relevant decision-making can be used for advancement in the healthcare system.

IoT with ML can reduce the patient's waiting time, reduce cost, provide timely and proper treatment, provide better and smarter services, early diagnosis, prediction of disease and body condition, better management of healthcare infrastructure, remote monitoring, and services, provide convenience to the users and so on, as compared to the traditional healthcare system.

Figure 17.4 represents the overall configuration of the smart healthcare system. In this figure, the whole system is partitioned into three sections for three different purposes. In the first section, various IoT devices generate and gather healthcare data. The second section communicates the generated healthcare data through various capillary communication technologies like Wi-Fi, ZigBee, Z-wave, and 5G. The IoT Healthcare data is stored in the cloud server. The third section uses various ML models for prediction and future decision-making.

4. 5G Network Enhancement for Smart-Healthcare

The IoT plays an essential role in overcoming healthcare-related challenges. Communication among IoT devices requires reliable, faster, and more efficient communication technology. Hospital management data can be stored and processed locally. 5G IoT enables continuous adaptation of advanced tools and techniques to automate patient monitoring and disease detection. Nowadays, researchers, academia, and industry are deeply engaged in healthcare-related advancement activities. In recent years, various healthcare management software and apps have designed and researched various AI/ML techniques for disease detection. With the 5G wireless communication network, we can look forward to designing and developing healthcare applications.

4.1 AR/VR for Patients' Appointments

One of the significant difficulties for ICU patients is visiting different medical departments for consultation. The design of AR/VR with high-quality audio will improve the service process and reduce patient movements. The same application can be used for telemedicine. Moreover, AR/VR technology can help medical students in training in critical operations.

4.2 Realtime Patient Monitoring Systems with Critical Alerts

This application will generate critical alerts when any patient's condition approaches critical situations. The system shall generate alerts by reading the patient's current health condition, pre-existing disease, doctor's suggestion, and previous training data.

4.3 Crowd Management and Hospital Management

Using image and video analysis and real-time digital guidance, hospitals can better manage crowds and monitor critical patients for instant treatment. Intelligent scheduling of hospital activity with touchless devices shall improve management activity while reducing contaminations.

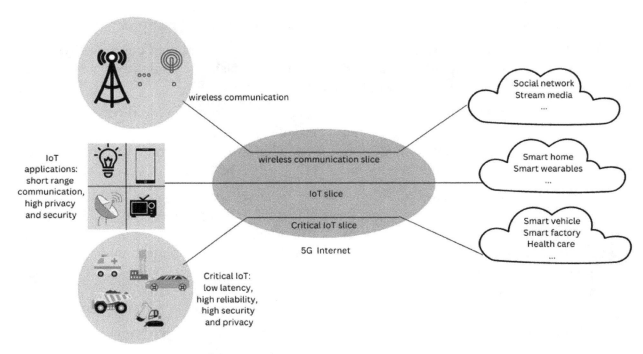

Fig. 17.5 5G network slicing with different IoT applications

5G wireless communication provides a high data rate, low latency, and uninterrupted, reliable connectivity. 5G is the first omnipresent connectivity solution that can provide more reliable, faster, and more efficient wireless communication services than existing wireless communication technologies like, LTE (4G) [16]. 5G networks have extremely low latency and high reliability to meet the performance needs of real-time, reliable, and secure communications, which can support some vertical industries, including healthcare, interconnected vehicles, and industrial production control [12]. Network slicing creates virtual networks based on different requirements like low latency, reliability, speed, and quality of service [13] used for faster, more reliable, and efficient communication services. It tries to maintain uninterrupted and consistent connectivity.

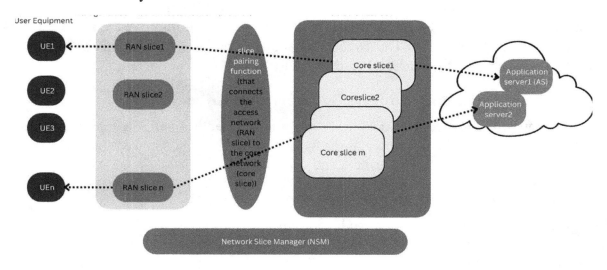

Fig. 17.6 Overall configuration in network slicing

This work focuses on network slice allocation to the applications as per their requirement specification. Proper allocation can improve the efficiency and performance of the network to handle IoT applications. There are various IoT applications, and their requirement also varies. We need to manage all these things so that interference, security, privacy, congestion, and latency can be reduced. In this work, the primary goal is network-slicing architecture and a network-slicing access selection method to optimize the network's performance, maximize resource utilization, and shorten the creation time of the slice. 5G network slicing is essential for delivering flexibility, scalability, and on-demand solutions for a diverse set of 5G network applications[14]. Separate slice allocation for different applications, combined with strong slice isolation mechanisms, can enable data communicated in different slices to be secure and private[15].

5. 5G Slicing in Smart Healthcare

In network slicing, multiple virtual networks of different characteristics can be deployed in parallel under the same network resource pool using Network Functions Virtualization (NFV) technology and Software Defined Network (SDN) separates the control plane from the data plane, allowing network operators to manage the network centrally and allocate network resources dynamically on demand [11]. Network slicing can satisfy the various networking demands of heterogeneous applications through dedicated slices and can improve the technical aspects. However, it faces numerous challenges in 5G networks, including resource allocation, orchestration, deployment, and service management [12]. We are focusing on,

- Network slice allocation to the healthcare applications as per their requirement specification.
- Optimal resource management for healthcare reduces interference and network congestion.
- Network-slicing access selection to optimize the network's performance, maximize resource utilization, and shorten slice creation time.
- Separate slice allocation with strong slice isolation enables privacy and security.

Different healthcare data have different levels of sensitivity based on latency, throughput, reliability, privacy, and security. For instance, remote surgery seeks ultra-low latency, remote health monitoring prefers accurate data transmission (i.e., high reliability), confidential healthcare data seeks privacy, and there is a demand for secure healthcare data communication, i.e., only authorized persons can access the data. In this work, remote healthcare data gathered from IoT-based healthcare infrastructure communicated to the healthcare center or healthcare practitioner through a 5G wireless communication network. Slices are created using 5G network slicing based on the application requirements. If we make more slices, the availability of resources will be reduced. So, in our work, we reduce the number of slices. By organizing the application requirements and analyzing past slicing data, the most availing requirements are considered, and the most required combinations of them are included to create slices. Dynamic slice creation can increase resource availability and efficient utilization.

6. Conclusion

Healthcare is very complex based on responsibility, sensitivity, timeliness, and dedication, i.e., proper diagnosis, timely treatment, proper medication, and drug referral at the patient's convenience. There are various categories of patients: some patients need immediate treatment, some patients are unable to afford the cost of treatment, some patients are unable to reach the hospital timely, in some cases, disease diagnosis is complex, and so forth. To handle these critical situations, the proposed model uses IoT technologies for autonomous configuration and management of healthcare systems for remote treatment, diagnosis, patient monitoring, surveillance, system management, and database management. AI/ML methods can be used on collected healthcare data for analysis and prediction. 5G slicing provides an efficient communication platform for different applications in the healthcare system. IoT automates the healthcare system, 5G wireless communication with network slicing provides uninterrupted communication, and again, for healthcare prediction, integration of ML improves efficiency. IoT, 5G, and ML together give better solutions to the difficulties of the traditional healthcare system. In the future, improvements in IoT technologies and ML methodologies improve the healthcare system, which could give society more benefits and more convenience. Enabling 5G-edge caching will play an important role in providing low latency. Although the generic network slicing shall improve the network's flexibility, using edge will add further new slicing dimensions.

REFERENCES

1. Aldahiri, Amani, Bashair Alrashed, and Walayat Hussain. "Trends in using IoT with machine learning in health prediction system." Forecasting 3.1 (2021): 181-206.
2. Al-Shargabi, Bassam, and Simak Abuarqoub. "IoT-Enabled Healthcare: Benefits, Issues and Challenges." The 4th International Conference on Future Networks and Distributed Systems (ICFNDS). 2020
3. H. Habibzadeh, K. Dinesh, O. Rajabi Shishvan, A. Boggio-Dandry, G. Sharma and T. Soyata, "A Survey of Healthcare Internet of Things (HIoT): A Clinical Perspective," in *IEEE Internet of Things Journal*, vol. 7, no. 1, pp. 53-71, Jan. 2020
4. Islam, S. R., Kwak, D., Kabir, M. H., Hossain, M., & Kwak, K. S. (2015). The Internet of things for health care: a comprehensive survey. IEEE access
5. Imran, M., Zaman, U., Imtiaz, J., Fayaz, M., & Gwak, J. (2021). Comprehensive Survey of IoT, Machine Learning, and Blockchain for Health Care Applications: A Topical Assessment for Pandemic Preparedness, Challenges, and Solutions.
6. I. Fathail and V. D. Bhagile, "Review: IoT Based Machine Learning Techniques for Healthcare Applications," *2020 International Conference on Smart Innovations in Design, Environment, Management, Planning and Computing (ICSIDEMPC)*, 2020, pp. 248-252
7. Jeong, Joon-Soo, Oakyoung Han, and Yen-You You. "A design characteristics of smart healthcare system as the IoT application." Indian Journal of Science and Technology 9.37 (2016).
8. Kashif Hameed, Imran Sarwar Bajwa, Shabana Ramzan, Waheed Anwar, Akmal Khan, "An Intelligent IoT Based Healthcare System Using Fuzzy Neural Networks", Scientific Programming, vol. 2020

9. Mohamed, R. M., Shahin, O. R., Hamed, N. O., Zahran, H. Y., &Abdellattif, M. H. "Analyzing the Patient Behavior for Improving the Medical Treatment Using Smart Healthcare and IoT-Based Deep Belief Network." Journal of Healthcare Engineering 2022 (2022).
10. Mohammed, Chnar Mustaf, and Shavan Askar. "Machine learning for IoT healthcare applications: a review." International Journal of Science and Business 5.3 (2021)
11. Sharma, Yogesh Kumar, and S. Khatal Sunil. "Health Care Patient Monitoring using IoT and Machine Learning." IOSR Journal of Engineering (IOSR JEN)-", ISSN (e) (2019)
12. Wijethilaka, S., & Liyanage, M. (2021). Survey on network slicing for Internet of Things realization in 5G networks. *IEEE Communications Surveys and Tutorials,* 23(2), 957–994. doi:10.1109/COMST.2021.3067807
13. Wei, H., Zhang, Z., & Fan, B. (2017, December). Network slice access selection scheme in 5G. In *2017 IEEE 2nd Information Technology, Networking, Electronic and Automation Control Conference (ITNEC)* (pp. 352-356). IEEE.
14. H. Zhang, N. Liu, X. Chu, K. Long, A. -H. Aghvami and V. C. M. Leung, "Network Slicing Based 5G and Future Mobile Networks: Mobility, Resource Management, and Challenges," in *IEEE Communications Magazine,* vol. 55, no. 8, pp. 138-145, Aug. 2017, doi: 10.1109/MCOM.2017.1600940
15. D. Sattar and A. Matrawy, "Optimal Slice Allocation in 5G Core Networks," in *IEEE Networking Letters*, vol. 1, no. 2, pp. 48-51, June 2019, doi: 10.1109/LNET.2019.2908351.
16. Roy, S. S., Dash, S., & Sahu, B. J. (2023). Internet of Things in the 5G Ecosystem and Beyond 5G Networks. In T. Murugan & N. E. (Eds.), *Handbook of Research on Data Science and Cybersecurity Innovations in Industry 4.0 Technologies* (pp. 476-504). IGI Global. https://doi.org/10.4018/978-1-6684-8145-5.ch024

Elderly Fall Detection Using Machine Learning

Amisha Sinha[1], Dezy Jha[2], Malaya Kumar Swain[3],
Soumya Sagar Rath[4], Nimisha Ghosh[5]

Department of Computer Science and Information Technology,
Institute Of Technical Education & Research, SOA University, Odisha, India

Abstract Elderly fall detection research is aimed at developing algorithms which can be used in different technologies to automatically detect falls and notify caregivers and emergency services. This work uses machine learning algorithms to detect falls. Such detection can greatly reduce the risk of injury in elderly populations. Ongoing research in this area is mainly focused on improving accuracy, reducing false results and developing more accurate detection algorithms. After applying different machine learning algorithms like LDA, CART, Random Forest, Logistic Regression and SVM to calculate average and standard deviation of accuracy, f1-score precision, recall and sensitivity, by running the algorithms 100 times to get the average. Consequently, with an accuracy of 100% the best performance is shown by LDA and SVM.

Keywords Artificial neural networks, Elderly fall detection, Machine learning, Sensor data

1. Introduction

Falling of elderly people around the world is a very common and notable problem which results in serious injuries, decreasing the quality of life which leads to a decrease in human age per year. The global population of elderly individuals, typically defined as those aged 60 years or older, is estimated to be around 962 million. According to the World Health Organisation (W.H.O), more than 6.5 Lakh fatal falls occur every year in the world, the majority of which are suffered by the aged people (W. H. O, 2018)[1], making it the second reason for injury related death, followed by road traffic injuries. The number of elderly individuals at risk of

---

[1]amishasinha59@gamil.com, [2]jhadezy101@gmai.com, [3]swainmalaya07@gmail.com,
[4]sagarsoumyarath07@gmail.com, [5]nimishaghosh@soa.ac.in

falls is expected to rise, making fall prevention and detection important for maintaining the health and well-being of elders. It is estimated that one fifth of the population will come under the age of 65 or above in the next 35 years. These groups of people fall very frequently which increases the risk of death and becomes life-threatening in life. Thus, it is imperative to improve our detection techniques which are associated with fall detection. Approximately 50% of the crowd of this age or above experiences fall every year. For this issue we use various sensors, machine learning algorithms, and data analytics to assess fall risk factors and detect fall in real-time.

This work intends to contribute to the field of elderly fall detection by applying machine learning techniques on data that are received from sensors to develop a predictive model for identifying elderly individuals at high risk of falling. Parameters analyzed for detection of elderly fall are accelerometer reading, distance, HRV, pressure (0-small, 1-medium and 2-high pressures), Sugar Levels and SpO2 levels.

The findings of this research have important implications for healthcare providers, caregivers, and elderly family members. By improving the accuracy of fall detection, we can potentially enable timely interventions, reduce fall-related injuries, and enhance the overall well-being of the elderly population. The proposed research will contribute to the advancement of knowledge in this area and provide valuable insights for future research and development of technology-based fall detection solutions.

In summary, this paper aims to address the extreme issue of elderly falls by proposing an approach for fall detection using sensors collected data and machine learning techniques. The research is significant in terms of its potential impact on improving the health and safety of elderly individuals and in the field of fall detection.

2. Literature Review

Fall detection is an important issue in healthcare, particularly for elderly people who are at an increased risk of falling. Many studies have been conducted to develop and evaluate fall detection systems for elderly people. In this literature review, we will discuss some of the research related to elderly fall detection. One study conducted by Ambrose et al. [2] devised a fall risk assessment tool using machine learning (ML) algorithms. Their model successfully predicted fall risk with a sensitivity of 75% and specificity of 69%. The tool incorporated five essential risk factors, namely age, sex, fall history, medication use, and mobility impairment. In a separate study by Lee et al. [3], a deep neural network (DNN) algorithm was employed to develop a fall prediction model. The DNN model achieved an impressive accuracy of 86%. The researchers utilized four key risk factors in their model, including age, sex, gait speed, and balance. Bagala et al. [4] presented a comprehensive evaluation of accelerometer-based fall detection algorithms. They implemented and compared a total of 13 published algorithms, analyzing various parameters, thresholds, and fall phases such as the beginning of the fall, falling velocity, fall impact, and post-fall orientation. Their evaluation was based on a real-world falls database. In Bourke et al. [5], 21 existing fall detection approaches have been evaluated for a waist-mounted accelerometer-based system. The researchers extracted three distinct features:

velocity, impact, and posture. Notably, the algorithm combining the aforementioned features achieved a remarkable sensitivity of 100%. Pierry et al. [6] conducted a survey and evaluation of different fall detection methods, specifically focusing on accelerometers. The methods were categorized into acceleration-based, acceleration combined with other sensor-based, and non-acceleration-based approaches. Aziz et al. [7] performed an accuracy comparison of ten accelerometer-based fall detection algorithms. Five of these algorithms utilized threshold-based methods, while the remaining five employed machine learning-based approaches for fall detection. Habib et al. [8] published a comprehensive survey on fall detection and fall prevention systems considering smartphones. The authors presented three taxonomies for the system's operation phases: sense, analyze, and communicate. Wang et al. [9] conducted an extensive review on low-power technologies used in wearable telecare and telehealth. They classified these technologies into two categories: hardware-based and firmware-based methods. Mubashir et al. [10] provided a comprehensive survey on fall detection systems and related algorithms. They categorized the approaches into wearable-based, ambience-based, and vision-based fall detection methods. Similarly, Vallabh et al. [11] classified fall detection systems into wearable-based, ambient-based, and camera-based systems. The researchers conducted a systematic analysis and summarized recent implementations for each classification.

3. Material and Methods

3.1 Dataset

In recent years, a significant number of fall detection datasets have emerged, encompassing various sensor types such as wearables, visual sensors, ambient sensors, and multimodal setups. Our extensive literature review focused on identifying publicly available datasets specifically based on wearable sensors. These datasets incorporate a range of sensor measurements, including accelerometers, distance sensors, Spo2 (blood oxygen saturation), HRV (heart rate variability), sugar level, and pressure, with the aim of predicting falls [12].

3.2 Pipeline of Work

This work has been executed following the pipeline as shown in Fig. 18.1. ML offers the system the ability to learn based on the dataset and data patterns. Sensors offer the data linked to various fall parameters

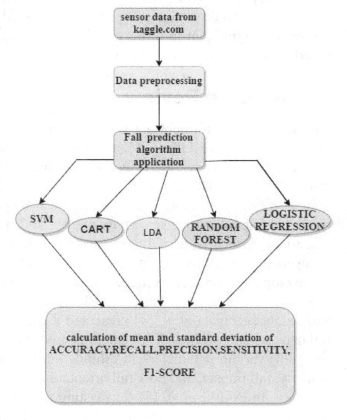

Fig. 18.1 Flow chart of elderly fall detection

during the data gathering procedure. In order to classify or detect fall activities based on the application requirements, the data is processed using ML algorithms. In general, all predictive models should be built using the training dataset and a separate test dataset should be used to evaluate them. The performance of predictive models should ideally be evaluated using data external to a comparable population. In this situation, it may be possible to use a split sampling strategy where the dataset is randomly split into a training/development set and a test/evaluation set. The size of the original data set determines the percentage of observations that belong to each data set. In this regard, we have split 80% of data for training and 20% of data for testing. However, there is a chance that each sample of the dataset will have unique valued features in the feature matrix. Lower performance is likely to occur in the classification result when the special valued output contains an error, or an infinite value. To keep all the feature values on the same scale, we have used z-score normalization. There is a high probability that the feature matrix will contain duplicate features or highly correlated and anti-correlated features. For this we have to calculate the correlation matrix. To represent the correlation matrix in graphic form we use the heat-map algorithm of machine learning.

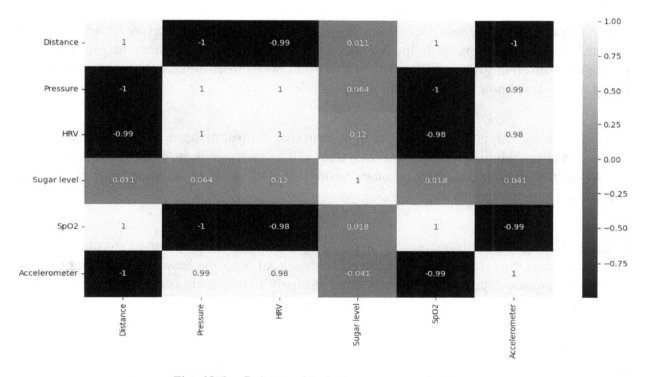

Fig. 18.2 Relationship between every feature

From this above Fig. 18.2 we can clearly see the relation between every feature. So we use PCA (principal analysis components) to reduce highly correlated features. Then, all the uncorrelated characteristics with correlation coefficients lower than 0.95 were selected. We also determined the feature relevance of each cluster's characteristics from the remaining associated clusters. The majority of each cluster feature was maintained, while the remainder were ignored. Thus, we have a feature set that is uncorrelated.

Five ML classifiers (SVM, CART, DT, RF, LDA) have been considered to separate autumn events from other regular activities. The hyper-parameters that these classifiers have employed.

SVM

SVM stands for Support Vector Machine, which is a dominating machine learning algorithm used for classification and regression analysis.

It looks at different features or characteristics of the things we want to classify or predict, and tries to find a line or curve that best separates or predicts them. SVM tries to make this line or curve as far away as possible from the closest points from each group, making it a powerful and accurate algorithm.

CART

CART (Classification and Regression Trees) is a machine learning algorithm that builds a tree-like model to classify or predict things based on certain features. It recursively splits the data into groups based on the most important feature until the data is fully classified or predicted.

LDA

LDA (Linear Discriminant Analysis) is a technique used for feature extraction and dimensionality reduction. It helps identify the most important features in our data that distinguish different groups or classes, and finds a lower-dimensional space that preserves as much of the discriminatory information as possible.

RF

RF(Random forest) is a popular machine learning algorithm used for classification and regression tasks. It builds multiple decision trees using random subsets of the data and features, and combines their predictions to make a final prediction. It is accurate, and can handle categorical and continuous data.

LR

LR(Logistic regression) is a statistical and machine learning technique used for classification tasks. It models the probability of a binary or categorical outcome based on one or more input variables. Logistic regression is widely used in many fields, such as healthcare, marketing, and finance, and is a simple and interpretable algorithm that can handle both categorical and continuous data.

Confusion matrix

A confusion matrix is a table used to evaluate the performance of a classification model. It provides a summary of the predicted and actual classes of a dataset. The matrix is particularly useful when dealing with binary classification problems, where the classes are typically labeled as "positive" and "negative." However, it can also be extended to handle multiclass classification problems.

The confusion matrix organizes the predictions into four categories:

TP: True Positive. This refers to the cases where the model correctly predicts the positive class (e.g. a disease is present) as positive.

TN: True Negative. This refers to the cases where the model correctly predicts the negative class (e.g. a disease is absent) as negative.

FN: False Negative. This refers to the cases where the model incorrectly predicts the negative class as positive (e.g. a disease is present but predicted as absent).

FP: False Positive. This refers to the cases where the model incorrectly predicts the positive class as negative (e.g. a disease is absent but predicted as present).

3.3 Accuracy

Accuracy is calculated as the ratio of accurately predicted values to the total number of predictions made.

$$\text{Accuracy} = (TN + TP) \div (TN + TP + FN + FP) \tag{1}$$

3.4 Precision

It is the total correctly predicted values against the total number of positive predicted values.

$$\text{Precision} = TP \div (TP + FP) \tag{2}$$

3.5 Recall

The Recall is calculated as the ratio of the total correctly predicted instances of the positive class to the total number of instances in the positive class.

$$\text{Recall} = TP \div (TP + FN) \tag{3}$$

3.6 F1-Score

It computes the harmonic mean of precision and recall to determine the F1 score.

$$\text{F1-Score} = (2 \times \text{recall} \times \text{precision}) \div (\text{recall} + \text{precision}) \tag{4}$$

Here,

TP: True Positive

TN: True Negative

FN: False Negative

FP: False Positive

4. Result

Table 18.1 presents the results obtained from various machine learning algorithms, including LDA, CART, random forest, logistic regression, and SVM. Each algorithm was executed 100 times, and the average and standard deviation of accuracy, precision, sensitivity, recall, and F1-score were calculated. The analysis reveals that LDA and SVM achieved the highest average accuracy of 100%, indicating their superior performance in fall detection.

Table 18.1 The outcomes of several machine learning algorithms, namely LDA (Linear Discriminant Analysis), CART (Classification and Regression Trees), RF (Random Forest), LR (Logistic Regression), and SVM (Support Vector Machine), were evaluated. The average value across these algorithms for various metrics was found to be 100%, indicating their excellent performance in fall detection

Algorithm	Accuracy (%) Average (%)	Accuracy (%) Standard deviation	Precision Average (%)	Precision Standard deviation	Recall Average (%)	Recall Standard deviation	Sensitivity Average (%)	Sensitivity Standard deviation	F1-Score Average (%)	F1-Score Standard Deviation
LDA	100	0.0	100	0.0	100	0.0	100	0.0	100	0.0
CART	99.9131	0.000809	99.9136	0.001208	99.9244	0.001052	99.9244	0.001052	99.9189	0.000858
RF	99.9931	0.000310	99.9933	0.000375	99.9903	0.000451	99.9903	0.000451	99.9918	0.000385
LR	99.9027	0.000838	99.8804	0.001307	99.9044	0.001100	99.9044	0.001100	99.8922	0.001023
SVM	100	0.0	100	0.0	100	0.0	100	0.0	100	0.0

4.1 Confusion Matrix

The confusion matrix is a widely used evaluation tool in fall detection systems and classification problems. It provides a concise summary of true positives, true negatives, false positives, and false negatives, allowing for the calculation of metrics like accuracy, precision, recall, specificity, and F1-score. It offers valuable insights into the system's performance in accurately detecting falls.

	Prediction Fall	**Prediction No Fall**
Actual Fall	TP=121	FN=0
Actual No Fall	FP=0	TN=287

	Training Set			
TARGET \ OUTPUT	Class0	Class1	Class2	SUM
Class0	121 / 29.66%	0 / 0.00%	0 / 0.00%	121 / 100.00% / 0.00%
Class1	0 / 0.00%	151 / 37.01%	0 / 0.00%	151 / 100.00% / 0.00%
Class2	0 / 0.00%	0 / 0.00%	136 / 33.33%	136 / 100.00% / 0.00%
SUM	121 / 100.00% / 0.00%	151 / 100.00% / 0.00%	136 / 100.00% / 0.00%	408 / 408 / 100.00% / 0.00%

Fig. 18.3 Confusion matrix

In this confusion matrix, the class labels are represented as Class 0, Class 1 and Class 2. These three classes represent-

Class 0: It generally refers to the reference class or baseline class. In our paper it represents 'Negative' sentiment i.e. the instances when fall don't occur. It includes activities like normal walking, sitting, standing or other non-fall-related movements.

Class 1: It generally represents the 'Positive' class. In our paper it refers to the instances where fall occurred. It may include activities such as slipping, abnormal running or any other motion which indicate a fall.

Class 2: It is an additional class that is used in multiclass classification scenarios. Here it refers to the activities that are neither falls (class 1) nor normal movements (class 0). It includes the activities such as lying down, crawling or any other non-fall activities that differ from typical movements.

We can measure Recall and Precision using the above confusion matrix as well as Accuracy and the F1-Score, which are used to evaluate the effectiveness of our ML model. We can use the formula in the equation (1), equation (2), equation (3) and equation (4) to calculate Accuracy, Precision, Recall and F1-Score simultaneously in Table 18.2.

Table 18.2 The evaluation of fall detection models involved assessing their accuracy, precision, recall, and F1-score

Class Name	Precision	1-Precision	Recall	1-Recall	f1-score
Class0	1.0000	0.0000	1.0000	0.0000	1.0000
Class1	1.0000	0.0000	1.0000	0.0000	1.0000
Class2	1.0000	0.0000	1.0000	0.0000	1.0000
Accuracy			1.0000		

The codes for this work can be found at: https://github.com/Sagarsoumya/ElderlyFallPrediction

5. Conclusion

In summary, the findings of this study indicate that the fall detection technology used had a relatively high accuracy in detecting falls among the elderly population. While further research is needed to confirm these findings, the technology may be a valuable tool for detecting falls and promoting safety among older adults.

In future, we intend to use a large sample size so that we may apply to real-world settings where falls are more unpredictable.

REFERENCES

1. W.H.O data: https://www.who.int/health-topics/ageing#tab=tab_1.
2. Lockhart, T.E., Soangra, R., Yoon, H. et al. Prediction of fall risk among community-dwelling older adults using a wearable system. Sci Rep 11, 20976 (2021). https://doi.org/10.1038/s41598-021-00458-5
3. AK Mishra, M Skubic, LA Despins, M Popescu, J Keller, M Rantz, C Abbott, M Enayati, S Shalini and S Miller (2022) Explainable Fall Risk Prediction in Older Adults Using Gait and Geriatric Assessments. Front. Digit. Health 4:869812. doi: 10.3389/fdgth.2022.869812
4. F. Bagalá, C. Becker, A. Cappello, L. Chiari, K. Aminian, J. M. Hausdorff, et al., "Evaluation of accelerometer-based fall detection algorithms on real-world falls", *PLoS ONE*, vol. 7, no. 5, May 2012.
5. A. Bourke, P. Van de Ven, M. Gamble, R. F. O'Connor, K. Murphy, E. Bogan, et al., "Evaluation of waist-mounted tri-axial accelerometer based fall-detection algorithms during scripted and continuous unscripted activities", *J. Biomech*, vol. 43, no. 15, pp. 3051–3057, Nov. 2010.
6. J. T. Perry, S. Kellog, S. M. Vaidya, J.-H. Youn, H. Ali and H. Sharif, "Survey and evaluation of real-time fall detection approaches", *Proc. 6th Int. Symp. High Capacity Opt. Netw. Enabling Technol. (HONET)*, pp. 158–164, Dec. 2009.
7. O. Aziz, M. Musngi, E. J. Park, G. Mori and S. N. Robinovitch, "A comparison of accuracy of fall detection algorithms (threshold-based vs. machine learning) using waist-mounted tri-axial accelerometer signals from a comprehensive set of falls and non-fall trials", *Med. Biol. Eng. Comput.*, vol. 55, no. 1, pp. 45–55, Apr. 2016.
8. M. A. Habib, M. S. Mohktar, S. B. Kamaruzzaman, K. S. Lim, T. M. Pin and F. Ibrahim, "Smartphone-based solutions for fall detection and prevention: Challenges and open issues", *Sensors*, vol. 14, no. 4, pp. 7181–7208, 2014.
9. C. Wang, W. Lu, M. R. Narayanan, S. J. Redmond and N. Lovell, "Low-power technologies for wearable telecare and telehealth systems: A review", *Biomed. Eng. Lett.*, vol. 5, no. 1, pp. 1–9, Mar. 2015.
10. M. Mubashir, L. Shao and L. Seed, "A survey on fall detection: Principles and approaches", *Neurocomputing*, vol. 100, pp. 144–152, Jan. 2013.
11. P. Vallabh and R. Malekian, "Fall detection monitoring systems: A comprehensive review", *J. Ambient Intell. Humanized Comput.*, vol. 9, no. 6, pp. 1809–1833, Nov. 2018.
12. Dataset used in this research paper: https://www.kaggle.com/datasets/laavanya/elderly-fall-prediction-and-detection?select=cStick.csv

2D CNN based Pituitary, Meningioma and Glioma Tumor Classification*

Reetichi Pattanaik[1], Suraj Sahu[2], Vishwas Kumar[3], Samrudhi Mohdiwale[4]

Department of CSIT, Institute of Technical Education and Research,
SOA (Deemed to be) University

Abstract Brain tumour is a conglomeration of anomalous cells in brain. It is bifurcated into sets benign (non-cancerous) and malignant (Cancerous). They can grow rapidly, feeding off of their own blood supply and compressing surrounding tissues. Despite extensive research, brain tumors remain a significant health threat. The main challenge involves designing an accurate deep learning architecture to classify brain tumors. The second challenge pertains to the scarcity of experts with the necessary experience and skill-set in classifying brain tumors using image-based deep learning models. These challenges represent significant obstacles in accurately classifying brain tumors and further emphasize the need for continued research in this area. To continue the research in this area the data-set is taken from Kaggle[1]. The proposed model gives the accuracy of 0.98 which is significantly higher than existing work. Further exploration and experimentation with these factors can lead to continued advancements in deep learning-based classification. The proposed model utilizes deep learning techniques, has the potential to improve the detection of minimal growth in brain tumors, a task that may be difficult for specialists to perform with complete accuracy. By reducing the work-load of specialists and improving the accuracy of classification, our model represents a significant advancement in the field of deep learning-based brain tumor detection.

Keywords 2D CNN, Deep learning, Glioma tumor, Meningioma tumor, Pituitary tumor

---

[1]reetichipattanaik@gmail.com, [2]surajkumarsahu@gmail.com, [3]vishwaskumar@soa.ac.in, [4]samrudhimohdiwale@soa.ac.in

*Supported by Department of CSIT, Institute of Technical Education and Research, SOA (Deemed to be university)

DOI: 10.1201/9781003489443-19

1. Introduction

The incentive behind this paper to overcome challenges of detection of brain tumor at early stage and play a role in betterment of people in a long term. The human brain being the central command with spinal cord makes up the central nervous system (CNS). Any threat to brain can cost a life. In spite of researches until recently, it has been most menacing disease. Brain tumor, otherwise known as intracranial tumor is a conglomeration of anomalous cells in brain. They grow faster on own blood vessels while compress its neighboring tissue. It can cause serious complications and even lead to death if they grow and exert pressure on nerves and eventually on the brain.

1.1 Issues and Challenges

Brain tumor can hamper brain functions, It is bifurcated into two sets benign (non-cancerous) and malignant (cancerous). Most common tumors are Glioma, Pituitary and Meningioma. Gliomas are brain tumors that start in glial cells. These are the supporting cells of the brain and the spinal cord. Pituitary gland tumors, pituitary neuroendocrine tumors (PitNETs) are brain tumors that start to grow in the pituitary gland. Meningiomas are tumors that start in the layers of tissue (meninges) that cover the brain and spinal cord.[11].[2]. The worldwide incidence rate of primary malignant brain and other CNS tumors in 2020, age-adjusted using the world standard population, was 3.5 per 100,000 populations. An estimated 168,346 males and 139,756 females who were diagnosed worldwide with a primary malignant brain tumor in 2020, an overall total of 308,102 individuals.[3] Scientists have developed sophisticated imaging tools include MRI (magnetic resonance imaging) scan, CT (computed tomography) scan and PET (positron emission tomography) scan to pin point location, shape and size, cellular makeup and normal nerve pathways. MRI scan is the most preferable scan among these three scans providing better differentiation between fat, water muscle and other soft tissue contrast. MRI scans of three planes, top to down (axial plane), from front to back (coronal plane), and side to side (sagittal plane) are considered.

Algorithmic approaches are used in machine-based diagnostic imaging equipment helping neuroscience physician in stratification of brain tumor. Deep learning is an artificial intelligence based model that uses multiple layers of artificial neural networks to extract high-level features from large data-sets. It has shown promising results in automating the analysis of medical images, including the detection and classification of brain tumors. Many Deep learning projects are imposed for well-being of the society like Drowsy Driver Detection [4], Fake News Detection [8] and Plant Disease Detection [14]. Deep learning has made significant advances in the medical field like Ultrasound analysis[10], physiological signals[5] and Decision making with EMRs[9]

In Turkey, Denizli [Abdullah Kavakli, 2023], a deep learning model for brain tumor classification from MRI was proposed and the proposed model is evaluated under several performance criteria. The Brain tumor images were fed to Convolutional Neural Network (CNN) for feature extraction. After feature extraction several layers would be added to classify the extracted features and give an approximate prediction of the output. There will be final

Soft-Max layer which will output the probability distribution of the extracted features. After adding layers AMSgrad optimizer which is an extension of ADAM, is utilized to update the weights of the network and train the model. During training the images will be fed in batches of 32 over several epochs and the weights will be adjusted iteratively to minimize the loss function.

1.2 Motivation and Objective

This paper upgrades changes that has been done in an existing model and enhance the performance of the model by making optimal changes.

This paper aims to detection brain tumor using different MRI images and providing optimal model for classification up to its best and provide favorable structure. Here firstly we use MRI images data-set and use different images techniques to process it. We have used an improved version of existing model which will give more optimal result with better accuracy. We have used different optimization techniques (Adam, SGD, Ada-grad, SGD (With momentum) AMSgrad), these are all well-known optimizer that are used on existing model with changes to batch size , epochs and CNN layers. As final result, we obtrained a model that classifies brain tumor using deep learning application with 2D CNN algorithm; with higher accuracy of 0.985. The presented model uses AMSgrad, which is an upgraded version of Adam optimizer. We can summarize the methodology followed in this paper as follows:

- Grading and fragmentation of MRI scans using an algorithm and combines for a result using 2D CNN layer.
- In this paper, we have used AMSgrad optimizer which is an improvement to Adam that converges to globally optimal solution and avoids immediate changes to learning rate.
- With changes to model with different parameters we observed that our proposed 2D CNN model with Deep learning approach gives a model with optimal functionality.

The incentive behind the paper to classify brain tumor with high accuracy rate by using improved model on the data-set. It will help specialist to detect minimal growth and make their work easy.

2 Background and Literature

2.1 Model

CNN: The model uses Convolutional Neural Networks (CNNs) which is a deep learning approach can be used to achieve desired goal. CNN in medical imaging has shown significant improvements in accuracy and speed compared to traditional methods, and has the potential to greatly enhance the diagnostic capabilities of healthcare professionals. Deep learning approach helps better analysis of complex patterns in data, with higher accuracy and with sturdy in nature. Its architecture is similar to connection of neurons in brain and inspired by visualization by occupational cortex. The main role is to reduce the image data but does not lose its features which will help when we have massive data-set. CNN broadly consists of three layers described as follows:

- Convolution Layer: In this layer input image is mapped with kernel and reforms output by different convolution operation and stores in featured map.
- Pooling Layer: Pooling layer helps in extraction of more relevant features and reducing dimensional. Which is further divided into Max pooling and Average Pooling.
- FC Layer (Fully Connected Layer): It is the last layer of CNN it performs relevant task with use of an activation function.

Optimizer: For improving the performance of a Deep Learning model accurate optimization method should be used. An optimizer is an algorithm used to reduce loss function and provide most accurate result. Optimizer are chose based on the need of the model in account of learning rate and weight.

2.2 Review

This section reviews about different studies that has been conducted on this field.

A Kapoor et al.[7] made a model using Firefly Algorithm (FA) and Traditional k-means algorithm enhanced using Particle Swarm Optimization (PSO) and achieved accuracy as high as 96%. Senan [12] hybrid models were trained using ml algorithms (AlexNet, AlexNet+SVM, ResNet-18, and ResNet-18+SVM) and on a later stage filtered images were trained using deep learning algorithms for distinguishing features. The model achieved an average accuracy of 94%. Shanthakumar [13] made a model using watershed segmentation to improve feature extraction process to maximize accuracy of tumour segmentation 93.51%. Gumaei [6] proposed a hybrid model for feature extraction using RELM (regularised extreme learning machine). The min-max normalization method was used to enhance prepossessing. This work obtained an accuracy rate of 94.23%. Swati [15] used a fine tuned VGG19 model, pretrained using Contrast-enhanced MRI to improve the results and obtained an average accuracy of 94.82%.

On viewing many related works based on this model are designed to only categorized tumor based only on Glioma, Pituitary and Meningioma where the accuracy goes as high as 0.96 and in such models due to its unique characteristics, meningioma tumors are challenging to detect, resulting in lower accuracy compared to the other two types. At the same time other models are only focused on just predicting if the MRI scans contain tumor or not. The Proposed model distinguishes the MRI images into four types (Glioma, Meningioma, No tumor and Pituitary tumor) which makes it convenient for machine to learn better features to train with and improve overall accuracy of the model.

3 Methodology

In this section we have the overview of the 2D CNN model, optimization method and changes that has been feed to the model. A brief overview of the algorithms used in this following paper.

3.1 Data-set

This paper performs experiment on Brain MRI data-set that was publicly accessible .It was downloaded from Kaggle website.[1]. The data-set contains 3270 gray-scale Brain MRI images which is further divided into four classes; 930 Glioma, 940 Meningioma, 900 Pituitary

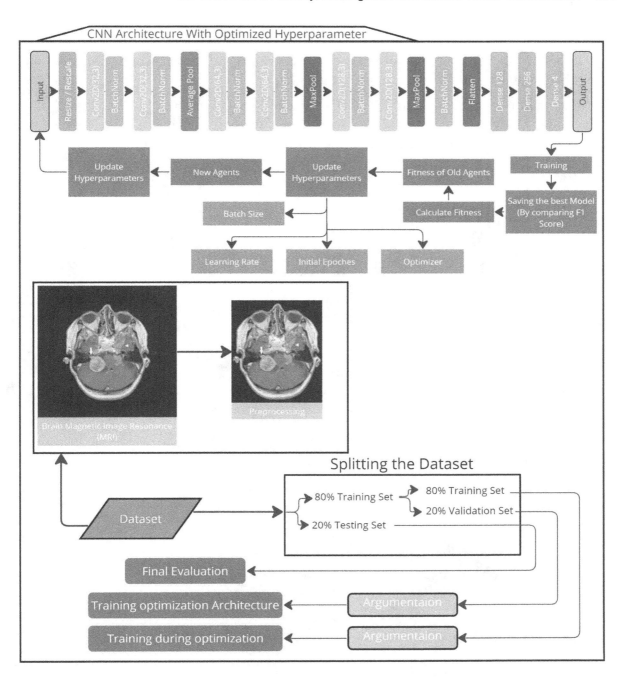

Fig. 19.1 Framework for the proposed model

and 500 No tumor. The data-set is split-ted into training set of 80 percent and testing for 20 percent of the total data-set. As data separation is uneven so data is augmented for achieving better accuracy and different version of images.

3.2 Image Preprocessing

The process of resizing images in to a dimension 160*160 as preprocessing step as images from data-set may have different dimensions.

182 Prospects of Science, Technology and Applications

3.3 2D CNN

The data-set uses 2D CNN model is based on dimension of kernel which leverage content across height and width of a segmented image. graphicx

The Flowchart depicts the 22 different CNN layer used in the proposed model with activation used as ReLU (Rectified Linear Unit) Activation Function which is a half wave rectified function which puts positive values in case of negative it simply takes zero.

3.4 AMSGrad

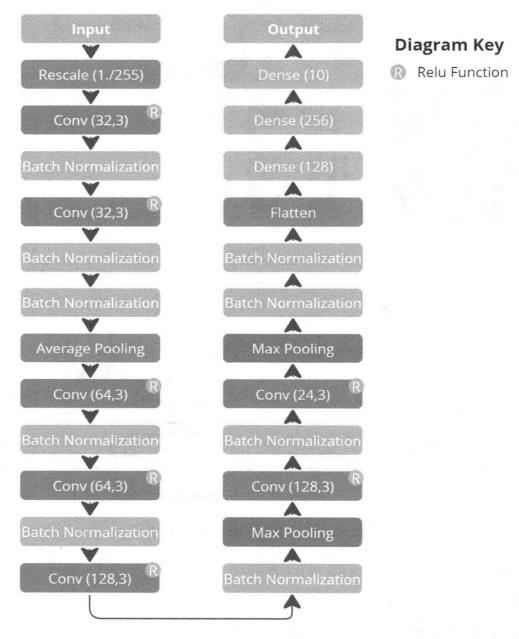

Fig. 19.2 Convolution Neural Network 22 Layers

The model uses AMSGrad algorithm which is an extension of Adam that resolves convergence issue faced on Adam. More broadly it is up gradation of gradient descent optimization algorithm and maintains maximum at second momentum. It uses maximum past squared gradients or second momentum V_t rather than exponential average updated iteratively over the time stamp t with the function of different parameter x via vector operation. First, the gradients (partial derivatives) are calculated for the current time step

$$g = f(x_{t-1}) \qquad (1)$$

Next, First momentum κ with the help of gradient and hyperparameter σ

$$\kappa t = \sigma \kappa_{t-1} + (1 - \sigma)g_t \qquad (2)$$

Hyperparameter can be decayed exponentially as such

$$\sigma_t = \sigma^t \qquad (3)$$

Second moment vector is updated using hyperparameter2 (is a decay function of infinity norms) ζ and square of gradient

$$v_t = \zeta v_{t-1} + (1 - \zeta)g^{2t} \qquad (4)$$

Maximum second momentum vector i is given as

$$\imath t = \max(\imath t - 1, v_t) \qquad (5)$$

Lastly, the parameter value can be calculated using learning rate ξ and a smoothing term that avoids division by zero ϱ

$$x_t = x_{t-1} - \xi * \kappa_t / i_t + \varrho \qquad (6)$$

4 Results and Discussion

4.1 Results

At Table 19.1 is the data collected of models run using the 5 optimizers with hyperparameters set to 16 batch size and a max of 64 epochs, trained multiple times to obtain Best run and Worst run results we can see that the accuracy obtained by the reference model using the different optimizers attain a accuracy of 92.8% using the Adam optimizer and a maximum of 96.34% accuracy using AMSgrad optimizer. Other optimizers could obtain better results if the model architecture is changed as seen in Table 19.2.

At Table 19.2 the hyperparameters remain unchanged and using the newer model architecture all optimizers perform significantly better than the reference model. Over the reference model the newer model is able to obtain a accuracy 95.1% using Adam and over 98% accuracy using the AMSgrad. The Selection of Optimizers becomes challenging since it can affect accuracy of the whole model rather than a particular type. In certain cases some optimizers can be useful where we need to classify a particular of Tumor for, instance AdaGrad has an accuracy close to 97.7% for Glioma Tumor and SGD(with momentum) is a better alternative to detect Meningioma Tumor as it achieves better accuracy than AMSgrad. The same can be said for reference model. In the reference model SGD(with momentum) has better accuracy for Meniglioma Tumor and Adam is a better optimizer to detect Pituitary Tumor with this Dataset.

Table 19.1 Performance matrix for reference model

Original Model		Accuracy			Glioma Tumor			Meningioma Tumor			No Tumor			Pituitary Tumor			Macro Avg		
Optimizer	Statistical metrics	Precision	Recall	f1 score	Precision	Recall	f1 score	Precision	Recall	f1 score	Precision	Recall	f1 score	Precision	Recall	f1 score	Precision	Recall	f1 score
Adam	Average	0.93568	0.91750	0.91191	0.89650	0.90925	0.91950	0.90865	0.88510	0.86150	0.95860	0.94150	0.91010	0.97895	0.95415	0.95655	0.91980	0.90430	0.95375
	Best Run	0.94493	0.92883	0.92960	0.9012	0.9173	0.9214	0.9126	0.8851	0.9073	0.9761	0.9513	0.9184	0.9698	0.9616	0.9713	0.9315	0.9174	0.9394
	Worst Run	0.92643	0.90618	0.89423	0.8918	0.9012	0.9176	0.9047	0.8451	0.8157	0.9411	0.9317	0.9018	0.9681	0.9467	0.9418	0.9081	0.8912	0.9681
SGD	Average	0.81555	0.87350	0.84500	0.72210	0.91530	0.85210	0.87560	0.64700	0.71075	0.76465	0.96510	0.84965	0.89985	0.96660	0.96750	0.79990	0.88425	0.83775
	Best Run	0.82618	0.88185	0.85675	0.7318	0.9183	0.8713	0.8871	0.6517	0.7197	0.7774	0.9861	0.8581	0.9034	0.9713	0.9779	0.8081	0.8971	0.8581
	Worst Run	0.80493	0.86515	0.83325	0.7124	0.9123	0.8329	0.8641	0.6423	0.7018	0.7519	0.9441	0.8412	0.8913	0.9619	0.9571	0.7917	0.8714	0.8174
Adagrad	Average	0.87975	0.86914	0.87506	0.73015	0.96015	0.84460	0.94265	0.64265	0.77690	0.90710	0.93100	0.91025	0.93910	0.92275	0.96850	0.84930	0.84245	0.86525
	Best Run	0.88460	0.87308	0.87675	0.7419	0.9812	0.8421	0.9541	0.6481	0.7819	0.9013	0.9312	0.9187	0.9411	0.9318	0.9723	0.8813	0.8712	0.8791
	Worst Run	0.87490	0.86520	0.87138	0.7184	0.9791	0.8471	0.9312	0.6372	0.7719	0.9129	0.9308	0.9018	0.9371	0.9137	0.9647	0.8173	0.8137	0.8514
SGD(With Momentum)	Average	0.96284	0.95078	0.93160	0.94650	0.96045	0.95100	0.97135	0.95330	0.95600	0.96715	0.97830	0.92630	0.96635	0.91105	0.89110	0.91180	0.95995	0.96285
	Best Run	0.96738	0.95440	0.93673	0.9512	0.9612	0.9541	0.9738	0.9578	0.9639	0.9731	0.9848	0.9328	0.9714	0.9138	0.8961	0.9149	0.9681	0.9639
	Worst Run	0.95830	0.94715	0.92648	0.9418	0.9597	0.9479	0.9689	0.9488	0.9521	0.9612	0.9718	0.9198	0.9613	0.9083	0.8861	0.9087	0.9518	0.9618
AMSgrad	Average	0.96794	0.96346	0.97095	0.97025	0.92865	0.95160	0.96650	0.96780	0.97180	0.97100	0.98295	0.97670	0.96400	0.97445	0.98370	0.97380	0.97875	0.96965
	Best Run	0.96565	0.97423	0.97643	0.9818	0.9489	0.9541	0.9712	0.9774	0.9751	0.9681	0.9918	0.9852	0.9415	0.9788	0.9913	0.9738	0.9832	0.9691
	Worst Run	0.97023	0.95270	0.96548	0.9587	0.9084	0.9491	0.9618	0.9582	0.9685	0.9739	0.9741	0.9682	0.9665	0.9701	0.9761	0.9738	0.9743	0.9682

Table 19.2 Performance matrix for proposed model

Modified Model		Accuracy			Glioma Tumor			Meningioma Tumor			No Tumor			Pituitary Tumor			Macro Avg		
Optimizer	Statistical metrics	Precision	Recall	f1 score	Precision	Recall	f1 score	Precision	Recall	f1 score	Precision	Recall	f1 score	Precision	Recall	f1 score	Precision	Recall	f1 score
Adam	Average	0.95104	0.93936	0.94989	0.91120	0.92105	0.92745	0.91835	0.88895	0.90880	0.96990	0.96815	0.88250	0.98470	0.96930	0.98080	0.94270	0.95365	0.96335
	Best Run	0.95813	0.94660	0.95388	0.9146	0.9334	0.9277	0.9241	0.9007	0.9124	0.9973	0.9749	0.9852	0.9985	0.9774	0.9902	0.9538	0.9592	0.9673
	Worst Run	0.94395	0.93213	0.94590	0.9078	0.9087	0.9272	0.9126	0.8972	0.9052	0.9825	0.9614	0.9798	0.9729	0.9612	0.9714	0.9316	0.9481	0.9394
SGD	Average	0.82564	0.87980	0.84453	0.77150	0.92115	0.86350	0.85195	0.64345	0.70715	0.77035	0.99130	0.83245	0.90955	0.96330	0.97500	0.80695	0.87910	0.84675
	Best Run	0.84470	0.88193	0.84935	0.7912	0.9251	0.8651	0.8914	0.6451	0.7126	0.7789	0.9914	0.8418	0.9173	0.9661	0.9779	0.8162	0.8861	0.8518
	Worst Run	0.80698	0.87756	0.83970	0.7518	0.9172	0.8619	0.8125	0.6418	0.7017	0.7618	0.9912	0.8231	0.9018	0.9605	0.9721	0.8017	0.8721	0.3417
Adagrad	Average	0.89040	0.85514	0.87275	0.73950	0.97920	0.84645	0.95125	0.58320	0.77020	0.91085	0.93625	0.91675	0.96000	0.92190	0.95760	0.84430	0.89115	0.86020
	Best Run	0.89550	0.87708	0.87945	0.7489	0.9871	0.8528	0.9594	0.6493	0.7723	0.9128	0.9408	0.9204	0.9619	0.9311	0.9723	0.8761	0.8948	0.8681
	Worst Run	0.88530	0.83320	0.86605	0.7301	0.9713	0.8401	0.9441	0.5171	0.7681	0.9089	0.9317	0.9131	0.9581	0.9127	0.9429	0.8125	0.8875	0.8523
SGD(With Momentum)	Average	0.95171	0.95895	0.96191	0.97625	0.91605	0.93685	0.90795	0.96160	0.95905	0.97450	0.99465	0.97000	0.94815	0.96350	0.98175	0.97490	0.97130	0.96660
	Best Run	0.96455	0.96880	0.97623	0.9333	0.9411	0.9524	0.9124	0.9661	0.9664	0.9813	0.9961	0.9967	0.9812	0.9699	0.9854	0.9755	0.9752	0.9861
	Worst Run	0.93888	0.94910	0.94560	0.9692	0.891	0.9213	0.9035	0.9571	0.9517	0.9677	0.9912	0.9413	0.9151	0.9571	0.9681	0.9743	0.9674	0.9671
AMSgrad	Average	0.97604	0.98251	0.98029	0.96150	0.98030	0.96970	0.98095	0.95005	0.96285	0.98095	0.99990	0.99615	0.98075	0.99980	0.99265	0.97675	0.97735	0.97035
	Best Run	0.98063	0.98638	0.98383	0.9643	0.9884	0.9798	0.9816	0.9571	0.9658	0.9875	1.0000	0.9989	0.9891	1.0000	0.9908	0.9772	0.9836	0.9766
	Worst Run	0.97145	0.97865	0.97675	0.9587	0.8722	0.9596	0.9603	0.943	0.9595	0.9744	0.9998	0.9934	0.9724	0.9996	0.9845	0.9763	0.9711	0.9639

Precision: It tells about all predicted values how many of them are actual true.

$$P = TP/TP + FP \tag{7}$$

Recall: It tells about all actual input how many is correct predicted.

$$R = TP/TP + FN \tag{8}$$

F1 Score: It is the combination of both recall and precision.

$$F1 = 2(P * R)/P + R \tag{9}$$

Confusion Matrix: A confusion matrix is an evaluation metric used to evaluate the performance of a proposed model. It is used to extract various metrics to show the classifier's performance for each section. Some important evaluation metrics include Precision, Recall and F1-score. article array

TP	FP
FN	TN

True Positive (TP) signifies how many positive class samples in model are predicted correctly.

True (TN) signifies how many negative class samples in model are predicted correctly.

False Positive (FP) signifies how many negative class samples in model are predicted incorrectly.

False Negative (FN) signifies how many positive class samples in model are predicted incorrectly.

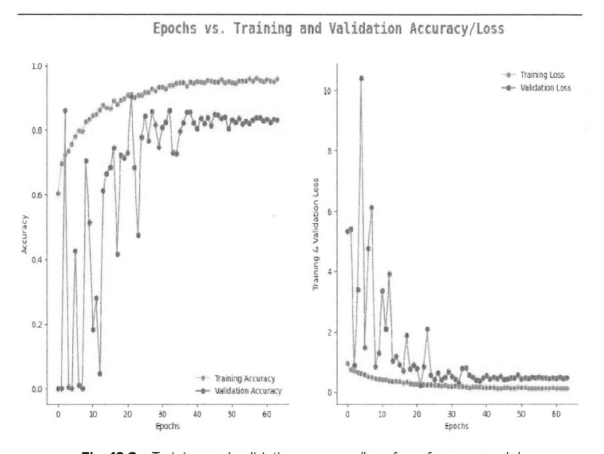

Fig. 19.3 Training and validation accuracy/loss for reference model

The training/validation accuracy and loss are common metrics used to evaluate the performance of a model. The Proposed Model has a accuracy of 98.25% where the reference model was only able to achieve a accuracy of 93.56% .It was also able to achieve better learning rate and had lower training loss. At Fig. 19.4 the difference between Training loss and Validation loss in the reference model indicates towards underfitting where it is unable to accurately capture the input and output variables. At Fig. 19.5 Proposed model has better and consistent Training/Validation accuracy than the reference model along with lower difference between Training loss and Validation loss than reference model.

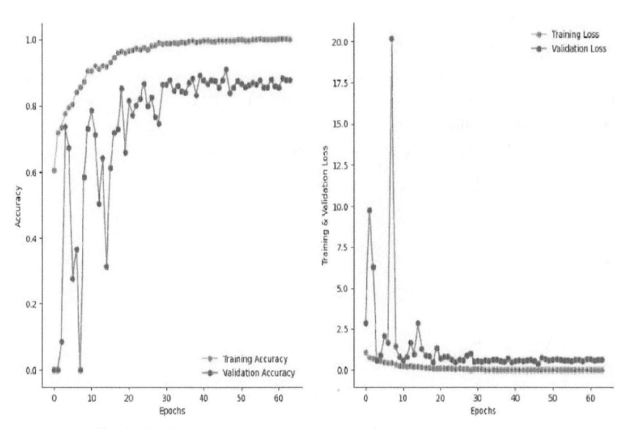

Fig. 19.4 Training and validation accuracy/loss for proposed Model

4.2 Discussion

In account of brain tumor classification we faced We came across many challenges for choosing desired optimizer. By testing different optimizers we found AMSGrad with best outcomes.

At first, the reference model was run in its base environment without any changes to any of its paraments and provides us with a base accuracy of 93% then the reference model was tested with multiple optimizers [Adam, SGD, Adagrad, SGD (with momentum) and AMSGrad]

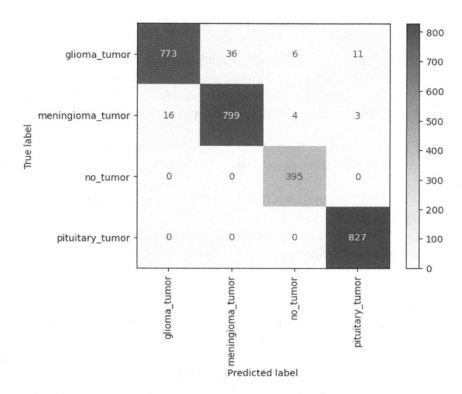

Fig. 19.5 Confusion matrix for proposed model

with a maximum of 64 epochs and AMSgrad optimizer obtained better results than Adam which was originally used. It could be noted some optimizers provides better accuracy for a particular types. For the reference model Higher accuracy for Pituitary Tumor was shown by Adam and Better Accuracy for Meniglioma Tumor was provided by SGD (with Momentum.) The similar tests were done for the proposed model and the over all accuracy for all optimizers was significantly better. Of all optimizers AMSgrad showed the highest accuracy of all tests done so far is around 98.3%. Similar to Reference model this also showed some exceptions where certain optimizers provided better accuracy for certain type using this Dataset.

5. Conclusion and Scope for Future

Hence we can derive a conclusion that Brain tumor is a life threatening disease early stage and accurate detection can save many lives. A computer aided system with artificial intelligence can make physician work more accurate and easier. This will also help non expert to have prerequisite knowledge about this field.

In future there is a scope for development not only in deep learning perspective but also computer driven programs. Which will help in healthcare field in emergency situation where a specialist may not be available but at that instant of time action need to be taken , this also creates a field for development of 3D imaging software.

REFERENCES

1. AbdullahKavakli and Sartaj. "Brain Tumor Classification with CNN". In: https://www.kaggle.com/code/abdullahkavakli/brain-tumor-classification-with-cnn/notebook 3.4 (2023).
2. Stephen R Ali et al. "Improving the effectiveness of multidisciplinary team meetings in skin cancer: Analysis of national Cancer Research UK survey responses". In: Journal of Plastic, Reconstructive & Aesthetic Surgery (2023).
3. Maria Antonietta Cerone et al. "The Cancer Research UK Stratified Medicine Programme as a model for delivering personalised cancer care". In: British Journal of Cancer 128.2 (2023), pp. 161–164.
4. Dhara Desai, Nisha Gandhi, and Shruti Dodani. "A Review of different Methods of Watermarking for Medical Images Authentication". In: ().
5. Oliver Faust et al. "Deep learning for healthcare applications based on physiological signals: A review". In: Computer methods and programs in biomedicine 161 (2018), pp. 1–13.
6. Abdu Gumaei et al. "A hybrid feature extraction method with regularized extreme learning machine for brain tumor classification". In: IEEE Access 7 (2019), pp. 36266–36273.
7. Anjali Kapoor and Rekha Agarwal. "Enhanced Brain Tumour MRI Segmentation using K-means with machine learning based PSO and Firefly Algorithm". In: EAI Endorsed Transactions on Pervasive Health and Technology 7.26 (2021), e2–e2.
8. Nei Kato et al. "Optimizing space-air-ground integrated networks by artificial intelligence". In: IEEE Wireless Communications 26.4 (2019), pp. 140–147.
9. Znaonui Liang et al. "Deep learning for healthcare decision making with EMRs". In: 2014 IEEE International Conference on Bioinformatics and Biomedicine (BIBM). IEEE. 2014, pp. 556–559.
10. Shengfeng Liu et al. "Deep learning in medical ultrasound analysis: a review". In: Engineering 5.2 (2019), pp. 261–275.
11. Timothy R Lukins et al. "Cervical spine immobilization following blunt trauma: a systematic review of recent literature and proposed treatment algorithm". In: ANZ journal of surgery 85.12 (2015), pp. 917–922.
12. Ebrahim Mohammed Senan et al. "Early diagnosis of brain tumour mri images using hybrid techniques between deep and machine learning". In: Computational and Mathematical Methods in Medicine 2022 (2022).
13. P Shanthakumar and P Ganesh Kumar. "Computer aided brain tumor detection system using watershed segmentation techniques". In: International Journal of Imaging Systems and Technology 25.4 (2015), pp. 297–301.
14. Gaurav Srivastava et al. "A Review on Mechanical properties of Aluminium 2024 alloy with various reinforcement metal matrix composite". In: J. Univ. Shanghai Sci. Technol 23 (2021), pp. 558–573.
15. Zar Nawab Khan Swati et al. "Brain tumor classification for MR images using transfer learning and fine-tuning". In: Computerized Medical Imaging and Graphics 75 (2019), pp. 34–46.

Dynamic Long Short-Term Memory Model for Stock Market Price Forecasting

Indrajit Sahu, Kiran Shankar Paira, Priti Rani Bhoi
Siksha 'O' Anusandhan University, Bhubaneswar, Odisha

Samrudhi Mohdiwale*
Supervisor, Siksha 'O' Anusandhan University, Bhubaneswar, Odisha

Abstract This research paper aims to explore the utilization of Stacked Long Short-Term Memory (LSTM) method for prediction the stock market prices using machine learning. The objective of the research is to determine the performance of LSTM model in predicting stock prices and to compare its accuracy with other similar existing prediction methods. Utilizing Stacked LSTM entails the ability to layer numerous LSTM units on one another. Each LSTM layer possesses the capability to capture various levels of abstraction, thereby enhancing the model's capacity to discern intricate patterns within the data. Historical stock data of NIFTY50 was collected for conducting this research, and an LSTM model was employed for the prediction of future prices of the stocks. The paper investigates different preprocessing techniques, such as data scaling and feature engineering. Data scaling is a preprocessing technique used to bring numerical features to a similar scale and Feature engineering associates building new features or transforming real features in the data to extract relevant information, such as creating technical indicators or incorporating time-based features. The LSTM model was successful to identify and capture fundamental trends and patterns of the NIFTY50 data, making it an efficient tool for stock price prediction. The work of the LSTM model for stock market price prediction was evaluated using several metrics, including accuracy, root mean squared error (RMSE), and mean absolute error (MAE). Accuracy was used as the primary metric for evaluating the performance of the model. The correctness of the model was decided by analyzing the predicted values with the real values in the testing dataset. Based on the findings, it is evident that employing an LSTM model can prove to be a valuable and effective approach for predicting stock prices. This can offer traders and investors valuable insights into market trends, aiding them in making well-informed decisions.

Keywords Stock price prediction, LSTM, Machine learning, Python programming language, Accuracy

---

*Corresponding author: samrudhimohdiwale@soa.ac.in

DOI: 10.1201/9781003489443-20

1. Introduction

1.1 Overview

The stock market is a vital economic indicator that has major impact on the global economy. With the rapid evolving technology and the increased availability of data, many researchers and analysts have turned to machine learning algorithms to forecast stock prices with greater accuracy [1]. In recent years, deep learning methods (popular method is long short-term memory models (LSTM)) [2], have attracted much attention due to their high performance and also able to capture the underlying patterns in your time series data.

Long Short-Term Memory also known as LSTM, is a popular recurrent neural network building that have demonstrated remarkable effectiveness in learning and recognizing long-term patterns and dependencies within time-series data [3]. Its proficiency in analyzing sequential information [4] makes it well-suited for predicting stock prices, leveraging its capability to capture long-term trends. By utilizing its unique memory cell structure, which includes forget gates, input gates, and output gates, LSTM models can retain relevant information and discard noise, enabling them to process historical stock market data and discern intricate relationships that impact future price movements. However, it is important to acknowledge that achieving complete accuracy in stock price prediction remains challenging because of the built-in volatility and randomness of monetary shares, emphasizing the need to supplement LSTM-based predictions with other tools and expertise for informed decision-making.

1.2 Technical Description

Forecasting stock market prices poses a significant challenge due to the intricate and ever-changing nature of the system. Nevertheless, deep learning techniques [5], particularly LSTM (Long Short-Term Memory) models, have emerged as a promising solution for achieving precise predictions. LSTM models are explicitly designed to capture temporal dependencies and long-term memory. By effectively addressing issues such as vanishing and exploding gradients commonly encountered in traditional RNNs, LSTM models excel at capturing extended trends in time-series data, including stock prices. The architecture of an LSTM model encompasses essential components such as memory cell, input gate, output gate, and forget gate. The memory cell retains the previous state of the model, while the input gate controls the flow of data from the input to the memory cell. The forget gate determines the relevance of data from the previous state to be discarded, and the output gate controls the information flow from the memory cell to the output. In recent years, LSTM models have demonstrated significant achievements in predicting stock prices, exhibiting state-of-the-art performance [6]. The capacity of LSTM models to com- prehend and model long-term trends in stock prices has empowered researchers to develop more accurate and dependable prediction models.

1.3 Issues and Challenges

The dynamic nature of the stock market presents a significant challenge in predicting stock prices [7]. Global economic conditions, geopolitical events, company news, and investor sentiments are some major factors that directly impact on stock markets. These factors are highly volatile and subject to frequent changes, posing a difficulty in accurately capturing and responding to market fluctuations. [8].

Another major challenge is the availability and quality of data. As there is a large amount of historical stock market data available on internet, the quality and quantity of data can be limited due to issues such as errors in data collection, missing data, and data inconsistencies. However, the quantity of data can be limited by factors such as data availability, storage capacity, and computational resources [9].

In addition, there is a risk of overfitting when using LSTM models for stock market prediction. Overfitting arises when a model becomes excessively intricate and begins to closely fit the training data, resulting in reduced accuracy when applied to testing data. This risk can be resolved by proper tuning of the model's hyperparameters and using other techniques.

Finally, concluding the results of LSTM models for stock market prediction can be a bit challenge. As these models can provide accurate predictions, understanding the basic factors that led to these predictions can be difficult. This can make it challenging to make a clear and precise trading strategy based on the model's predictions.

1.4 Motivation

Stock market prediction is extensive research due to its potential for providing valuable insights to investors and financial institutions. Accurately forecasting stock prices aid in making informed decisions and maximizing profits. LSTM (Long Short-Term Memory) has emerged as a popular artificial neural network for analyzing and predicting time series data, exhibiting promising results in stock price prediction when compared to traditional statistical models. [10]. This research aims to assess the effectiveness of LSTM in stock price prediction and compare its performance against other widely used prediction models. Additionally, the study will explore the impact of factors like economic indicators, news sentiment, and social media sentiment on stock prices. The outcomes of this research is expected to yield significant implications for investors and financial institutions, contributing to the advancement of more precise and reliable prediction models.

1.5 Objectives

The primary objective of this research is to examine the utilization of LSTM models for predicting stock prices. Several key objectives have been outlined to accomplish this aim. Firstly, a performance evaluation will be conducted to compare LSTM models with traditional time-series forecasting models in terms of stock price prediction accuracy [11]. The research will also focus on identifying the significant factors that influence the precision of LSTM models in predicting stock prices. Furthermore, the paper will propose potential solutions to address the challenges and limitations associated with using LSTM models in stock price prediction. Real-world trading scenarios will be examined to provide insights into the potential and limitations of LSTM models in this context. Additionally, aclear and concise trading strategy will be developed based on the predictions generated by the LSTM model. The research findings are expected to offer benificial vision into the effectiveness and potential of LSTM models for stock price prediction, benefiting investors, traders, and financial analysts in making well-informed decisions in the stock market.

2. Literature Review

2.1 Existing Models

The paper named as Machine Learning with Applications and the title of the journal is Predicting stock market index using LSTM in the year 2022 has proposed a method based on LSTM Algorithm. Their conclusions are explained in such a way that the proposed model can be easily applied to other broad market indices featuring comparable data patterns. This customization process ensures that the model's application remains versatile while maintaining the originality of its design [12].

In recent paper, Dilhan et al, proposed a hybrid RNN (LSTM + GRU) for prediction banking stock. The authors introduced a novel approach which utilizes the Secondary Decomposition algorithm, Attention-based LSTM, Multi-Factor Analysis, and Improved VMD. The study's conclusion highlights that the hybrid model proposed in this research demonstrates notable advancements in prediction accuracy when compared to other existing models [13].

Similarly, Beak et al. introduced a new modularized forecasting-based framework named as ModAugNet for stock market prediction. In the paper, the authors employed various techniques such as Long short-term memory (LSTM), data augmentation for addressing the overfitting through deep learning. Their research findings suggest the introduction of a unique augmentation technique for financial time-series forecasting. This approach enhances the model's resilience against overfitting, eliminating the necessity of generating artificial time-series data to supplement the existing training dataset.

More paper like Knowledge-Based Systems, Applied Soft Computing, Decision Analytics Journal, Journal of King Saud University – Computer and Information Sciences, ScienceDirect, International Journal of Computer Engineering Technology and Wseas transactions on Computer research etc. have proposed many methods in the year between 2017-2023 [14].

Many papers have used methods like Multi-Layer Sequential Long Short Term Memory Algorithm, RNN, Forecasting, Artificial Neural Network, Support Vector Machines, Predictive analytic, Deep learning, Generative Adversarial Network (GAN), Convolutional Neural Network (CNN), Random forest, prediction, time series analysis, Date Mining, Machine learning, Stock Prediction, Date Cleaning, Data Prediction, Data Normalization, Support

Vector Machine (SVM), Genetic Algorithms and so many. All paper shows good results and the proposed model demonstrate promising performance in the prediction of stock prices, showcasing its competitiveness when compared to the current state-of-the-art methodologies in this domain. [15].

Based on the results presented in the research paper, it can be concluded that the LSTM model showcased superior performance compared to other similar models, exhibiting minimal tolerance for errors. The paper proves the point by critically analyzing by different Machine Learning models (LSTM, ANN, and SVM) and proving the best in their fields. It shows an advancement in the correctness measures on bench marking with Moving Average. It is also concluded as the random forest algorithm demonstrated higher efficiency in comparison to logistic regression. Lastly, the papers have concluded as the Stock prediction using deep learning-based ML can increase the prediction accuracy [16].

3. Methodology

3.1 Dataset Description

The dataset contains information on the daily performance and it is taken from the Nifty50 index over a period of time. The dataset has 4 tuples and it includes the opening and closing prices, highest and lowest prices of each stock. In the provided dataset, the column having name "high" represents the maximum price by the stock during the trading day. On the other hand, the column having name "low" indicates the minimum price by the stock within the same trading day. The "close" column represents the final price at which the stock was traded by the end of the day, while the "open" column denotes the price at which the stock commenced trading at the beginning of the trading day. These columns provide crucial information regarding the price fluctuations and trading dynamics of the stock throughout the day. There are nearly 5800 rows in the dataset, which suggests that it contains multiple years of data from January 2000 to April 2023 of financial data. The dataset is branched into training and testing sets to measure the performance of the model. Overall, the Nifty50 dataset provides a valuable resource for analyzing and understanding the Indian stock market[17].

3.2 Proposed Methodology - Flow Chart

The flowchart describes the process of creating a stock market price prediction machine learning application using a dataset of Nifty50 from 2000 to 2023. The raw data is preprocessed and then used to train a stacked LSTM model with 3 layers containing 256, 128, and 64 nodes respectively. The model validation accuracy is checked and the model is fine-tuned until it satisfies the desired accuracy. If the accuracy is not satisfactory, the model is sent back to the validation accuracy section where the loop continues until it satisfies the desired accuracy. Finally, the model is deployed for use in predicting stock market prices.

194 Prospects of Science, Technology and Applications

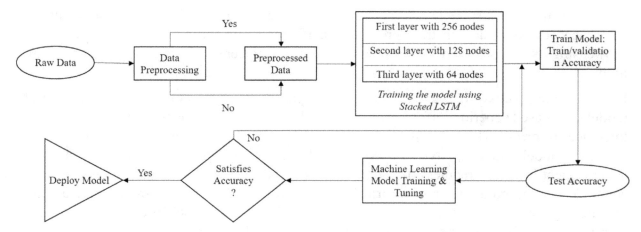

Fig. 20.1 Schematic layout/model diagram

3.3 Evaluation - Accuracy

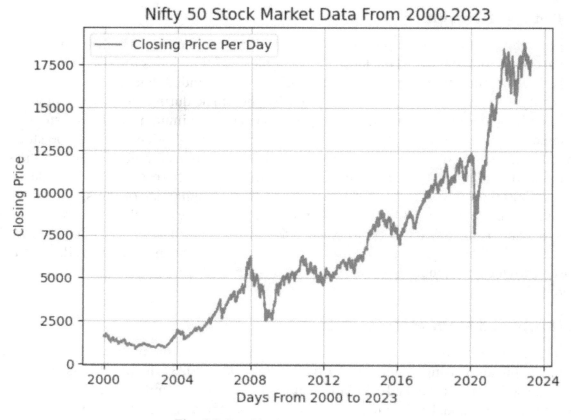

Fig. 20.2 Closing price year wise

The delicacy evaluation of the stock request price vatication operation using machine literacy was performed by assessing the performance of the model in the terms of RMSE and MSE. RMSE is a generally used metric that measures the average magnitude of vatication crimes. It calculates the square root of the normal of the squared differences between the prognosticated

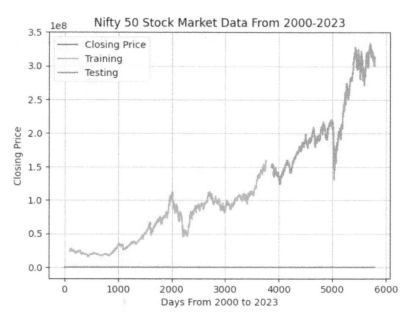

Fig. 20.3 Training and testing of the model using stacked LSTM model

values and the factual values. A less RMSE value indicates an advanced position of delicacy, as it signifies a lower average vaticination error.

Furthermore, MSE quantifies the average squared disparity between predicted values and actual values. It gauges the comprehensive prediction error, with a lower MSE signifying a higher precision in forecasting stock market prices.

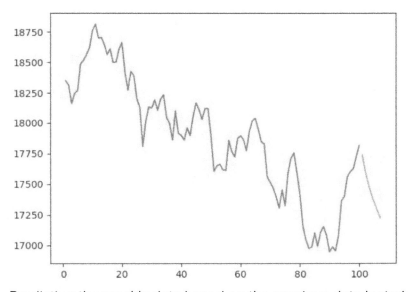

Fig. 20.4 Predicting the weekly data based on the previous data by trained model

During the evaluation process, the developed model achieved a promising performance with a low RMSE and MSE.

These results indicate that the model's prognostications were close to the factual stock prices, demonstrating its capability to capture the underpinning patterns and trends in the request data.

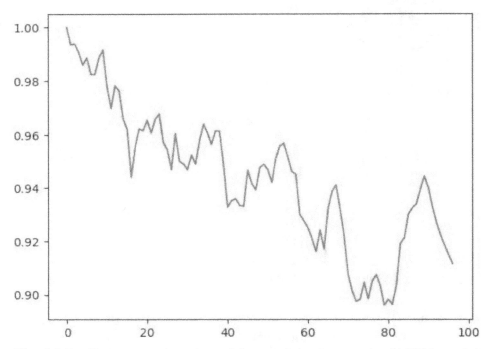

Fig. 20.5 Training and testing of the model using stacked LSTM model

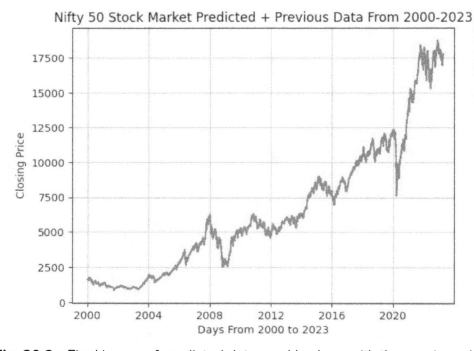

Fig. 20.6 Final image of predicted data weekly along with the previous data

4. Conclusion

The research aims to assess the effectiveness of incorporating the Stacked Long Short-Term Memory (LSTM) method in machine learning for the prediction of stock prices. The accuracy of the model is evaluated by comparing its predicted values against the actual values derived from the testing dataset. To further enhance the accuracy of LSTM models in stock price prediction, the study recommends integrating ensemble methods and feature engineering techniques. The LSTM model is trained to forecast stock prices across various time horizons, spanning from day 1 to day 30. Its ability to capture intricate trends and patterns within the NIFTY50 dataset establishes it as a reliable and valuable tool for stock price prediction. Performance evaluation incorporates metrics such as accuracy, mean absolute error (MAE), and root mean squared error (RMSE). The research findings conclude that LSTM surpasses other existing pre diction models, exhibiting superior accuracy and reliability. As a result, the LSTM model holds substantial potential as a valuable resource for investors and traders, enabling them to make well-informed decisions regarding their stock market investments.

REFERENCES

1. Huang, J. Y., Liu, J. H. (2020). Using social media mining technology to improve stock price forecast accuracy. Journal of Forecasting, 39(1), 104–116.
2. Roondiwala, M., Patel, H., Varma, S. (2017). Predicting stock prices using LSTM. International Journal of Science and Research (IJSR), 6(4), 1754–1756.
3. Roh, T. H. (2007). Forecasting the volatility of stock price index. Expert Systems with Applications, 33(4), 916–922.
4. Kim, M. J., Min, S. H., Han, I. (2006). An evolutionary approach to the combination of multiple classifiers to predict a stock price index. Expert Systems with Applications, 31(2), 241–247.
5. Kamalov, F., Smail, L., Gurrib, I. (2020, November). Stock price forecast with deep learning. In 2020 International Conference on Decision Aid Sciences and Application (DASA) (pp. 1098–1102). IEEE.
6. Zarandi, M. F., Rezaee, B., Turksen, I. B., Neshat, E. (2009). A type-2 fuzzy rule-based expert system model for stock price analysis. Expert systems with Applications, 36(1), 139–154.
7. Kara, Y., Boyacioglu, M. A., Baykan, O . K. (2011). Predicting direction of stock price index movement using artificial neural networks and support vector machines: The sample of the Istanbul Stock Exchange. Expert systems with Applications, 38(5), 5311–5319.
8. Wang, C., Chen, Y., Zhang, S., Zhang, Q. (2022). Stock market index prediction using deep Transformer model. Expert Systems with Applications, 208, 118–128.
9. Kumar, D., Sarangi, P. K., Verma, R. (2022). A systematic review of stock market prediction using machine learning and statistical techniques. Materials Today: Proceedings, 49, 3187–3191.
10. Adebiyi, A. A., Ayo, C. K., Adebiyi, M., Otokiti, S. O. (2012). Stock price prediction using neural network with hybridized market indicators. Journal of Emerging Trends in Computing and Information Sciences, 3(1).
11. Hassan, M. R., Nath, B. (2005, September). Stock market forecasting using hidden Markov model: a new approach. In 5th International Conference on Intelligent Systems Design and Applications (ISDA'05) (pp. 192–196). IEEE.

12. Bhandari, H. N., Rimal, B., Pokhrel, N. R., Rimal, R., Dahal, K. R., Khatri, R. K. (2022). Predicting stock market index using LSTM. Machine Learning with Applications, 9, 100320.
13. Dilhan, M. S., Wagarachchi, N. M. (2022, September). Stock Market Prediction using Artificial Intelligence. In 2022 International Research Conference on Smart Computing and Systems Engineering (SCSE) (Vol. 5, pp. 35–41). IEEE.
14. Baek, Y., Kim, H. Y. (2018). ModAugNet: A new forecasting framework for stock market index value with an overfitting prevention LSTM module and a prediction LSTM module. Expert Systems with Applications, 113, 457–480.
15. Shi, Y., Song, X., Song, G. (2021). Productivity prediction of a multilateral-well geothermal system based on a long short-term memory and multi-layer perceptron combinational neural network. Applied Energy, 282, 116046.
16. Rouf, N., Malik, M. B., Arif, T., Sharma, S., Singh, S., Aich, S., Kim, H. C. (2021). Stock market prediction using machine learning techniques: a decade survey on methodologies, recent developments, and future directions. Electronics, 10(21),
170 Sisodia, P. S., Gupta, A., Kumar, Y., Ameta, G. K. (2022, February). Stock market analysis and prediction for NIFTY50 using LSTM Deep Learning Approach. In 2022 2nd International Conference on Innovative Practices in Technology and Management (ICIPTM) (Vol. 2, pp. 156–161). IEEE.

Health is Wealth: Menu Driven Health Monitoring System for Improved Quality of Life

Shreeja Mahapatra[1], Ayushi Pradhan[2], Amlanjyoti Satapathy[3], Nimisha Ghosh[4]

Siksha O Anusandhan (Deemed to be) University, Bhubaneswar, India

Abstract Everyone's health is a top priority, and therapeutic decisions are often based more on doctors' expertise and intuition than on the knowledge concealed in the data. Thus, the main objective here is to build a dynamic website to help doctors in diagnosing health problems with preliminary symptoms based on test results to minimise errors and save time and money. We have focused on the prediction of mainly 4 diseases that are heart disease, diabetes, mental health issues and breast cancer. Comparative analysis of results for all the 4 health problems has been done using machine learning models like Support Vector Machine (SVM), Random Forest Classifier, AdaBoost Classifier, Gradient Boost Classifier, Logistic Regression and deep learning model using Keras python library. We have also implemented hyperparameter tunings such as grid search, random search and Bayesian optimisation in order to get more precise results. Additionally, Wilcoxon signed-rank test has been performed for the confirmation that the chosen model is best. The best results are as follows; heart disease: gradient boost classifier using random search (99.85%), diabetes: logistic regression using grid search (77.30%), breast cancer: SVC using grid search (92.96%), mental health issues: gradient boost classifier using grid search (81.82%).

Keywords Deep neural network, Health, Hyperparameter tuning, Machine learning, Web application

1. Introduction

It is undeniably true that one's health should come first in all spheres of life, and the widespread adoption of the computer-based technologies in the healthcare industry has led

---

[1]b.shreejamahapatra@gmail.com, [2]b.ayushipradhan@gmail.com, [3]b.amlanjyotisatapathy@gmail.com, [4]nimishaghosh@soa.ac.in

to the accumulation of electronic data. Due to the considerable amounts of data, doctors are facing challenges to analyse symptoms accurately and identify diseases at an early stage. In this regard, using analytical tools and data modelling can help in enhancing the clinical decisions. Machine learning is such an analytical tool that plays a significant role by carrying out data analysis and pattern discovery. It eventually provides decisions and predictions based on the same [1]. On the other hand, Deep Learning, a subset of machine learning, produces high-level abstraction in data by using several processing layers [2]. Deep learning methods are less susceptible to misclassification and noise than other methods. The architecture of neural networks incorporates the idea of deep learning. Deep Neural Network (DNN) is a version of deep learning which employs neural network with more layers and units in a layer. Advantage of DNN over neural networks is the rise in precision and generalisation after learning new examples because of deeper architecture [2]. Moreover, there are many works in the literature which have used the results obtained from the models to optimise them using hyperparameter tuning such as random search [3], grid search [4] and Bayesian optimization [5]. Hyperparameter tuning is an approach that impartially searches for different values for model hyperparameters and results in a model that is best by choosing a subset that achieves the best performance. A search space in Random Search is described as a bounded domain of hyperparameter values, and sample points are randomly selected for testing within that domain [3], while a search space in Grid Search is described as a grid of hyperparameter values, and all grid points are evaluated [4]. In order to undertake an effective and efficient search of a global optimisation issue, Bayesian Optimisation offers a systematic technique based on the Bayes theorem. It operates by creating an objective function's probabilistic surrogate function. Prior to selecting candidate samples for assessment on the actual objective function, the surrogate function is effectively searched utilising an acquisition function [5]. Additionally, we have performed a non-parametric statistical hypothesis test that is the Wilcoxon signed-rank test. It is used to confirm that the chosen sample (in our case model) is the best.

Furthermore, in today's busy world where people prefer instant results rather than standing in long queues and waiting for their turn to get diagnosed in hospitals, we have thought of helping people by creating a time-saving dynamic website which is a practical implementation of our desired best models. In this regard, Flask provides tools, libraries and technologies that allow us to build a web application which is user-friendly and provides wide functionalities [6]. Moreover, Flask is majorly preferred to deploy machine learning models because it gives freedom to implement one's own rules and does not require much setup. It also contains various extensions which result in a dynamic look [6]. When a user (doctor) is accessing this API, he/she has to provide as input the basic symptoms of the disease based on the preliminary tests done. Flask API will invoke the corresponding model and return the status of the patient's probability of suffering from the disease. Taking this into consideration, for 4 very common diseases viz. heart disease, diabetes, mental health issues and breast cancer (particularly in women), we have implemented various machine learning and DNN models to get the most optimised result. These aforementioned diseases are taken into account because they have become very common among middle-aged and even young people and if not diagnosed at an early stage may become incurable for life. Also, misdiagnosis can result in other serious issues as well. To achieve the desired result, 4 different datasets have been used for the 4 diseases and a comparative analysis has been performed between various models for prediction of each disease. Motivated by the literature, among the multiple machine learning models that we have applied, Gradient Boost classifier using random search has given the best performance for heart disease: accuracy: 99.85%, precision: 99.70%, recall: 100%, f1-score: 99.85%, AUC score: 0.99. Logistic regression using grid search has given the best result for diabetes (accuracy: 77.30%, precision: 56.20%, recall: 72.40%, f1-score: 62.79%, AUC score: 0.75), breast cancer is predicted using SVC with grid search(accuracy: 92.96%, precision: 96.38%, recall: 92.75%, f1-score: 94.44%, AUC score: 0.93) and mental health issues is giving the best result in gradient boost using grid search (accuracy: 81.82%, precision: 95.66%, recall: 81.50%, f1-score: 87.97%, AUC score: 0.82).

2. Literature Review

Doctors rely on manual methods to predict diseases. These methods are based on their training, experience, and observations of their patients. One of the most common manual methods used by doctors to predict diseases is physical examination of the patient. For example, they look for abnormalities in the patient's skin, eyes, mouth, and other organs, as well as examine the patient's vital signs such as blood pressure, heart rate, and respiratory rate. Now as modern technology has come into effect, errors in clinical decisions have reduced exponentially. Many researchers have developed various machine learning and deep learning models for the prediction of various diseases. These models have not only minimised clinical errors but have also helped in saving time and money.

In [9], heart disease prediction is done using three machine learning algorithms that are Decision Tree (DT), Random Forest (RF) classifier and Logistic Regression (LR). It is found that the DT provides the best result of 94.7% accuracy. Also during our study we found that the Cleveland heart disease dataset from the UCI machine learning repository was used in this paper and it had only 303 patient's data. Author et. al. [2], heart disease was predicted using DNN and hyper parameter tuning was implemented such as grid search, random search and Bayesian optimisation on the above mentioned dataset. In this study it was found that Bayesian optimization gave the best accuracy that is 91.67%.

In [7], diabetes prediction has been done using six machine learning models that are Naïve Bayes (NB), SVM, DT, K-nearest neighbour (KNN), LR and RF. LR gave 74% accuracy, SVM gave 77% accuracy, 74% accuracy was achieved by using Naïve Bayes, DT and RF achieved 71% accuracy and KNN achieved 77%. As a result, SVM and KNN had the highest accuracy (77%). Therefore, SVM and KNN methods are suitable for forecasting a patient's diabetes state, according to the experimental results. Author et. al. [8], diabetes has been predicted using one dimensional convolutional neural network (1DCNN). An accuracy of 97.02% is obtained using a dataset from the Sylhet Diabetes Hospital in Sylhet, Bangladesh containing data of 520 cases.

Researchers in [10] have provided a study on mental health at work in which they have proved that neural network techniques are best for predicting mental health. They have applied Artificial Neural Network (ANN), Convolution Neural Network (CNN) and Recurrent Neural Network (RNN) on the dataset. The accuracies obtained are 92.2%, 89.1% and 86.9% respectively. Author et. al. [11], the proposed model has used eight mainstream machine learning calculation methods, namely DT, RF, SVM, Naïve Bayes, LR, XGBoost (XGB), Gradient Boosting Classifier (GBC) and ANN to build up the expectation models utilising a huge dataset (1429 individual's survey). The final outcome received was 87.38 percent, which was using SVM.

In study of [12], a comparative analysis of machine learning algorithms is performed. The accuracies obtained are KNN: 95.61%, SVM: 96.49%, Kernel SVM: 97.36%, NB: 94.73%, DT: 94.74%, RF: 94.73%, LR: 97.36%, Linear Regression: 96.49%. The study in [13] has suggested a deep neural network with feature selection techniques to predict breast cancer. Accuracy of the proposed method is found to be 99.42%.

During the analysis of existing systems in health care, we found that most of them have considered only one disease at a time. We have analysed the above drawback and developed a comprehensive project that would predict any of the aforementioned diseases in a single platform.

3. Materials and Methods

3.1 Data Preparation

Initially, for our work we have collected 4 different databases for the 4 diseases viz. Heart disease, diabetes, breast cancer (particularly in women) and mental health issues. For heart disease we have taken reference from [15]. Although there are 76 attributes in this database, all published experiments only mention using a portion of 14. It has data of 1025 patients. The dataset for prediction of diabetes is originally from the National Institute of Diabetes and Digestive and Kidney Diseases. It holds 8 attributes and data of 768 cases. The OSMI (Open Source Mental Illness) community runs large scale surveys in the Tech industry. We have used one of its dataset for prediction of mental health based on what provisions and resources are made available by the employer and how effective the entire system is. It contains 6 features and data of 1386 employees. For breast cancer, the data is referred from [16]. It holds data of 569 patients and their details over 5 attributes. Fig. 21.1 shows the pair-plot of 3 features (mean perimeter, mean area, mean smoothness) of breast cancer with the outcome (result). Fig. 21.2 shows the pair-plot of 3 features (age, diabetes function, insulin) of diabetes with the outcome (result).

Health is Wealth: Menu Driven Health Monitoring System for Improved Quality of Life **203**

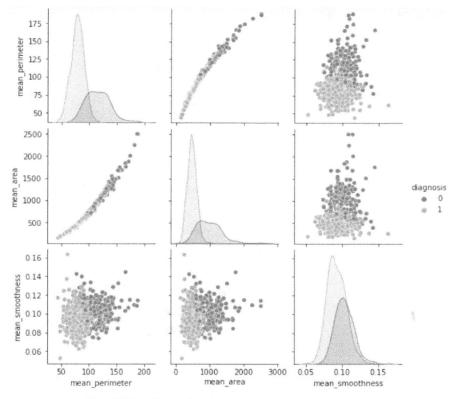

Fig. 21.1 Pair-plot for breast cancer dataset

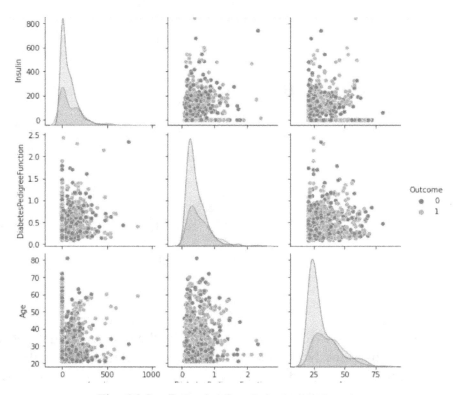

Fig. 21.2 Pair-plot for diabetes dataset

3.2 Methodology: Pipeline of the Work

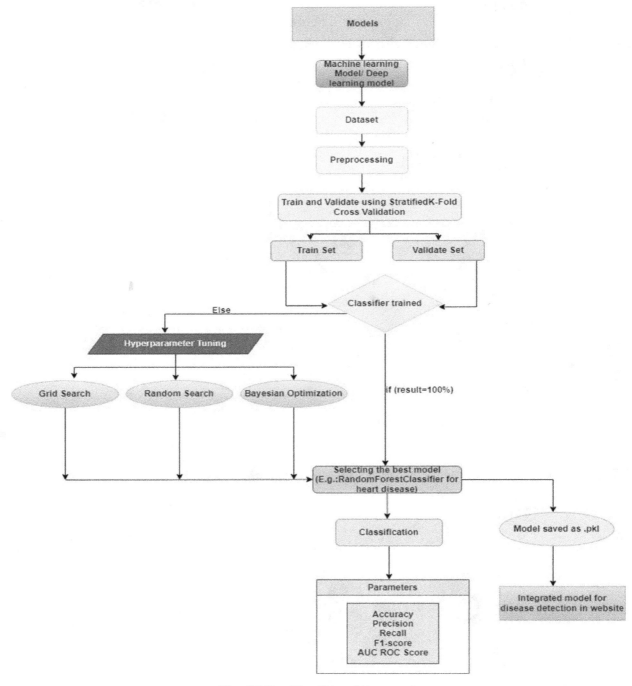

Fig. 21.3 Pipeline of the work

The pipeline of the work is shown in Fig. 21.3. Initially, the considered datasets undergo preprocessing (data normalisation). Normalisation is a method for dissecting tables to remove data redundancy (repetition) and standardise the information for better data workflow by reducing the range of all the attribute values between 0.001 to 1. After normalisation, repeated k-fold cross validation (k=3, n-repeats=10) is applied on the datasets for training and validation using an appropriate value of k such that the accuracy is maximum. Thereafter, getting the best sets for training and validation we apply various machine learning models to the dataset. We have implemented five machine learning models namely, SVM, RF, AdaBoost Classifier, Gradient Boost Classifier (GBC) and LR. We have implemented each of the above-mentioned models for each disease. If 100% accuracy is gained from any model we have gone for the classification. Otherwise we have implemented hyper-parameter tuning using grid search, random search and Bayesian optimisation for all the previously executed models following which we have finalised the results. Not only this, but we have also implemented neural network using Keras classifier for the aforementioned diseases and have also applied hyper-parameter tuning to it. Additionally, we have performed Wilcoxon signed-rank test to ensure that the null hypothesis is rejected. After getting all the required results we have selected the best model for each disease based upon parameters like accuracy, precision, recall, f1-score and AUC-ROC curve. The selected model for heart disease is GBC using random search while for diabetes it is LR using grid search, mental health issues are predicted from GBC using grid search and for breast cancer SVC using grid search is selected as the best model.

These selected models are then saved to the disk as pickle files and loaded using python unpickling so that every time a user runs the web application the models do not retrain themselves from scratch. Thereafter the saved models are integrated with the web application for prediction of each disease. Model integration with the website has been achieved by using Flask. It is a micro web framework written in python which makes complex web designing simpler. The web application is dynamic in nature and has been designed using JSX and ReactJS.

4. Results

The results are obtained on various parameters such as accuracy, precision, recall, f1-score, AUC and time (for hyper-parameter tuning). These values define the performance of the models. The percentage of correctly classified data instances over all data instances is known as Accuracy [2].

$$\text{Accuracy} = (\text{True Positive} + \text{True Negative}) / (\text{Total Sample Size}) \tag{1}$$

The exactness with which a model can classify is referred to as Precision [2]. False positives will be lower the greater the Precision, and higher the lower the Precision.

$$\text{Precision} = \text{True Positive} / (\text{True Positive} + \text{False Positive}) \tag{2}$$

The sensitivity and comprehensiveness of the model are described by Recall [2]. A higher recall results in fewer false negatives, whereas a lesser recall results in more false negatives.

$$\text{Recall} = \text{True Positive} / (\text{True Positive} + \text{False Negative}) \tag{3}$$

The F1 score [2] is the weighted harmonic mean of precision and recall, which combines the two, making it just as significant as accuracy.

$$\text{F1 Score} = 2 * (\text{Precision} * \text{Recall}) / (\text{Precision} + \text{Recall}) \tag{4}$$

Table 21.1 Results of applied models along with hyper-parameter tuning for heart disease

Heart Disease			
Model	Accuracy	Best Hyperparameter	Method
Random Forest Classifier	99.55+-0.007	Criterion: entropy; max depth: 10; max features: 10; n-estimators: 150	Grid search
Gradient Boost Classifier	**99.85+-0.006**	**n_estimators: 50, max features: 1, max depth: 10, learning rate: 1, criterion: friedman mse**	**Random Search**
AdaBoost Classifier	88.82+-0.018	Algorithm: SAMME.R; n-estimators: 10	Grid search
Logistic Regression	84.55+-0.025	C:10; max iter: 500; intercept scaling: 1.0	Grid Search
Support Vector Classifier	95.30+-0.014	C: 10; degree: 5	Grid search
Neural Network using Keras library	79.64+-0.033	Activation: softsign,batch_size: 961, dropout: 1, dropout_rate: 1, epochs: 32, layers1: 1, layers2: 1, learning_rate: 0.86, neurons: 8, normalisation: 1, optimizer: Adam	Bayesian Optimisation

Each model (with hyperparameter tuning) has been executed 100 times and the mean of the obtained results have been considered. The results are represented in a tabular format.

Tables 21.1, 21.2, 21.3, and 21.4 report results of applied models (RF classifier, SVC, AdaBoost classifier, GBC, LR and neural network using Keras) with hyper-parameter tuning for heart disease, diabetes, mental health issues and breast cancer respectively. Table 21.5 contains the result of best models (integrated in website) after hyper-parameter tuning for each dataset.

Table 21.2 Results of applied models along with hyper-parameter tuning for diabetes

Diabetes			
Model	Accuracy	Best Hyperparameter	Method
Random Forest Classifier	76.27+-0.032	max depth: 5; max features: 8; n-estimators: 70	Bayesian Optimisation
Gradient Boost Classifier	75.63+-0.019	n_estimators: 50, max_features: 1, max_depth: 1, learning_rate: 1, criterion: friedman_mse	Random Search
AdaBoost Classifier	74.69+-0.034	Algorithm: SAMME.R; n-estimators: 10	Grid search
Logistic Regression	77.30+-0.044	C:10; max iteration: 500; intercept scaling: 1.0	Grid Search
Support Vector Classifier	76.55+-0.032	C: 3; degree: 5	Random search
Neural Network using Keras library	66.97+-0.033	Activation: softsign, batch size: 961, dropout: 1, dropout rate: 1, epochs: 32, layers1: 1, layers 2: 1, learning rate: 0.86, neurons: 61, normalisation: 1, optimizer: Adam	Bayesian Optimisation

Table 21.3 Results of applied models along with hyper-parameter tuning for mental health issues

Mental Health Issues			
Model	Accuracy	Best Hyperparameter	Method
Random Forest Classifier	81.40+-0.008	max depth: 4; max features: 10; n-estimators: 123	Bayesian Optimisation
Gradient Boost Classifier	**81.82+-0.028**	**n-estimators: 150, max features: 1, max depth: 10, learning rate: 0.01, criterion: friedman mse**	**Grid Search**
AdaBoost Classifier	79.45+-0.025	Algorithm: SAMME.R; n-estimators: 3	Grid search
Logistic Regression	79.33+-0.021	C:10; max iteration: 500; intercept scaling: 1.0	Grid Search
Support Vector Classifier	81.41+-0.008	C: 6; degree: 19	Bayesian
Neural Network using Keras library	79.80+-0.032	Batch size: 200, epochs: 100, optimizer: 0, optimizer activation': 0, optimizer dropout rate: 0, optimizer learning rate: 0.001, optimizer neurons: 6, optimizer weight constraint: 1.0	Grid Search

Table 21.4 Results of applied models along with hyper-parameter tuning for breast cancer

Breast Cancer			
Model	Accuracy	Best Hyperparameter	Method
Random Forest Classifier	92.27+-0.026	Criterion: entropy; max depth: 10; max features: 1; n-estimators: 50	Random Search
Gradient Boost Classifier	92.91+-0.023	n_estimators: 150, max_features: 10, max_depth: 1, learning_rate: 1, criterion: friedman_mse	Grid Search
AdaBoost Classifier	92.38+-0.026	Algorithm: SAMME; n-estimators: 10	Random Search
Logistic Regression	92.28+-0.016	C:4; max iteration: 866; intercept scaling: 3	Bayesian
Support Vector Classifier	**92.96+-0.027**	**C: 3; degree: 5**	**Grid search**
Neural Network using Keras library	68.24+-0.093	Batch size: 200, epochs: 100, optimizer: 7, optimizer activation: 0, optimizer dropout rate: 0.3, optimizer learning rate: 0.001, optimizer neurons: 1, optimizer weight constraint: 2.0	Grid Search

Table 21.5 Results of best models after hyperparameter tuning for each dataset

	Best Model	Accuracy	Precision	Recall	F1-score	AUC
Heart Disease	GBC using random search	99.85+-0.0	99.70+-0.012	100	99.85+-0.006	0.99
Diabetes	LR using grid search	77.30+-0.04	56.20+-0.097	72.40+-0.096	62.79+-0.081	0.75
Mental Health	GBC using grid search	81.82+-0.02	95.66+-0.019	81.50+-0.031	87.97+-0.020	0.82
Breast Cancer	SVC using grid search	92.96+-0.02	96.38+-0.031	92.75+-0.039	94.44+-0.022	0.93

When we need to assess or visualise the performance of the multi-class classification problem, we use the ROC (Receiver Operating Characteristics) curve and its area under the curve (AUC) parameter. These curves demonstrate the degree to which a model can differentiate between two or more distinct classes [2]. The model is more accurate at classifying 0 classes as 0, and classifying 1 classes as 1, the higher the AUC. The true positive rate is displayed against the false positive rate on the ROC curve [2]. For better visualisations, we have plotted and compared the ROC curve for all models of all the 4 datasets (heart disease, diabetes, breast cancer and mental health issues).

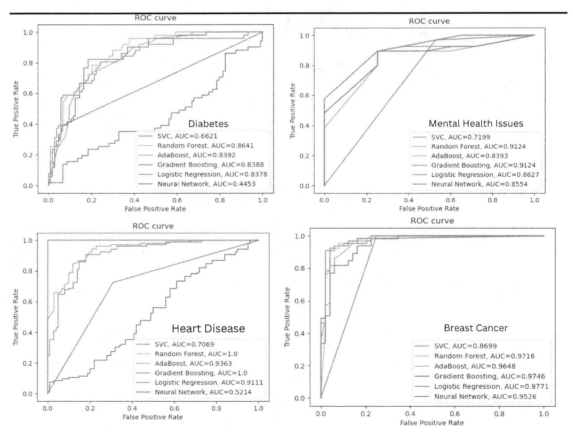

Fig. 21.4 Comparative analysis of ROC curves for the 4 diseases

Additionally, Wilcoxon signed-rank test is performed to confirm that the selected models for each dataset is the best. It is a non-parametric statistical test that is used when the sample size is small and the data does not have a normal distribution. We have executed the test for 5000 random samples. It is observed that the null hypothesis is rejected for all compared models. Hence, we conclude that the chosen models are the best.

5. Discussion

After the completion of the models and successfully getting all the results, we have designed a menu driven web application named 'Health is Wealth'. This application provides real time predictions to the users. The application is primarily designed using JSX and ReactJS and the selected models for each disease run in the backend with the help of Flask. Following figures are a glance at how the web application runs.

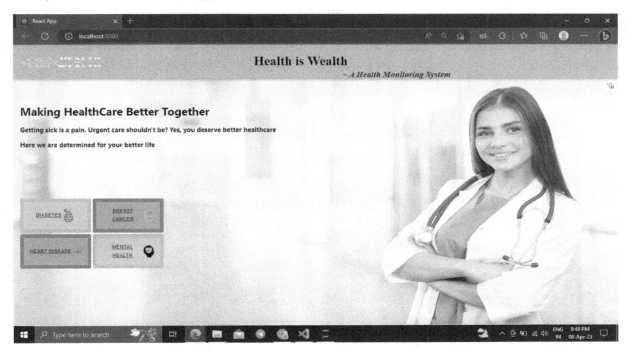

Fig. 21.5 Homepage of the website, where we can see 4 buttons representing 4 diseases viz. heart disease, diabetes, breast cancer and mental health issues

Similarly, a user can predict any of the 4 diseases by clicking on the specified button and later entering the details of the preliminary tests in the form that will open. The prediction will be displayed once the user clicks on the 'Result' button in the form. Moreover, the entire code of the project with datasets can be found on: Health is Wealth: Menu driven health monitoring system for improved quality of life

6. Conclusion

Lack of proper care for health has very severe consequences on one's physical as well as mental wellness. With increasing population and hustle in daily life people tend to neglect

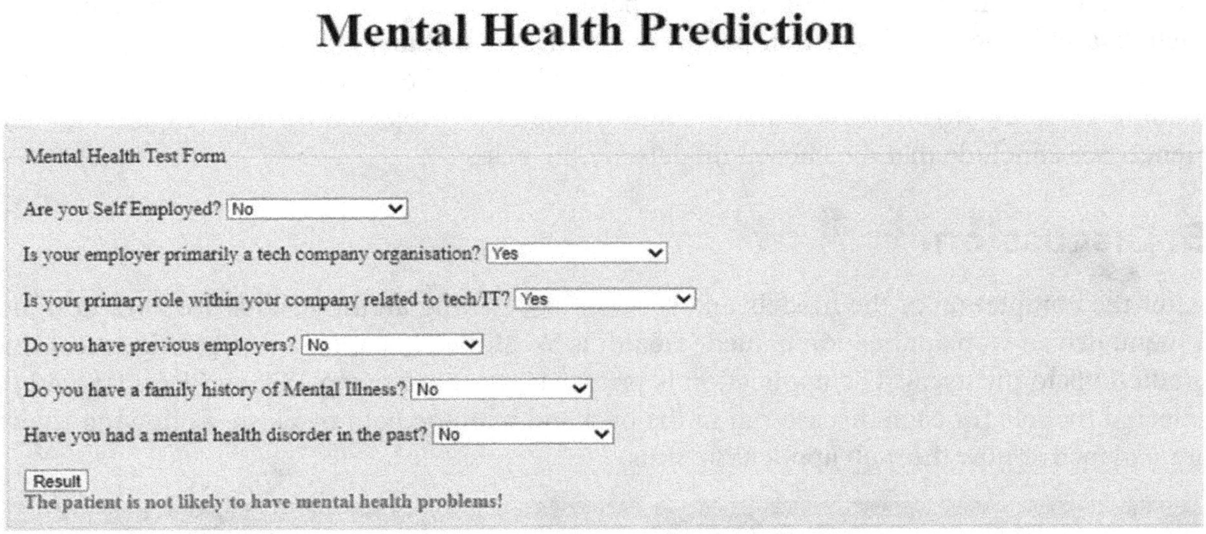

Mental Health Prediction

```
Mental Health Test Form
Are you Self Employed? [No ▾]
Is your employer primarily a tech company organisation? [Yes ▾]
Is your primary role within your company related to tech/IT? [Yes ▾]
Do you have previous employers? [No ▾]
Do you have a family history of Mental Illness? [No ▾]
Have you had a mental health disorder in the past? [No ▾]
[Result]
The patient is not likely to have mental health problems!
```

Fig. 21.6 Shows the tab that opens when a user clicks on the mental health button. After entering the details in the form, if the result is negative it is displayed in red colour at the bottom of the screen as shown

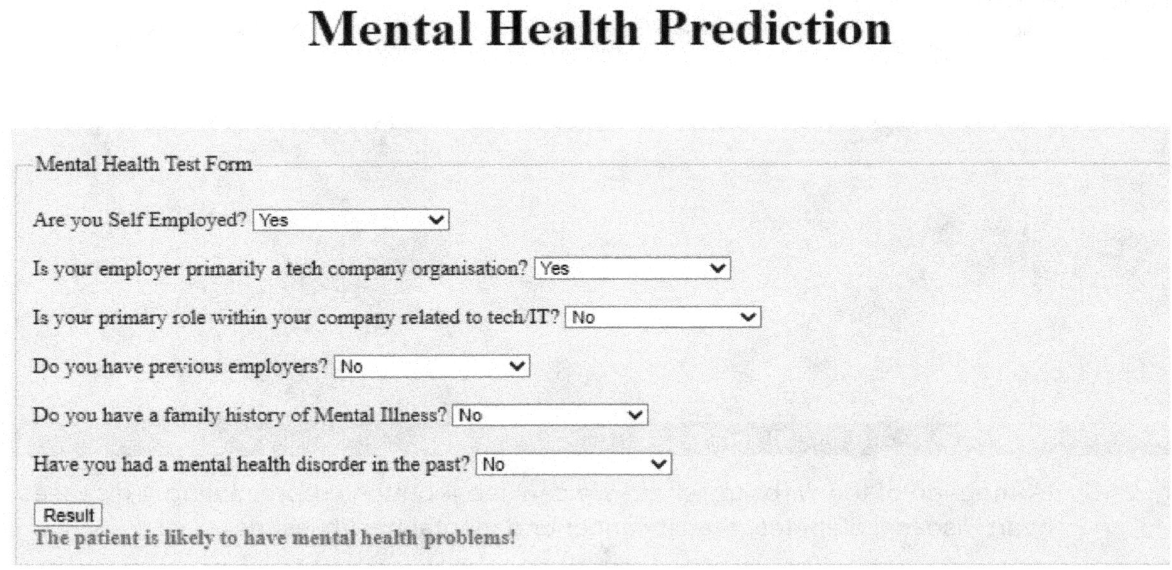

Mental Health Prediction

```
Mental Health Test Form
Are you Self Employed? [Yes ▾]
Is your employer primarily a tech company organisation? [Yes ▾]
Is your primary role within your company related to tech/IT? [No ▾]
Do you have previous employers? [No ▾]
Do you have a family history of Mental Illness? [No ▾]
Have you had a mental health disorder in the past? [No ▾]
[Result]
The patient is likely to have mental health problems!
```

Fig. 21.7 Shows the positive result after entering the details

their health. This project provides a time saving as well as free platform for the prediction of 4 very common diseases that are heart disease, diabetes, breast cancer (in women) and mental health issues. With the help of various machine learning models we have successfully found out the best models for the prediction of the above-mentioned diseases. The best results are obtained for heart disease by GBC using random search (99.85%), LR using grid search has provided best performance for diabetes (77.30%), SVC using grid search for breast cancer (92.96%) and GBC for grid search for mental health issues (81.82%). These models are integrated with the web application for real-time use. The web application is designed in such a way that both doctors and patients can use it without facing any difficulty. To summarise, we have made a comprehensive platform for the prediction of the aforementioned diseases using machine learning models and deep learning models with hyper-parameter tuning. In the future, we intend to work on more health related curated datasets to provide the user with a comprehensive website.

REFERENCES

1. N. Kosarkar, P. Basuri, P. Karamore, P. Gawali, P. Badole and P. Jumle, "Disease Prediction using Machine Learning," 2022 10th International Conference on Emerging Trends in Engineering and Technology - Signal and Information Processing (ICETET-SIP-22), Nagpur, India, 2022, pp. 1–4, doi: 10.1109/ICETET-SIP-2254415.2022.9791739.
2. F. F. Firdaus, H. A. Nugroho and I. Soesanti, "Deep Neural Network with Hyperparameter Tuning for Detection of Heart Disease," 2021 IEEE Asia Pacific Conference on Wireless and Mobile (APWiMob), Bandung, Indonesia, 2021, pp. 59–65, doi: 10.1109/APWiMob51111.2021.9435250.
3. R. G. Mantovani, A. L. D. Rossi, J. Vanschoren, B. Bischl and A. C. P. L. F. de Carvalho, "Effectiveness of Random Search in SVM hyper-parameter tuning," 2015 International Joint Conference on Neural Networks (IJCNN), Killarney, Ireland, 2015, pp. 1–8, doi: 10.1109/IJCNN.2015.7280664.
4. Li Yang, Abdallah Shami, "On hyperparameter optimization of machine learning algorithms: Theory and practice",Neurocomputing, Volume 415, 2020, Pages 295–316, ISSN 0925-2312.
5. V. Nguyen, "Bayesian Optimization for Accelerating Hyper-Parameter Tuning," 2019 IEEE Second International Conference on Artificial Intelligence and Knowledge Engineering (AIKE), Sardinia, Italy, 2019, pp. 302–305, doi: 10.1109/AIKE.2019.00060.
6. A. Yaganteeswarudu, "Multi Disease Prediction Model by using Machine Learning and Flask API," 2020 5th International Conference on Communication and Electronics Systems (ICCES), Coimbatore, India, 2020, pp. 1242–1246, doi: 10.1109/ICCES48766.2020.9137896.
7. M. A. Sarwar, N. Kamal, W. Hamid and M. A. Shah, "Prediction of Diabetes Using Machine Learning Algorithms in Healthcare," 2018 24th International Conference on Automation and Computing (ICAC), Newcastle Upon Tyne, UK, 2018, pp. 1–6, doi: 10.23919/IConAC.2018.8748992.
8. L. Xu, J. He and Y. Hu, "Early Diabetes Risk Prediction Based on Deep Learning Methods," 2021 4th International Conference on Pattern Recognition and Artificial Intelligence (PRAI), Yibin, China, 2021, pp. 282–286, doi: 10.1109/PRAI53619.2021.9551074.
9. A. Bhowmick, K. D. Mahato, C. Azad and U. Kumar, "Heart Disease Prediction Using Different Machine Learning Algorithms," 2022 IEEE World Conference on Applied Intelligence and Computing (AIC), Sonbhadra, India, 2022, pp. 60–65, doi: 10.1109/AIC55036.2022.9848885.
10. A. Chahar, V. M. Sanjay, A. Basrur, S. A. Kyalkond, A. Ajit and N. Patil, "Mental Health At Work Prediction Using Neural Networks," 2022 8th International Conference on Advanced Computing and Communication Systems (ICACCS), Coimbatore, India, 2022, pp. 458–461, doi: 10.1109/ICACCS54159.2022.9785283.
11. T. Jain, A. Jain, P. S. Hada, H. Kumar, V. K. Verma and A. Patni, "Machine Learning Techniques for Prediction of Mental Health," 2021 Third International Conference on Inventive Research in Computing Applications (ICIRCA), Coimbatore, India, 2021, pp. 1606–1613, doi: 10.1109/ICIRCA51532.2021.9545061.

12. K. Srivastava, P. Garg, V. Sharma and N. Gupta, "Comparative Analysis of Various Machine Learning Techniques to Predict Breast Cancer," 2022 3rd International Conference on Issues and Challenges in Intelligent Computing Techniques (ICICT), Ghaziabad, India, 2022, pp. 1–6, doi: 10.1109/ICICT55121.2022.10064517.
13. M. O. F. Goni, F. M. S. Hasnain, M. A. I. Siddique, O. Jyoti and M. H. Rahaman, "Breast Cancer Detection using Deep Neural Network," 2020 23rd International Conference on Computer and Information Technology (ICCIT), DHAKA, Bangladesh, 2020, pp. 1–5, doi: 10.1109/ICCIT51783.2020.9392705.
14. M. A. Rahman, S. M. Shoaib, M. A. Amin, R. N. Toma, M. A. Moni and M. A. Awal, "A Bayesian Optimization Framework for the Prediction of Diabetes Mellitus," 2019 5th International Conference on Advances in Electrical Engineering (ICAEE), Dhaka, Bangladesh, 2019, pp. 357–362, doi: 10.1109/ICAEE48663.2019.8975480.
15. A. Janosi, W. Steinbrunn, R. Detrano, et. al., International application of a new probability algorithm for the diagnosis of coronary artery disease. (1989), American Journal of Cardiology, 64, 304–310.
16. W.N. Street, W.H. Wolberg and O.L. Mangasarian, Nuclear feature extraction for breast tumour diagnosis. IS&T/SPIE 1993 International Symposium on Electronic Imaging: Science and Technology, volume 190.

Fine-Tuning and Comparing XLSR Models on Odia Speech

Pragyan Prusty, Ajit K Nayak*

Dept. of CSIT, ITER, SOA, BBSR

Abstract Automatic Speech Recognition or ASR is where a machine listens to speech input, understands it and outputs what it recognizes. There are many such technologies that rely on speech such as smart home devices, voice assistants and translation services that answer your queries and carry out your requests, making your life comfortable. However, speech technology is only available for a few languages spoken around the globe. In order to be accessible by each and every one, it must be available in every language such as Odia. Odia is one of the official languages of India and is designated as the sixth classical language of India. It has a rich and long literary history. Facebook AI has released XLSR [1] (cross-lingual speech representations), a multi–lingual pre-trained model for a variety of speech tasks. In this research we focus on fine-tuning XLSR-53 and its descendant XLS-R [2] on Odia dataset and then comparing the two. This direction will ignite further research to make ASR more accessible across the world in every language.

Keywords Speech-recognition, Odia, ASR, XLSR, wav2vec, wav2vec2.0

1. Introduction

ASR started with 'Audrey', the Automatic Digit Recognizer that was the first speech recognition system developed by Bell Labs in 1952. It could recognize the digits from 0 to 9. In 1962, IBM introduced 'Shoebox' that could understand 16 English words. 1970s and 80s saw the era of Hidden Markov Model that used probability functions to guess the right word. In late 1980s, the addition of neural networks developed ASR research. Currently the speech recognition systems with best accuracy are using Deep Neural Networks on top of HMM. [3] Wav2Vec2 [4] is a pretrained model for Automatic Speech Recognition (ASR) and was released in 2020.

*Corresponding author: ajitnayak@soa.ac.in

DOI: 10.1201/9781003489443-22

After its excellent performance, in the same year Facebook AI released XLSR (short for cross-lingual speech representations), a multilingual version of Wav2Vec2 pretrained on 53 languages. In 2021, XLS-R was released that was pretrained on 128 languages. When these models are adapted to a particular task or dataset it is called as fine-tuning the model. It means weights of a pre trained network are used as the beginning values for training a new network as shown in Fig. 22.1. We will be fine-tuning both the models on the dataset of Odia.

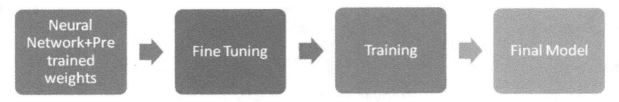

Fig. 22.1 Flow diagram of fine tuning a model

2. Approach

Automatic Speech Recognition models interpret speech to text for which we require a feature extractor and a tokenizer. According to the model's input format the speech signal is processed like a feature vector by the feature extractor and the model's output format is converted to text by the tokenizer. Hence XLSR models are accompanied by a feature extractor called Wav2Vec2FeatureExtractor and a tokenizer called Wav2Vec2CTCTokenizer.

The signal is mapped to a series of context representations by the pre-trained XLSR. This series of context representations is then mapped to its relevant transcription for speech recognition. For this a linear layer is appended on top of the transformer block as shown in Fig. 22.2 that categorizes each context representation to a token class. The number of tokens in the vocabulary is correlated to the output size of this layer. So at first, we defined a vocabulary based on the transcriptions and loaded it as a json file into an instance of the Wav2Vec2CTCTokenizer class.

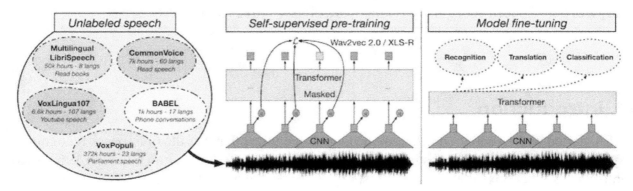

Fig. 22.2 The XLSR approach

Using Connectionist Temporal Classification (CTC) [5] XLS-R is fine-tuned. To train neural networks for sequence-to-sequence problems like ASR, this algorithm is used. For a given input sequence, this algorithm gives us an output distribution over all possible transcriptions. Then we define the feature extraction pipeline. For more accessibility, we wrap the feature extractor and tokenizer into a single Wav2Vec2Processor class.

In order to fine-tune, two strategies are used that are combined in practice. First is to freeze the initial layers because in the early stages the network is learning fundamental features. The initial component of XLS-R consists of a stack of CNN (Convolutional Neural Network) layers that are used to extract acoustic and meaningful features from the raw speech signal. This part has been trained already during pre-training and need not be fine-tuned anymore. Hence we freeze those layers. The second strategy is to decrease the learning rate in our current training. We use a value up to 10 times smaller than usual for the learning rate during fine-tuning. As a result, the model will then try to adapt itself to the new dataset in small steps. This way we avoid losing the features it has already learned.

3. Experimental Setup

3.1 Datasets

For our experiment we have considered the Common Voice dataset [6]. It is a multilingual corpus of read speech of various languages. We have used the Odia dataset containing eight recorded hours and one validated hour. It has been recorded in 34 voices. XLS-R was pretrained on audio data with a sampling rate of 16 kHz. The sampling rate is 48 kHz for Common Voice. Hence we down sample the fine-tuning data to 16 kHz.

3.2 Deep Learning Frameworks

Trained on 53 languages of Multilingual LibriSpeech, Common Voice and BABEL, XLSR-53 consists of 56,000 hours of speech data and comprises about 300M parameters. XLS-R is trained on more than 436,000 hours of speech recordings. It is almost 10 times more hours of speech than XLSR-53. It has covered up to 128 different languages, approximately twice more languages than its predecessor. It comes in sizes scaling from 300 million to 2 billion parameters. In this research we fine-tuned the pre-trained checkpoint Wav2Vec2-XLS-R-300M [7].

3.3 Parameters Table

Table 22.1 Parameters table for XLSR-53 and XLS-R-300M

Models	Batch Size	Gradient Accumulation Steps	Epochs	Learning Rate
XLSR-53	16	2	150	3e-4
XLS-R-300M	16	2	150	3e-4

The number of samples processed before the model is updated is the batch size. We have taken the batch size as 16. The number of complete passes through the training dataset is the number of epochs. We have taken it 150. Learning rate refers to the rate at which an algorithm converges to a solution and it is taken 3e-4. The Gradient Accumulation steps is taken 2.

4. Result and Discussion

After the fine-tuning and training of the models we got the training loss, validation loss and WER (Word Error Rate) as follows:

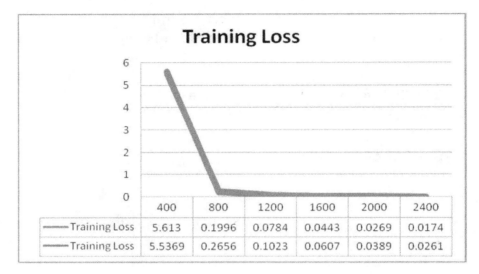

Fig. 22.3 Training Loss of XLSR-53 and XLSR-300-M

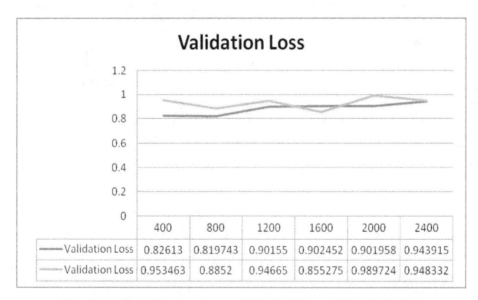

Fig. 22.4 Validation loss of XLSR-53 and XLS-R-300M

5. Conclusion

The Training Loss has come same for both the models as seen in Fig. 22.3. From Fig. 22.5 the WER of XLSR-53 is comparatively less than the WER of XLS-R-300M Similarly from Fig. 22.4 the Validation Loss of XLSR-53 is comparatively less than the WER of XLS-R-300M. The reason may be due to the many languages the XLS-R model is pre-trained

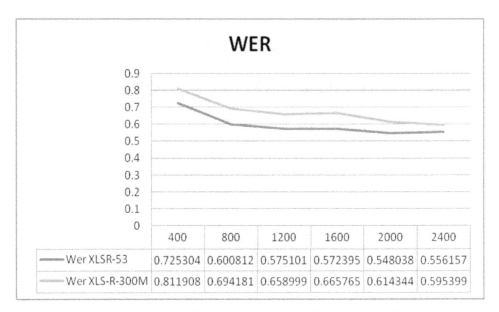

Fig. 22.5 WER of XLSR-53 and XLS-R-300M

on. XLSR-53 is pre-trained on 53 languages whereas XLS-R-300M is pre-trained on 128 languages. More the languages any model is pre-trained on, fewer is the model capacity available to learn representations for each language.

From Fig. 22.4 we see that the validation loss is nearly remaining constant which likely means that the model is overfitting to the data. A smaller model or an increased dataset or more likely both can be used to get better results.

Humans acquire speech skills just by listening to others around them. This suggests that there is a finer way to develop speech recognition techniques and models, that will not require huge amounts of labeled data. This will further help in providing speech technology to each and every one in their native language.

REFERENCES

1. Conneau, A., Baevski, A., Collobert, R., Mohamed, A. and Auli, M: Unsupervised cross-lingual representation learning for speech recognition, *arXiv preprint arXiv:2006.13979*(2020).
2. Babu, A., Wang, C., Tjandra, A., Lakhotia, K., Xu, Q., Goyal, N., Singh, K., von Platen, P., Saraf, Y., Pino, J. and Baevski, A.: XLS-R: Self-supervised cross-lingual speech representation learning at scale. *arXiv preprint arXiv:2111.09296*(2021).
3. https://blog.deepgram.com/the-history-of-automatic-speech-recognition/
4. https://ai.facebook.com/blog/wav2vec-20-learning-the-structure-of-speech-from-raw-audio/
5. https://distill.pub/2017/ctc/
6. https://commonvoice.mozilla.org/en/datasets
7. https://huggingface.co/facebook/wav2vec2-xls-r-300m

Chronic Kidney Disease Detection Using Machine Learning Model

Rishav Pandit[1], Amlan Mahapatra[2], Afraa Jumaa[3], Sushant Kumar[4], Suman Sau[5]

ITER SOA University, Bhubaneswar

Abstract Chronic kidney disease diagnosis is often expensive, time demanding, and incursive, and hence dangerous. That is why, especially in countries with limited resources, many patients approach the end stage without receiving curative treatment. Therefore, early diagnosis and treatment of chronic kidney disease help the patient to stop the illness's progression. Machine learning techniques such as Logistic Regression, XGboost, and Random Forest classifiers with hyper-parameter optimization are used to predict chronic renal disease in this work. The UCI ML Repository's data set was obtained. The comparison of different machine learning algorithms is done using performance metrics such as accuracy, precision, and sensitivity. The XGboost classifier has a success rate of over 99 percent in categorizing patients with chronic renal disease.

Keywords CKD, Logistic Regression, Random Forest, XGBoost

1. Introduction

A condition known as chronic kidney disease (CKD) causes a slow decline in kidney function. This occurrence can be observed over a period of months or years of several living circumstances of the patient. Currently, kidney disease is a major problem as from 1990 to 2017 it has increased by 29.3%, there is no such age-standardized that is changing significantly. There are approximately 1.23 million people died worldwide in 2017. Ho et al. [4] proposed a computer-assisted diagnosis based on K-clustering detection following image processing. A chronic kidney disease patient in a severe stage acknowledges to treatment can be revived.

---

[1]1941017081.c.rishavpandit@gmail.com, [2]1941017046.amlanmahapatra@gmail.com,
[3]1941017182.c.afraajumaa@gmail.com, [4]1941017079.c.sushantkumar@gmail.com, [5]sumansau@soa.ac.in

DOI: 10.1201/9781003489443-23

This severe stage of the patient, which ranges from stage 1 to stage 4, might be considered Non-dialysis Dependent CKD. The last phase. End-stage renal disease is referred to as stage 5. The patient at the last stage needs to get intensive care and requires dialysis.

Miguel et al. [2] demonstrated the viability of a study using a distributed approach to handle warnings from CKD patients, the management of the alarm systems is done by monitoring the CKD patient within the eNefro project. Hsieh et al. [5] proposed that by employing ultrasound pictures, it could be possible to create a real-time system to analyze chronic renal disease. The linked learning has also been utilized to forecast, allocate, and categorize chronic kidney disease stages from the supplied dataset by creating powerful classifiers using XG-boost, Random forest, and logistic regression. Singh et al. [8] demonstrated numerous approaches for understanding the hierarchical relationship of ICD-9 codes to enhance the prediction capability of the model. They [8] et al. have presented and tested a unique attribute engineering approach that takes advantage of this hierarchy while lowering feature dimensionality. Chiu et al. [1] conferred an intuitive model for the detection of determining the severity of a patient's chronic kidney illness, these intuitive models use Artificial Neural Networks. Asif Salekin and John Stankovic [7] the effectiveness of models KNN, neural network, and Random forest to detect CKD. Using the IBK algorithm, authors in [7] values were used to fill in the data set's missing data. In [7], the Random Forest approach produced RMSE of 0.0184 and an F1 score of 0.993, outperforming the other two techniques further both the authors reduced the model's characteristics and chose its key characteristics, these attributes were designed to make model effective and simple for Chronic Kidney disease.

In this paper, we have formulated the attribute reduction, learning algorithm selection, and tuning for Chronic Kidney Disease. Our contributions are:

1. Comparing the several Machine learning technique on the CKD data set with hyper-parameter optimization, the XG-boost model is found to perform best. The performance of this model is then optimized based on the entire statistics, and it is significantly better than the previous model.
2. Other machine learning methods like logistic regression and random forest classifier are used for comparing the results with the other issued models.
3. A potent non-linear tree-based machine learning method is XGBoost. The XGBoost hyper-parameter combination is designed to minimize the overhead encountered during grid-searching for optimal values.

2. Related Work

Adeola and Qing et al. [7] applied the application of artificial intelligence technique, and for a quick and precise diagnosis of chronic kidney disease with pertinent symptoms, a method using the Extreme Gradient Boosting (XGBoost) that combines three characteristics has been presented. This model was created with a 97 percent accuracy rate. Rahul et al. [3] focuses on the classification technique that with an accuracy of 98.48 percent, 94.16 percent, and 99.24 percent, tree-based DT, RF, and Logit Regression were developed respectively. Weilum et al. [9] focused on the 3 machine learning models: XGBoost, a boosting tree model, and a bagging tree model called Random Forest, they came up with the accuracy of 53.43%, 55.23% respectively. Pedrao A. Moreno-Sanchez [6] has used cross-validation to achieve accuracy of 98.9 and 97.5 percent with the XGBoost classifier across a set of 4 features. Pedrao [6] also mentioned that taking a small number of features will reduce the cost of diagnosis of Chronic Kidney Disease (CKD) with a promising solution for a developing country.

3. Background

3.1 Dataset

In this work, we used data from the UCI ML Repository. The data collection contains information from 400 adults aged 19 to 85 who underwent routine health checks in 2015. There are several missing values in the collected data set that also need to have been included. There are 26 attributes gathered, 25 of which are clinical aspects and the others are target qualities. Clinical history, physical examination, and lab tests are the three aspects of the highlighted qualities. The targeted traits were divided into positive and negative categories based on their properties which means if positive then it will be expressed by "Presence of Disease" and if it is negative then it will be expressed by "No Disease".

3.2 Logistic Regression

The most fundamental method, the logistic function, often known as the sigmoid function, is a component of logistic regression. Another type of regression that forecasts a continuous dependent parameter is linear regression. Additionally, logistic regression forecasts categorical dependent variables using a set of independent variables. Logistic The linear regression is regression. Logistic regression is used to determine the relationship between the example alpha and the Boolean class flag beta as follows:

$$(\beta = 1 \mid \alpha) = \frac{1}{1 + \exp(x_0 + \sum_{i=1}^{n} x_i \alpha_i)} \tag{1}$$

$$(\beta = 0 \mid \alpha) = \frac{\exp(x_0 + \sum_{i=1}^{n} x_i \alpha_i)}{1 + \exp(x_0 + \sum_{i=1}^{n} x_i \alpha_i)} \tag{2}$$

3.3 Random Forest

A decision tree forest is created by using a random forest. It's also known as a supervised learning algorithm. The forest's toughness is shown by the trees that grow there. The method is accurate to within α of the total amount of trees in the forest. It is represented in Fig. 23.1.

Algorithm

Step 1 Where (ab), choose 'a' features at random from the entire 'b' feature.

Step 2 Using the optimal split point approach, determine the node known as "n" from the feature "a."

Step 3 The nodes should be split into daughter nodes using the best split.

Step 4 Repeat the instructions above till you get to '1'.

Step 5 The processes outlined above are repeated a total of 'S' times to create a forest with an 'S' number of trees.

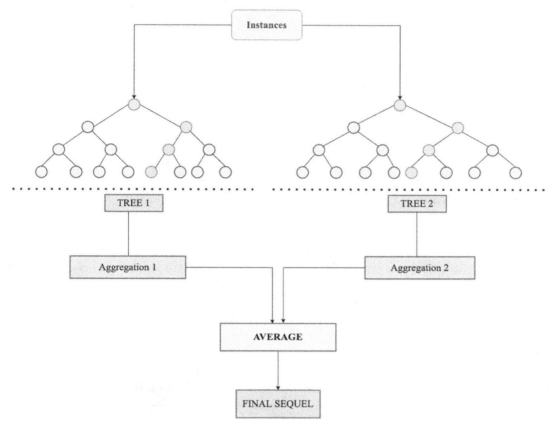

Fig. 23.1 Specified structure of random forest

3.4 XGBoost

To improve speed and performance, XGBoost adds a variety of optimizations to the gradient boosting technique. The Gradient Decision Tree framework uses a machine learning method to construct it. Fig. 23.2 depicts the situation. Unlike Random forest, each XGBoost tree model minimizes the residual from the preceding tree model. The Gradient Boosting Decision Tree uses just the first derivative of error information. XGBoost uses both the 1st and 2nd derivatives. Moreover, the XGBoost supports a contrived cost function.

This Python-based approach includes a new regularisation technique. The model must be fine-tuned step by step to attain optimal performance. Because of the large number of hyperparameters, tuning XGBoost can be a difficult task. These settings are divided into three categories: general, booster, and learning task; XGBoost can be tuned using grid and random search methods. Prediction accuracy cannot be the only criterion for classifier selection. Other criteria should be taken into account; the criteria will serve as the standard definition for all performance metrics. Here are some basic terminologies:

	Prediction	Actual
True Positive (TP)	positive	true
False Negative (TN)	negative	false
False Positive (FP)	positive	false
True Negative (FN)	negative	true

Various performance measures are formed using the aforementioned basic language.

- Accuracy on Machine learning, which is used to determine the amount of particular class that is predicted over a total number of samples collected.

$$\text{Accuracy} = \frac{TP + TN}{TP + FP + FN + TN} \tag{3}$$

- The proportion of accurately anticipated caused instances to all predicted caused cases is known as precision.

$$\text{Precision} = \frac{TP}{TP + FP} \tag{4}$$

- The ratio of accurately predicted well cases to all positive instances is known as sensitivity.

$$\text{Sensitivity} = \frac{TP}{TP + FN} \tag{5}$$

4. Results

Comparative measures for the above algorithm summary are given in Table 23.1.

Algorithm\Measure	Accuracy	Precision	Sensitivity
Random Forest	97%	97%	100%
Logistic Regression	90%	97%	100%
XGBoost	99%	97%	100%

The three classification methods utilized in the research had ultimate accuracy of 90%, 97%, and 99%, respectively.

The three classification algorithms' ultimate sensitivity levels are 100%, 100%, and 100%, respectively.

The ultimate precision of the three classification systems is 97%, 97%, and 7%, respectively.

5. Conclusion

In this paper, the XGBoost with hyper-parameter Random Search model developed for Chronic Kidney Disease performed after expectation. We conducted a detailed evaluation of machine learning techniques on several classifiers, but XGBoost outperforms them on average. To discover characteristics from the dataset, we used the chi-squared selection, correlation matrix, and feature significance filter techniques for feature selection. The suggested entire model

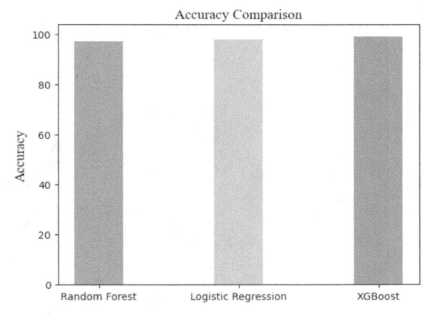

Fig. 23.2 Comparative study in terms of accuracy

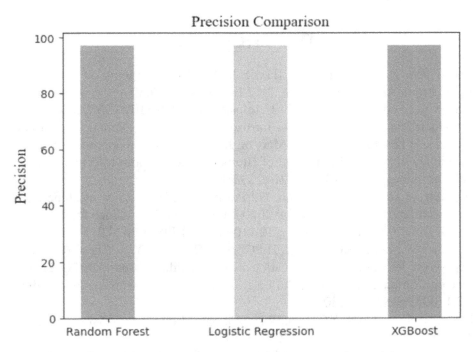

Fig. 23.3 Comparative study in terms of precision

receptively obtained 99%, 97%, and 100%, in accuracy, precision, and sensitivity. Random forest and logistic regression were also applied in the suggested model, with an accuracy of 97% and 90%, respectively. The proposed methodology can help diagnoses take less time and money.

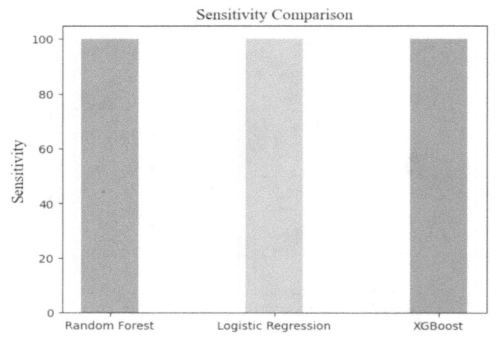

Fig. 23.4 Comparative study in terms of sensitivity

REFERENCES

1. Chiu, R.K., Chen, R.Y., Wang, S.A., Jian, S.J.: Intelligent systems on the cloud for the early detection of chronic kidney disease. In: 2012 International Conference on Machine Learning and Cybernetics. vol. 5, pp. 1737–1742 (2012). https://doi.org/10.1109/ICMLC.2012.6359637
2. Estudillo-Valderrama, M.A., Talaminos-Barroso, A., Roa, L.M., Naranjo-Hernández, D., Reina-Tosina, J., Areste-Fosalba, N., Milán-Martín, J.A.: A distributed approach to alarm management in chronic kidney disease. IEEE Journal of Biomedical and Health Informatics 18(6), 1796–1803 (2014). https://doi.org/10.1109/JBHI.2014.2333880
3. Gupta, R., Koli, N., Mahor, N., Tejashri, N.: Performance analysis of machine learning classifier for predicting chronic kidney disease. In: 2020 International Conference for Emerging Technology (INCET). pp. 1–4 (2020). https://doi.org/10.1109/INCET49848.2020.9154147
4. Ho, C.Y., Pai, T.W., Peng, Y.C., Lee, C.H., Chen, Y.C., Chen, Y.T., Chen, K.S.: Ultrasonography image analysis for detection and classification of chronic kidney disease. In: 2012 Sixth International Conference on Complex, Intelligent, and Software Intensive Systems. pp. 624–629 (2012). https://doi.org/10.1109/CISIS.2012.180
5. Hsieh, J.W., Lee, C.H., Chen, Y.C., Lee, W.S., Chiang, H.F.: Stage classification in chronic kidney disease by ultrasound image. In: IVCNZ '14 (2014)
6. Moreno-Sanchez, P.A.: Development and evaluation of an explainable prediction model for chronic kidney disease patients based on ensemble trees (2021)
7. Ogunleye, A., Wang, Q.G.: Enhanced xgboost-based automatic diagnosis system for chronic kidney disease. In: 2018 IEEE 14th International Conference on Control and Automation (ICCA). pp. 805–810 (2018). https://doi.org/10.1109/ICCA.2018.8444167

8. Singh, A., Nadkarni, G.N., Guttag, J.V., Bottinger, E.P.: Leveraging hierarchy in medical codes for predictive modeling. Proceedings of the 5th ACM Conference on Bioinformatics, Computational Biology, and Health Informatics (2014)
9. Wang, W., Chakraborty, G., Chakraborty, B.: Predicting the risk of chronic kidney disease (ckd) using machine learning algorithm. Applied Sciences 11, 202 (12 2020). https://doi.org/10.3390/app11010202

WhatsApp Chat Analysis Application

Sarthak Kumar Behera[1], Zarit Ahmed[2], Sahil Saswat Jena[3], Abhijit Pattanaik[4]

Students of Department of Computer Science & Information Technology,
Institute of Technical Education & Research, SOA University, Bhubaneswar, Odisha, India

Suman Sau[5]

Assistant Professor, Department of Computer Science & Information Technology,
Institute of Technical Education & Research, SOA University, Bhubaneswar, Odisha, India

Abstract The WhatsApp chat analyzer is a novel application that uses machine learning and NLP techniques to provide insights into hidden facts present in WhatsApp group chats. By importing the chat file, the tool generates visualizations using libraries such as pandas, seaborn, matplotlib, and emoji. Although WhatsApp chats can be challenging to analyze due to their dynamic nature, the WhatsApp chat analyzer can analyze any topic with ease and provide accurate results in a visually appealing format. This project's ability to extract valuable insights from WhatsApp chats can be highly beneficial in various scenarios, making it an important addition to the field of data analysis.

Keywords Whatsapp conversations, Streamlit, Pandas, Matplotlib, NLP

1. Introduction

WhatsApp is a chatting app which provides users to communicate by text, images, audio and video files, and voice calls. [1] It supports group chats with multiple users and broadcast messages to up to 256 users. The application is free to use but requires an internet connection to function. It functions on various operating systems including Android, Windows and iOS.

---

[1]1941017023.b.sarthakkumarbehera@gmail.com, [2]1941017119.b.zaritahmed@gmail.com, [3]1941017084.b.sahilsaswatjena@gmail.com, [4]1941017007.b.abhijitpattanaik@gmail.com, [5]sumansau@soa.ac.in

DOI: 10.1201/9781003489443-24

WhatsApp is widely used across different demographics, [2] and according to data from 2014, it had up to 450 million monthly average users. In 2015, [3] it was found that among all smartphone usages, 19.83% were only of this app, while Facebook had only 9.38%. The application is popular due to its cost-free messaging function, cross-platform compatibility, and international functionality. Persons must individually ban persons they choose not to interact with in the encrypted application. Individuals need a compatible mobile device or tablet, a stable network connection, and a phone number for using the app. The application allows users to share their location in real-time, organize lists of contacts, and make international calls without incurring additional charges. The application has added additional capabilities, including the ability to react immediately in a private conversation and connect a message back to the group from the sender's personal profile. The application is exceptionally beneficial for group conversations, enabling members to track discussions more conveniently and respond to particular questions or comments.

1.1 Problem Statement

A robust statistical assessment tool, WhatsApp Chat Analysis Application [5], creates illuminating charts from WhatsApp chat files. It helps users understand their communication patterns and make better decisions. With data manipulation techniques, it's an essential addition to any user's toolkit, for both personal and professional purposes. Our suggestion is to utilize advanced dataset manipulation techniques to enhance our comprehension of the WhatsApp conversations saved on our mobile devices. By applying these techniques, we can extract valuable insights from the chat data, enabling us to gain a deeper understanding of our communication patterns and improve our messaging skills.

1.2 Existing System

The current system has undergone significant development, with a range of new features that were not available in older versions. In previous versions, users were unable to display their status, share documents, or send their location. Moreover, sharing images through doc's format was not possible. However, the current version of the system offers all these features and more. One such feature is the ability to access WhatsApp on a desktop computer through the web-based WhatsApp that can be accessed with a QR code. Additionally, users can use the "export chat" feature to obtain chat details for data analysis, which can be sent or shared through email, Facebook, or other messenger applications. These updates have significantly improved the user experience and expanded the functionality of the platform [5].

1.3 Proposed System

Data pre-processing is essential in the early phases of the project to comprehend the implementation and application of numerous built-in modules. Instead of building functionality from start, this technique enables engineers to understand why various modules are beneficial. Developers can enhance coding readability and client accessibility by adopting these modules. Numerous libraries are used, including SciPy, NumPy, and other libraries Pandas, Matplotlib, CSV, and Seaborn.

The next stage involves exploratory data analysis, starting with the application of a sentiment analysis algorithm that provides a breakdown of chat messages into positive, negative, and neutral categories, which are then plotted on a pie chart. Line graphs are then plotted, showcasing the author and message count of each date, as well as the author and message count of each author. Other graphs include a count of media sent by writers and a structured graph on the same, messages without authors, and a graph of hour versus message count. These analysis techniques provide valuable insights into the WhatsApp chat and help improve communication and decision-making skills [5].

1.4 Objective

In the current era, data is the driving force behind technological advancements. However, obtaining relevant data requires extensive research and analysis based on the specific requirements of the tool. As machine learning enthusiasts create models to solve complex problems, the necessity for suitable data grows increasingly vital. Through careful examination of information, this initiative seeks to offer an in-depth knowledge of the numerous WhatsApp discussion kinds. Machine learning algorithms that investigate chat data may find this analysis to be a useful input. Proper learning instances are necessary for these models to be accurate and reliable. In order to improve the effectiveness of machine learning models, our project focuses on delivering an extensive investigation of data on different kinds of WhatsApp discussions [5].

2. Literature Review

The impact of instant messaging apps like WhatsApp on society has been a popular topic of research in recent years. A study was conducted in southern India [6] to investigate the usage patterns of WhatsApp among young adults aged 18 to 23 years. The study found that students spent an average of sixteen hours every day online, including a total of eight hours each day on WhatsApp. WhatsApp was found to be the primary mode of communication among youth, with images, audio, and video files being frequently exchanged. Another study [7] was carried out in Karachi, Pakistan, with the goal of assessing WhatsApp's performance as a texting service. WhatsApp has become Pakistan's top messaging service thanks to the rising popularity of smartphones and social media sites. A database of a WhatsApp group chat consisting of 55,563 records [8] was analyzed using RStudio to identify patterns in the messages sent, duration of use, and response levels. Forensic analysis of WhatsApp Messenger has also been conducted to understand the various forms of communication and services provided by the platform. These studies provide insights into the impact and usage patterns of WhatsApp among different age groups and regions.

3. Methodology

Organising, converting, analysing, and simulating the information used in the framework are all part of the data analysis step. The main goal of this procedure is to derive useful conclusions from the raw data that can be applied to decision-making. In this particular study, we analyzed the data from a WhatsApp group called Unizik Staff Community. The group consisted of three sub-groups, each having 256 members [9]. Our objective was to identify the most active users in the group, determine the periods of each week that are busiest, find the best 10 to 20 users that are highly engaged, and analysis of word count of each user. We also sought to understand each user's activities and the number of messages they sent on the platform. By conducting this analysis, we wanted to learn as much as we could about the productive and beneficial members of the WhatsApp community.

3.1 Design of System Framework

The general system architecture is depicted in Fig. 24.1, which provides an overview of the different stages involved in the system's operation. These stages include data acquisition, data entry, data conversion, data analysis and investigation, and data presentation and display [9]. Specifically, Fig. 24.2 illustrates the data acquisition stage, while Python and its libraries are responsible for handling the remaining stages of data input, data conversion, data analysis and investigation, and data presentation and display.

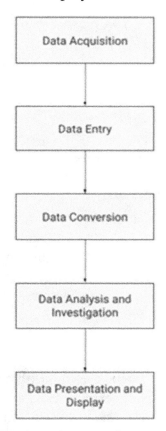

Fig. 24.1 Design of system framework diagram

3.2 Data Acquisition Stage

The data collection phase entails accessing the WhatsApp group that requires analysis in order to export the WhatsApp data file [9]. To achieve this, we followed the necessary procedures, which includes entering to the chat page, selecting "Export Chat" option from the settings, and choosing whether to include media or not. It is crucial to acknowledge that exporting data containing media files can result in larger file sizes, which may prolong the data collection process and result in unnecessary data waste. To illustrate the steps involved in the data collection process, please refer to Fig. 24.2.

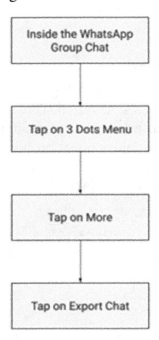

Fig. 24.2 Data acquisition stage flow chart

3.3 Tools used for Implementation

Programming with Python. For this undertaking, Python was employed as the programming language. Python is an open-source language for programming accessible without charge. It is a sophisticated language that endorses both structured and object-oriented programming. Python boasts compatibility with diverse platforms and operating systems as one of its advantages. Moreover, Python offers an extensive standard library, rendering it a favored selection for data science ventures. In our project, we made use of assorted Python libraries to manage the distinct phases of the data analysis procedure, comprising input, alteration, discovery, and visual representation of data [9].

Numpy. NumPy stands as a widely utilized Python library renowned for its application in scientific computing and data analysis. Its extensive capabilities in handling multidimensional arrays render it an optimal choice for managing vast volumes of data. The rich assortment of functions and methods accessible within NumPy facilitates efficient and effective data manipulation, an imperative aspect of our data analysis endeavor. Through the utilization of NumPy, we achieved successful classification of WhatsApp chat data based on specific time intervals [10].

Pandas. Pandas, a widely renowned Python library acclaimed for its expertise in data manipulation and analysis, assumes a pivotal and indispensable role. Its user-friendly interface empowers users to interact seamlessly with tabular data, streamlining the process of extracting, transforming, and loading data from a diverse array of file formats, such as CSV, Excel, and JSON. By harnessing the inherent intuitive data structures offered by Pandas, users are bestowed with a remarkable ability to effortlessly execute complex data operations. In the specific context of this project, the strategic employment of Pandas materialized as an effective approach to extract and manipulate 13 the dataset, thereby facilitating the execution of a diverse range of data manipulations [9].

Matplotlib. The utilization of the Matplotlib library played a crucial role in visualizing the data within this system. Matplotlib, a widely acclaimed Python library, empowers users to effortlessly generate 2D plots and graphs. In the specific context of this project, Matplotlib was imported to generate a graph that effectively depicted the dataset, enabling a comprehensive visual representation of the underlying information [11].

Seaborn. This is based on the foundations of Matplotlib, was also utilized in this work for data visualization purposes. As a higher-level library, Seaborn provides expanded plot options and enhanced aesthetics. However, it relies on the core functionality of Matplotlib [9].

Streamlit. The Streamlit library was utilized in this project to generate visually appealing web-based elements and components that showcase the analysis of WhatsApp chats. With Streamlit, various types of charts and visualizations can be created to enhance the presentation of the analyzed data [11].

NLP. For this project, various features of natural language processing (NLP) were employed, such as text parsing, stop word elimination, and text analysis. Text parsing was used to break down the messages into individual words, enabling analysis of metrics like total words and commonly used words. A stop word file was utilized in the Python program to eliminate all stop words and only display meaningful words. Text analysis was conducted to determine the count of links as well as media shared within the WhatsApp chat [10].

4. Analyzing the Results

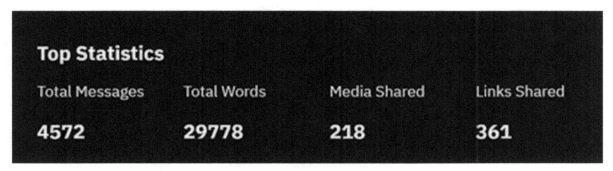

Fig. 24.3 Total number of data transfers in the chat

The figure (Fig. 24.3) displays an overview of the overall volume of shared messages, speech content, and links in a specific WhatsApp group or individual chat.

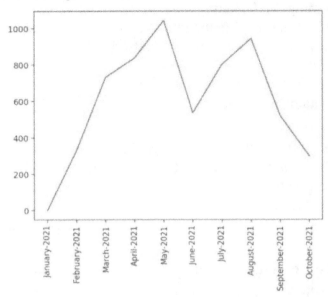

Fig. 24.4 Monthly count of messages over a certain period

The graph in Fig. 24.4 shows the monthly message count for the analyzed WhatsApp group or individual chat. The results indicate that the months of April and August had the highest message transfer activity, while January had the lowest. Also, we can see a decline in messages after August.

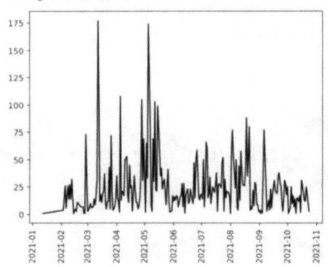

Fig. 24.5 Daily count of messages over a certain period

The graph in Fig. 24.5 represents the daily message activity, where we can observe the count of messages for each day. According to the graph, the months of March and May show the highest count of messages compared to other months.

Fig. 24.6 Most active day of the week and most active month of the year

The bar graph shown in Fig. 24.6 displays the most active day and month in terms of message transfers. According to the graph, Wednesday and May had the highest message activity, while Sunday and January had the lowest.

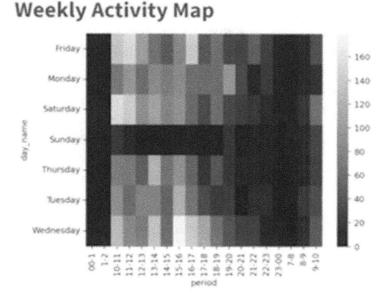

Fig. 24.7 Total no of messages on a per-hour basis in the week

The provided visual representation (Fig. 24.7) illustrates a heat map that displays the overall number of messages received during each hour of the day, for each of the seven days of the week. The analysis of the heat map reveals that the message activity is relatively low during the hours of 12am to 2am, whereas there is a noticeable increase in activity during the hours of 10am to 6pm.

Furthermore, it can be observed that Sunday has the least amount of message activity compared to the other days of the week. In summary, based on the heatmap analysis, it can be inferred that the message activity follows a predictable pattern throughout the week, with certain hours and days experiencing more significant activity than others.

234 Prospects of Science, Technology and Applications

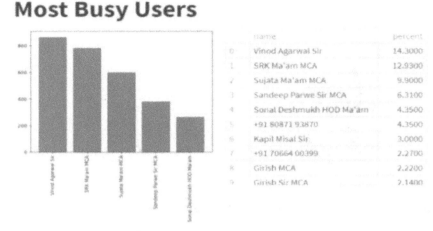

Fig. 24.8 Most active users and their contribution percentage in the chat

The bar chart in Fig. 24.11 presents the quantity of messages sent by the most active users in the chat. Moreover, it includes the proportion of contribution by each user located on the right.

Considering the chart's analysis, it can be concluded that Vinod Agarwal is the most active user in the chat, having sent more than 800 messages, which amounts to 14.3% of the total messages sent in the chat. It is noteworthy that the chart effectively conveys the message distribution among the users, highlighting the most active ones and their contributions.

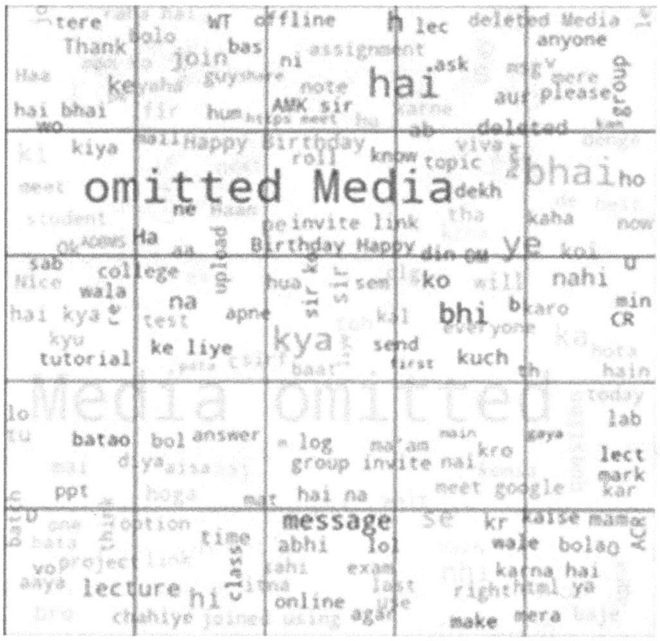

Fig. 24.9 Wordcloud of all words used in the chat

The presented visual aid (Fig. 24.9) exhibits a comprehensive analysis of the words used in a particular chat, depicting their usage frequency with the help of varying sizes. Upon examining the chart, it is evident that certain words, such as "media", "omitted" and "hai" are used more frequently than others, as they appear to be the largest in size.

Conversely, words like "student", "min", "best", and "now" are the least used, being the smallest in size. The chart successfully displays the distribution of word usage in the chat, highlighting the commonly used words and the less frequently used ones.

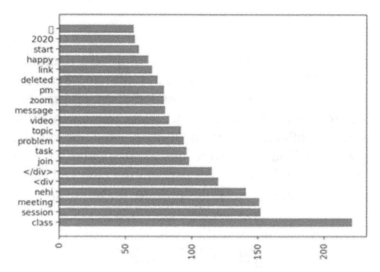

Fig. 24.10 Frequency of top 20 words of a chat

The provided visual aid (Fig. 24.10) showcases a bar graph that portrays the top 20 frequently used words in a chat and their respective frequencies. Based on the analysis of the graph, it can be concluded that the word "class" has the highest frequency, with a count of over 200. Conversely, words like "2020" and "start" were used less frequently, as their respective counts are relatively lower. The bar graph effectively represents the distribution of the most commonly used words in the chat, highlighting the ones that were used frequently and the ones that were not used as often.

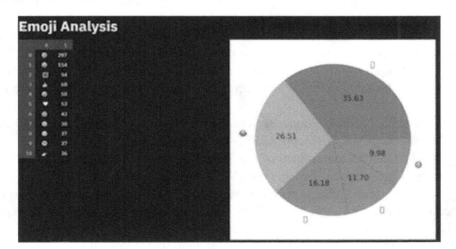

Fig. 24.11 Emoji count and usage percentage

The presented visual representation (Fig. 24.11) exhibits a bar chart that displays the frequency and usage percentage of the top 10 emojis used in a particular chat. The analysis of the chart reveals that the laughing emoji has been used the most, with a frequency of 207 times, accounting for 35.63% of the total emoji usage in the chat. It is noteworthy that the chart effectively conveys the distribution of the most commonly used emojis in the chat, highlighting the frequency and usage percentage of each emoji.

5. Final Thoughts

Upon conducting an extensive analysis of the WhatsApp chat data utilizing diverse Python libraries, including Streamlit, NumPy, Pandas, Matplotlib, and Seaborn, we can confidently ascertain that both the WhatsApp application and Python programming language exhibit formidable capabilities for data analysis. The successful implementation of our data analysis project demonstrates the potential of Python and its associated libraries in effectively analyzing and presenting complex data. The obtained results are particularly useful for chat organizers, as they can identify the most and least active users in the chat and make informed decisions based on the analysis. Overall, the project holds promise for future applications in data analysis and visualization.

REFERENCES

1. N. Thakur, "Forensic Analysis of WhatsApp on Android Smartphones," University of New Orleans Theses and Dissertations., pp. 1–33, 2013.
2. C. Steele, "what-is-whatsapp-an-explainer," 20 February 2014. [Online]. Available: https://www.pcmag.com/news/320871/what-is-whatsapp-an-explainer.
3. C. Montag, K. Błaszkiewicz, R. Sariyska, B. Lachmann, I. Andone, B. Trendafilov, M. Eibes and A. Markowetz, "Smartphone usage in the 21st century: who is active on WhatsApp?," 4 August 2015. [Online]. Available: https://bmcresnotes.biomedcentral.com/articles/10.1186/s13104-015-1280-z. [Accessed 12 March 2019].
4. WhatsApp, "About WhatsApp," 20 April 2019. [Online]. Available: https://www.whatsapp.com/about/.
5. K, Ravishankara & Dhanush, & Vaisakh, & S, Srajan. (2020). WhatsApp Chat Analyzer. International Journal of Engineering Research and. V9.10.17577/IJERTV9IS050676.
6. Available from: http://www. statista.com/statistics/260819/numberof-monthly-active-WhatsApp-users. Number of monthly active WhatsApp users worldwide from April 2013 to February 2016 (in millions).
7. Ahmed, I., Fiaz, T., "Mobile phone to youngsters: Necessity or addiction", African Journal of Business Management Vol.5 (32), pp. 12512–12519, Aijaz, K. (2011).
8. Aharony, N., T., G., The Importance of the WhatsApp Family Group: An Exploratory Analysis. "Aslib Journal of Information Management, Vol. 68, Issue 2, pp.1–37" (2016).
9. Blessing Nwamaka Iduh, WhatsApp Network Group Chat Analysis Using Python Programming. International Journal of Latest Technology in Engineering, Management & Applied Science (IJLTEMAS) Volume IX, Issue II, February 2020.
10. Shaikh Mohd Saqib. Whatsapp Chat Analyzer. International Research Journal of Modernization in Engineering Technology and Science. Volume: 04/Issue:05/May-2022.
11. Marada Pallavi, Meesala Nirmala, Modugaparapu Sravani, Mohammad Shameem. WhatsApp Chat Analysis. International Research Journal of Modernization in Engineering Technology and Science. Volume: 04/Issue:05/May-2022.

Prospects of Science, Technology and Applications – Prof. Renu Sharma (eds)
© *2024 Taylor & Francis Group, London, ISBN 978-1-032-78833-3*

Characterization of Antimony Chalcogenide-based PIN Photodiode

25

Vedika Pandey and Sumanshu Agarwal*

Department of Electronics and Communication Engineering,
Institute of Technical Education and Research, Siksha 'O' Anusandhan
(Deemed to be University), Bhubaneswar, India –751030

Abstract   Calibrated detection of photons by using photo-detectors has become an integral part of various engineering solutions. A photo-detector is also a type of photo-detector that converts the optical signal to an electric current and measures across its terminals. In this paper, we designed and discussed the potential of an antimony chalcogenide based photodiode device model. The proposed design is a PIN hetero-junction photodiode, where intrinsic antimony chalcogenide $Sb_2(S, Se)_3$ and its compound form $(Sb_2(S_{1-x}Se_x)_3)$ is used as an active layer and sandwiched between P-type hole transport layer (Spiro-MeOTAD) and N-type electron transport layer (TiO_2). The influence of physical photodiode qualities on photodiode functionality as well as photodiode design elements were investigated using numerical simulations. In this research work, we employed the SCAPS software which simulates solar cell and is well used in scientific community. As a consequence, numerical simulations were run for the suggested design which self consistently solves drift-diffusion equations linked with Poisson's equation in one-dimensional geometry to assess current-voltage characteristic properties, responsiveness and quantum efficiency. Using SCAPS-1D software and MATLAB, we also examined the impact of Mole fraction on the performance of photodiode. In addition, we considered 100 mW/cm² incident optical power and calculated the responsivity and quantum efficiency values using short circuit current obtained from simulation results of SCAPS-1D. Our results indicate that the proposed design is promising and provide good responsivity for a large range of wavelengths.

Keywords   Photodiode, Numerical simulation, Optimization, Drift-diffusion, Antimony chalcogenide

---

*Corresponding author: sumanshuagarwal@soa.ac.in

DOI: 10.1201/9781003489443-25

1. Introduction

Sensors are a critical component of any optoelectronic device and a component of an optical fiber communication system that affects system performance as a whole. Thus, the core intellectual for the recipient constitutes a photodiode. The photodiode is basically a type of photo-detector that detects the signals whenever incident light falls into it and roughly translates it into an optical power signal back into corresponding variations in electric current. The Photodiodes often feature PN or PIN constructions. Due to their greater applications and quick response time PIN configurations are often employed for the design of various photodiode devices. During earlier times, light-emitting was developed utilizing a wide range of substances that are inorganic, like silicon and germanium, amongst others. However, photodiodes are devices called semiconductors that are produced employing semiconductor elements. And one of these substances can be used to make photodiode detectors. Those are elements from the chalcogen family and antimony.

Antimony chalcogenides Sb_2S_3, Sb_2Se_3 and $Sb_2(S,Se)_3$ are currently highly desirable substances as a result of having outstanding strength, sufficient chemical preservation, nontoxic properties, affordability, appropriate adaptable band gap that could be utilized for producing any optical devices, and substantial absorbing efficiency. Materials having a combination of group IV and group V elements (such as Si, Ge, As, etc.) and chalcogen components (such as S, Se, and Te) are known as chalcogenide semiconductor components. A significant advantage of those substances lies in the fact they quickly generate crystalline chalcogenides with a variety of intriguing attributes, which include a wide band gap energy, which depends on the ratio of the composition of the chalcogen element to the various elements that compose it. As a result, these mixtures can be used to research physical science and a wide range of material properties depending on the varied chemical constituent ratios. Because of the changes in their characteristics brought on by light, amorphous chalcogenide electronic components have been the subject of intense research in recent years.

Additionally, two elements within an identical family may come together, as in the case of $Sb_2(S, Se)_3$ which is created by mixing Sb_2S_3 and Sb_2Se_3. Sb_2S_3 and Sb_2Se_3 are combined to form Antimony Sulfide Selenide $(Sb_2S, Se)_3$ as well as a semiconductor combination $(Sb_2(S_{1-x}Se_x)_3)$ which is a member of the antimony chalcogenide family and a sustainable substance with adjustable band gaps and an exceptionally high absorption coefficient of variation. Here x defines the Mole fraction of the material. Because of their excellent physical features, antimony chalcogenide semiconductors are considered to be attractive materials for a variety of optical devices such as photodiodes, and so on. In this paper, we address the possibility and potential of an antimony chalcogenide-based photodiode. The suggested design is a PIN hetero-junction photodiode with intrinsic antimony chalcogenide $(Sb_2(S_{1-x}Se_x)_3)$ as the active layer positioned between a P-type HTL having (Spiro-MeOTAD) material. The ETL is made of TiO2 material. The most attractive feature of choosing this material is its adjustable band gaps and its environmental friendly nature. The device simulator SCAPS-1D, were employed for investigating the current-voltage characteristic under darkness and illumination, as well as the optical properties are also evaluated using this tool like- responsivity, and quantum efficiency. The design of the proposed device has also gone through several changes to observe whether varying the mole fraction of the semiconductor composition impacts functionality as well as investigate whether those changes contribute to another efficient photodiode. The simulated results can be used for developing an advance photo-detector for optoelectronic devices.

2. Device Structure

The proposed schematic design of a PIN hetero-junction photodiode based on antimony chalcogenide is shown in Fig. 25.1. With a thickness of 300 nm, the intrinsic ($Sb_2(S_{1-x}Se_x)_3$) is employed as an active layer and is positioned between the p-type HTL (Spiro-MeOTAD) and N-type ETL (TiO_2) for a doping concentration of 10^{18} cm^{-3}. The P-type and N-type layers are 50 nm thick.

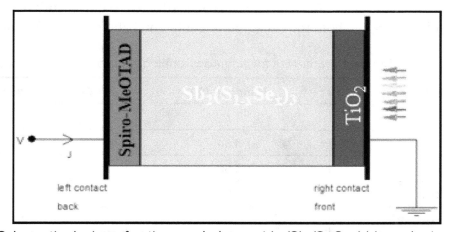

Fig. 25.1 Schematic design of antimony chalcogenide ($Sb_2(S_{1-x}Se_x)_3$) based p-i-n photodiode

2.1 Software Used for Simulation

These days, simulation and modeling are two of the easiest, most accessible and well-liked methods for a full examination and optimization of photovoltaic as well as photodiode response. Regarding the mathematical evaluation for the devices, there are numerous freely accessible applications along with teaching method tools capable of calculating the basic semiconductor calculations. Considering the literature contains an abundance of information concerning SCAPS and its prospective applications for investigating the behavior of a photodiode, SCAPS has been selected among the available applications for this work. Therefore, we employ SCAPS-1D for simulating our proposed photodiode model. SCAPS-1D is a software simulation tool that was constructed by Burgle man et al. from the Electronics and Information Systems (ELIS) division of the University of Gent in Belgium. This simulator tool is an instance of free, open-source applications technology that was initially designed to serve Copper indium selenide and (Cadmium telluride) compound's cellular structures. This software are able to forecast device properties like the current density vs. voltage curve (JV curve), quantum efficiency, variation in responsivity, and other particular electrical and optical properties of the photodiode device structure.

3. Simulation Methodology and Numerical Values

The simulations were carried out using SCAPS-1D, and we used a linear relationship to compute the mole fraction dependent band gap and uniform photo-generation rates of the charge carriers in the active layer (equal to the integrated value obtained from Beer-Lambert's equation). Furthermore, we use 100 mW/cm² incidents optical and determine responsivity and quantum efficiency values using SCAPS-1D simulation results. Furthermore, it provides various advanced physical models that describe the behavior of several properties of the device model that could be employed for designing optical devices.

The band gap (Eg) of our device model is calculated using the empirical equation:

$$Eg(x) = (1 - x) Eg(1) + xEg(0) \qquad (1)$$

Where x is the mole fraction of antimony selenide and $1 - x$ is the mole fraction of antimony sulfide. $E_g(1)$ is the band gap of Sb_2S_3 (i.e. 1.7eV) and $E_g(0)$ is the band gap of Sb_2Se_3 (i.e. 1.1eV). The corresponding values of both the cutoff wavelength and band gap are tabulated in Table 1.

Table 25.1 Typical Parameters for antimony chalcogenide based photodiode

Mole Fraction (x)	Bandgap (E_g)	Cutt-off wavelength (λ_c)
0	1.7	729
0.1	1.64	756
0.2	1.58	785
0.3	1.52	816
0.4	1.46	819
0.5	1.4	886

The device's functionality is also impacted by the fault density located in the operative layer (Active). As the defect density increases, more photo-generated carrier recombination occurs, significantly decreasing the device's efficiency. When the defect density is low, the carrier diffusion length increases and the recombination process fails, resulting in improved photodiode performance. The defect density in the active layer is set to 10^{15} cm³. For further investigating the unique properties of the photodiode including Dark and light J-V characteristics, responsivity, and quantum efficiency, numerical simulations for the proposed layout were performed employing the device Simulation tool, and these self-consistently solve drift-diffusion equations coupled with Poisson's equation in one-dimensional dimensional geometry. The software enabling mathematical analysis which examines photodiodes should be capable of solving essential semiconductor mathematical equations. The one-dimensional semiconductor equations are solved by SCAPS-1D. The above formulas serve as vital for assessing photodiode efficiency and potential production. The Poison equation, as well as the drift-diffusion equation, constitutes the equations that regulate these. The subsequent combination of formulas within the one-dimensional constitutes the building blocks for the totality of the drift-diffusion equations structure:

Table 25.2 Typical Parameters for antimony chalcogenide based photodiode

Parameters	unit	(Sb$_2$(S$_{1-x}$Se$_x$)$_3$)	ETL	HTL
Thickness, W	nm	300	50	50
Band gap, Eg	eV	Eq. (1)	3.2	3.2
Electron affinity, χ	eV	3.8	4.0	1.9
Effective DOS for electron, N_c	cm^{-3}	1 × 10^9	1 × 10^9	1 × 10^9
Effective DOS for hole, N_v	cm^{-3}	1 × 10^9	1 × 10^9	1 × 10^9
Mobility of electron, μ_e	cm²/V-s	50	50	50
Mobility of hole, μ_h	cm²/V-s	50	50	50
Radiative recombination coefficient, B	cm³/s	1 × 10^{-10}	1 × 10^{-10}	1 × 10^{-10}
Dielectric constant	–	10	10	10
Thermal velocity of electron	cm/s	1 × 10^7	1 × 10^7	1 × 10^7
Thermal velocity of hole	cm/s	1 × 10^7	1 × 10^7	1 × 10^7

4. Result and Discussion

The simulated results were compared with the calculated data for mole fracion values (from 0 to 0.5) to highlight how comparable the outcomes are in both scenarios where the highlighted dotted circles are the values of current-voltage obtained from simulation and the highlighted lines are the values of current-voltage values obtained from calculation under darkness. The simulated results corroborates well with the analytical calculations for dark current (see Fig. 25.2). That establishes the modeling methodology. Further we plot responsivity vs wavelength for different mole fractions in Fig. 25.3. For a given incident optical power of 100 mW/cm², we computed the responsivity and quantum efficiency using short circuit current density data from the SCAPS simulation results. We find that responsivity is linear as expected until cutoff wavelength is reached. The cutoff wavelength was identified using $E_g = hc/\lambda_c$, where h is the planck's constant, c denotes light speed, and λ_c defines the cutoff wavelength. This relation follows the principle that no photon of energy less than the bandgap energy generates free e-h pair. Furthermore, we do not observe any drop in the quantum efficiency and it remain almost unity for all the mole fractions studied here.

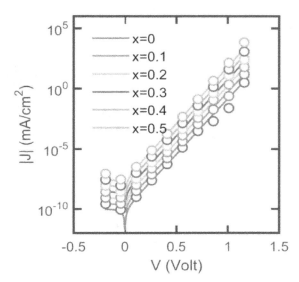

Fig. 25.2 Comparision between (a) simulated and (b) Analytical result of current-voltage characteristics of photodiode under darkness with mole fraction

242 Prospects of Science, Technology and Applications

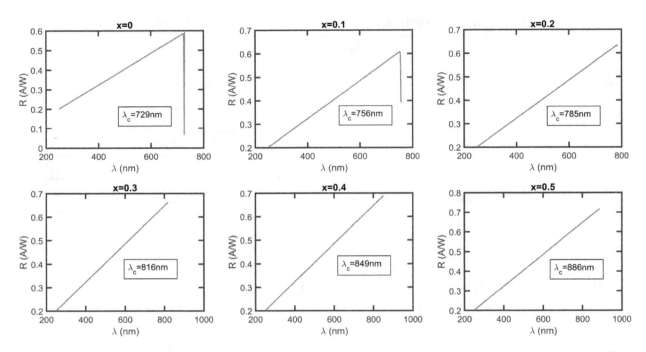

Fig. 25.3 Variation of responsivity with wavelength for various mole fractions (x = 0 to 0.5)

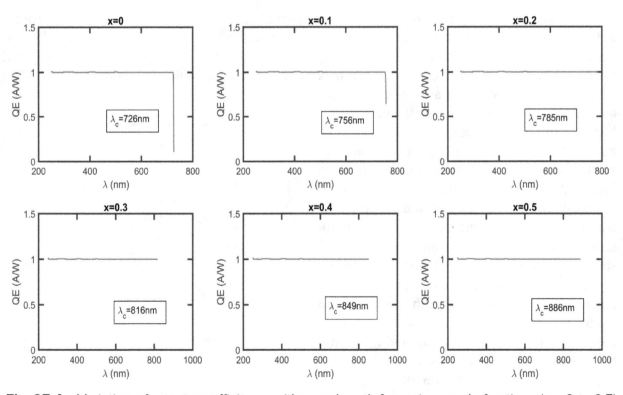

Fig. 25.4 Variation of quantum efficiency with wavelength for various mole fractions (x = 0 to 0.5)

5. Conclusion

The Antimony chalcogenide (Sb$_2$(S$_{1-x}$Se$_x$)$_3$) based PIN photodiode was simulated using the SCAPS-1D Simulator Tool A comparison between simulated and analytical data was given to assess the dependability of the model and simulation parameters. While SCAPS Tool is designed for solar cell simulation, coupling it with other computation tools like Matlab or Python we can use it for characterizing and predicting the output of a photodiode. Our results indicate that the proposed photodiode design is promising and provides good responsivity for a large range of wavelengths. While antimony chalcogenides are seen as promising candidate for optoelectronic device, their use in photodiodes has not been explored yet. To the best of our knowledge, this is the first report discussing the photodiode based on antimony chalcogenides. And our results also suggests a new photodiode structure, which is both less complicated and more effective, could potentially be developed in the future.

5.1 Implication

The active layer used in this research work is still in research phase and therefore material parameters are approximate values. More confident results can be obtained by using more reliable material parameters.

REFERENCE

1. Agarwal, S., and Nair, P. R.: Device engineering of perovskite solar cells to achieve near ideal efficiency. Applied physics letters 107(12), 12390(2015).
2. Elbar, M., Alshehri, B., Tobbeche, S., and Dogheche, E.: Design and simulation of InGaN/GaN p-i-n photodiodes. Physica Status Solidi (a) 215(9),1700521(2018).
3. Green, M. A.: Self-consistent optical parameters of intrinsic silicon at 300 K including temperature coefficients. Solar Energy Materials and Solar Cells 92(11),1305–1310, 2008.
4. Pierret, Robert. F.: Semiconductor Device Fundamentals. 2nd edition.
5. Lu, Y., Zhang, Y., and Li, X. Y.: Advanced Optical Manufacturing and Testing Technologies. Optoelectronics Materials and Devices for Sensing and Imaging 928401, 2014.
6. Liu, X., Chen, C., Wang, L., Zhong, J., Luo, M., Chen, J., Xue, D. J., Li, D., Zhou, Y., and Tang, J.: Improving the performance of Sb2Se3 thin film solar cells over 4% by controlled addition of oxygen during film deposition. Progress in Photovoltaics: Research and Applications 2015, 23 (12), 1828-1836.
7. Messina, S., Nair, M., and Nair, P.: Antimony selenide absorber thin films in all-chemically deposited solar cells. Journal of The Electrochemical Society 2009, 156(5), H327-H332.
8. Bandic, Z. Z., Bridger, P. M., Piquette, E. C., and McGill, T. C.: Appl. Phys. Lett 1998, 72, 3166.
9. Shen, Y. C., Mueller, G. O., Watanabe, S., Gardner, N. F., Munkholm, A. M., Krames R., Appl. Phys. Lett. 2007, 91, 141101.
10. Boroditsky, M., Gontijo, I., Jackson, M., Vrijen, R., Yablonovitch, E., Krauss, T., Cheng, C.C., Scherer, A., Bhat, R., and Krames, M.: J. Appl. Phys. 2000, 87, 3497.

Customer Churn Prediction in Banking Sector Using Machine Learning Techniques

Anshul Srivastava, Abhigyan Bhadra and Laxmipriya Moharana*

Ece Dept., Siksha 'O' Anusandhan (Deemed to be University),
Bhubaneswar, Odisha, India

Abstract The primary assets of a banking sector are the investors or the customers of that bank. The stability of customers in a particular bank decides the economic growth rate of that bank. It is more important to identify the customers, whose are at menace of churn. With this estimation banking sectors can plan to implement new terms and facilities for the existing customers to deviate them from churning. Bank officers always keep in track a record of leaving and newly joining customers. Evaluation of customers leaving from a sector or organization known as customer churn rate prediction. In this modern era machine learning techniques have enormously used for this work.

In this paper, we have discussed some of the supervised machine learning algorithms such as xgboost, random forest, Logistic regression, SVM (support vector machine), KNN (K-nearest neighbour) and decision tree, which have been used to predict the churn rate of customers in a set of banking data. The data set consists of some customer parameters like age, gender and area of residence. These parameters have taken as the predictors to our learning model. When all three parameters taken together for prediction, they give us good accuracy scores with acceptable computational time in comparison to individual parameters as predictors. Among all the machine learning algorithms, decision tree and random forest provides more than 72% accuracy score with less computational time in prediction process.

Keywords   Churn rate, Machine learning analysis, Accuracy score, Computational time

---

[1]laxmimoharana@soa.ac.in

DOI: 10.1201/9781003489443-26

1. Introduction

The backbone of the world of economy are the banking sectors. The main stake holders in any financial sector is the customers. Taking care of ccustomer turnover is an important aspect for any financial sectors and firms. Customer churn is a very basic and serious problem that hampers the growth rate of any sector. Customer churn in a banking sector describes the situation in which the customer leaves the bank i.e., closes his/her account from the bank or switches his/her account to another bank. Customer churn degrades the revenue and reputation of a bank. The potential customers those who are in the edge of churn should be identified and necessary steps should be taken by the bank to keep their subscription intact. Generally, customers end their subscription from a bank to get more interest on their saving money, to make any of their financial dealing in a simple way and to get more security for their accounts and lockers. To decrease the churn rate the banking officials should take proper care of their customers by introducing new plans and new laws to the investment schemes as well as they should council the potential investors in a very effective way. First of all, identifying the people who are likely to unsubscribe or leave is the most important work to accomplish.

Now a days application of machine learning is the very effective way to solve many problems. In the process of prediction, it plays a great role in all the technical and non-technical fields. In the banking industry, machine learning has become a potent tool for anticipating customer attrition. Large volumes of customer data can be analyzed by machine learning algorithms to find trends that can point to future churners. Banks can create precise churn prediction models using machine learning, allowing them to take preventative action and keep customers.

This paper basically focusses the effectiveness of the churn prediction algorithms with computational time and accuracy scores. The paper is organized as 1. Introduction 2. Literature Survey 3. Methodology 4. Result discussion 5 Conclusion

2. Literature Survey

Banking sectors should identify the factors that influence the customer to churn, then they can prepare a well applicable frame work to make the customers' association with the bank for a longer period. Researchers found some of the factors that heavily influence their churn are the length of customer association, customer's age, customer's gender and the number of mobile banking transactions [1]. Customer churn rate can be controlled by satisfying the customers with proper utilization of company resources and provide them the optimum financial benefits. A well explained discussion about this is mentioned by researchers in [2]. Customer churn also heavily affects the telecommunication companies. The reasons of churning are mentioned by authors in [3]. Customer transactions and their billing information gives sufficient clues of their churn in a telecom sector [4]. Categorize the customers as per the possibility of customer churn is a fundamental step to retain them in the same bank. Researchers has formed a road map to label the customer in [5].

Many methods have been implemented by the researchers to predict the churn rate in banking sectors and telecom industries. Hybrid data mining techniques have been implemented by the authors to predict churn rate [6]. Churn rate prediction has been successfully implemented by the researchers by using a series of Monte Carlo simulation [7]. To get better performance in churn prediction process, researchers integrate sampling techniques and cost-sensitive learning to the existing random forest algorithm [8]. Ant colony algorithm, effectively implemented by the researchers to predict the churn rate [9].

In this paper we have predicted customer churn rate by considering their age, gender and domicile from an existing bank customer's database. We have implemented some supervised machine learning algorithms like xgboost, random forest, Logistic regression, SVM, K neighbors and Decision Tree. We have calculated the accuracy score and computational time for each machine learning algorithms to get a better prediction algorithm.

3. Methodology

This work mainly focuses to predict the churn rate in banking sector. The frame work of the work has been represented in Fig. 26.1

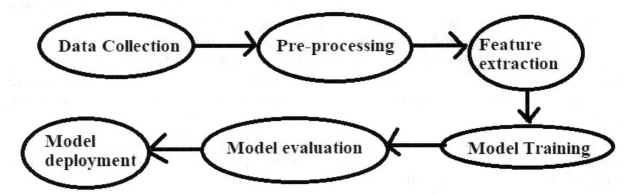

Fig. 26.1 Framework for bank churn prediction using machine learning techniques

The first step involves the data collection process. Data set regarding customers has been retrieved from Kaggle web-site [10]. The data set contains the customer information such as gender, customer Id, age, basic income, domicile information etc. In the pre-processing state we have discarded the parameters which are not required for our analysis and handling the missing values, outliers, and other data quality issues. In Feature extraction phase we have only considered the features which are used as the predictors for our learning model and other features are ignored. In training stage different machine learning algorithms such as logistic regression, decision trees, KNN, xg-boost etc. have been used to train our model. After training the testing phase initiated. Where our model worked upon the new phase of the given bank data set. From the data set we have chosen the 70:30 proportion of data as training and testing data set respectively. In this stage, it is being verified that our model has provided a good accuracy score with less computational time, which is elaborately discussed in result section.

4. Result and Discussion

We have predicted the churn rate of the customers for banking sectors. After collecting the data set from Kaggle, the pre-processing has been performed. After pre-processing the selected parameters i.e., age, gender and area of residence have been used to train the model by using different machine learning algorithms. We selected these three parameters as the predictors to our learning model. Fig. 26.2 represents the training process of our model with the predictors and response variable. We have taken Age, gender and area of residence as three predictors to the model. Also, by considering each predictor individually, the model was trained. We have applied Logistic regression, KNN, SVM, Decision tree, Random Forest and XG boost machine learning algorithms to train our model. The accuracy score and prediction time of each algorithm for all cases has been presented in Table 26.1–26.4. By considering all three parameters such as age, gender and area of residence of customers, we got quite good accuracy scores in comparison to the other cases, where age, gender and area of residence were taken as predictors individually. While considering each of the machine learning algorithms in Table 26.1, decision tree and random forest provides more than 70% accuracy score with manageable computational time. In case of age as predictor which is shown in Table 26.2, the results are better than gender and area of residence cases shown in Table 26.3 and 26.4. In case of age as predictor, the decision tree and Xg boost algorithm provides 70% of accuracy scores with acceptable computational time. When we have taken gender and area of residence as predictors the results are not so good i.e., below 60%. These results signifies that the age, gender and area of residence of the customers can provide sufficient clue whether they are in the edge of churn or not. All the works we have done using Python environment.

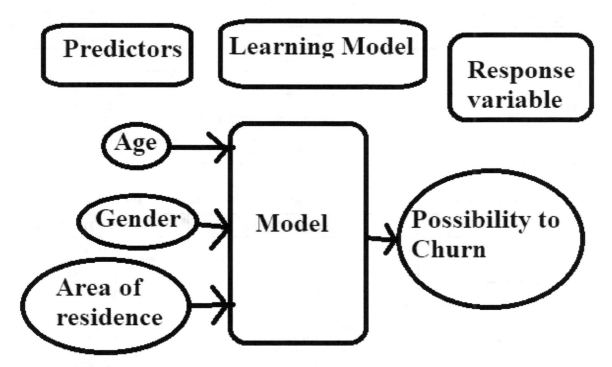

Fig. 26.2 Projection of learning model to churn rate calculation

Table 26.1 Prediction results for the predictors age, gender and area of residence

Type of algorithm	Accuracy in %age	Prediction Time in Sec
Logistic Regression	67.76	0.002672
KNN	71.05	0.00321
SVM	67.14	0.00273
Decision Tree	72.35	0.00291
Random Forest	72.33	0.00256
Xg boost	71.95	0.00318

Table 26.2 Prediction results for the age as predictor

Type of algorithm	Accuracy in %age	Prediction Time in Sec
Logistic Regression	69.44	0.00251
KNN	69.94	0.00257
SVM	66.03	0.00298
Decision Tree	70.53	0.00344
Random Forest	70.50	0.00261
Xg boost	70.53	0.00300

Table 26.3 Prediction results for the gender as predictor

Type of algorithm	Accuracy in %age	Prediction Time in Sec
Logistic Regression	56.55	0.002749
KNN	56.55	0.002608
SVM	56.55	0.00273
Decision Tree	55.51	0.002563
Random Forest	57.50	0.002942
Xg boost	58.53	0.002745

Table 26.4 Prediction results for the area of residence as predictor

Type of algorithm	Accuracy in %age	Prediction Time in Sec
Logistic Regression	55.06	0.00264
KNN	58.41	0.00259
SVM	53.45	0.002482
Decision Tree	58.41	0.002470
Random Forest	58.41	0.002680
Xg boost	58.41	0.002599

5. Conclusion

To increase the revenue of a bank, the officials should identify the customers who are in the edge of churn. Once these customers have been identified new plans should be introduced to make them satisfied as well as different and variable profit incurred terms should be offered to the existing and long-term customers. Machine learning algorithms are proved to be very helpful in the churn detection process. In our work we have shown that by considering the customer parameters such as age, gender and their area of residence the churn rate can be predicted successfully with more than 70% accuracy. Also, it was observed that the churn rate is not dependent with the customer's gender and their area of residence as single parameters. But by considering the age of the customer as a parameter to calculate churn rate, we got 70% accuracy in prediction. So, we can say age of the customer may be a clue to get the information regarding the closing or switching of their account from a bank.

REFERENCES

1. Keramati, Abbas, Hajar Ghaneei, and Seyed Mohammad Mirmohammadi. Investigating factors affecting customer churn in electronic banking and developing solutions for retention. International Journal of Electronic Banking 2(3), 185-204 (2020)
2. McDonald, Lynette M., and Sharyn Rundle-Thiele.Corporate social responsibility and bank customer satisfaction: a research agenda. International Journal of Bank Marketing (2008).
3. Bharti, Akanksha.Customer churn management. ACADEMICIA: An International Multidisciplinary Research Journal 7(5), 96-102(2017)
4. Ahn, Jae-Hyeon, Sang-Pil Han, and Yung-Seop Lee.Customer churn analysis: Churn determinants and mediation effects of partial defection in the Korean mobile telecommunications service industry. Telecommunications policy 30(10-11), 552-568 (2006)
5. Chayjan, Mahdiyeh Rezaei, Tina Bagheri, Ahmad Kianian, and Niloufar Ghafari Someh. Using data mining for prediction of retail banking customer's churn behaviour. International Journal of Electronic Banking 2(4), 303-320, (2020)
6. Tsai, Chih-Fong, and Yu-Hsin Lu. Customer churn prediction by hybrid neural networks. Expert Systems with Applications 36(10), 12547-12553 (2009)
7. Vafeiadis, Thanasis, Konstantinos I. Diamantaras, George Sarigiannidis, and K. Ch Chatzisavvas. A comparison of machine learning techniques for customer churn prediction. Simulation Modelling Practice and Theory .55, 1-9 (2015)
8. Xie, Yaya, Xiu Li, E. W. T. Ngai, and Weiyun Ying. Customer churn prediction using improved balanced random forests. Expert Systems with Applications 36(3), 5445-5449 (2009)
9. Verbeke, Wouter, David Martens, Christophe Mues, and Bart Baesens. Building comprehensible customer churn prediction models with advanced rule induction techniques. Expert systems with applications.38(3), 2354-2364 (2011)
10. Kaggle data set for bank churn prediction Available at https://www.kaggle.com/datasets/gauravtopre/bank-customer-churn-dataset (2021)

Plant Disease Detection

Tirthankar Biswas*, Swastik Kumar Pati and Sunita Sarangi

Siksha 'O' Anusandhan, Bhubneswar, Odisha

Abstract Plant disease identification is crucial for crop health and yield maximization. The application of machine learning methods for the automated diagnosis of plant diseases has attracted growing attention in recent years. In this thesis, we investigate the application of pooling model and convolutional neural network (CNN) for plant disease diagnosis.

The suggested method uses pooling to minimize the dimensionality of the feature maps after CNN is used to extract features from pictures of damaged plants. After that, a classification model for identifying diseases receives the reduced feature maps. The efficiency of the suggested approach is shown by evaluation results on a collection of photos of plants with various illnesses.

The effectiveness of the suggested approach is also examined, along with the effects of various factors on it, including the size of the convolutional kernels, the quantity of pooling layers, and the size of the feature maps. The findings suggest that the proposed method detects plant illnesses with high accuracy and has the potential to be employed in real-world applications for automated disease detection and prevention in agriculture.

Keywords Machine learning, CNN, Classification, Disease identification

1. Introduction

1.1 Background Study

One of the biggest problems in agriculture is plant disease, which can significantly reduce crop quality and productivity. Plant diseases can be stopped in their tracks and the associated losses reduced by early detection and prompt treatment. Inaccuracies in disease diagnosis might result from the use of time-consuming and subjective traditional methods of plant disease

---

*Corresponding author: tirthankarbiswas98@gmail.com

detection, such as visual inspection by human experts. Therefore, approaches for accurately and automatically detecting plant diseases are required. For the automatic diagnosis of plant diseases, machine learning approaches have recently become more and more popular in the agricultural sector. CNNs, one of many machine learning algorithms, have demonstrated promising outcomes in the detection of plant diseases. A particular kind of deep learning neural network called a CNN can automatically extract features from incoming photos without the requirement for feature extraction explicitly.

Pooling is a method frequently combined with CNNs to decrease the dimensionality of the feature maps and improve the model's resistance to minute image fluctuations. The model becomes more resilient to different input conditions as a result of pooling, which aids in establishing invariance to translation, rotation, and scale of the input image.

On the use of CNNs and pooling for plant disease detection, numerous studies have been done. The majority of this research has demonstrated that CNNs with pooling layers can accurately and efficiently identify a wide range of plant illnesses. Nevertheless, there is still potential for development in terms of lowering false positives and raising the rare disease detection accuracy.

In order to achieve high accuracy and robustness to a variety of input conditions, this thesis intends to investigate the application of CNNs and pooling models. The findings of this work may provide light on the automated identification and prevention of plant diseases using deep learning techniques, which will support sustainable agricultural practises.

1.2 Motivation

As india mainly a rural area agriculture is the main source of income in many families. India being the second largest patron of wheat and rice which is world's major food masses. This sector is the major contributor of the country's GDP and hence it's the most pivotal sector in Indian Economy. Now-a-days with the proliferation of pollution the conditions in humans as well as crops are affected. Besides this the operation of fungicides and chemical which will mixed with the ground water can also beget of several conditions. The planter cannot guess the complaint by the symptoms in the earlier stages of it, which is directly commensurate with a loss and a peril of epidemic. That's why we're end to make a medium which can descry if there's any kind of complaint in the crops at early stages by the image processing of the splint of the crops by Machine Learning algorithm.

2. Methodology

The technique for the suggested approach to plant disease detection utilizing a CNN (CNN) and pooling model is separated into four major stages: data collection and preparation, model design, model training, and evaluation.

Data collection and preparation: The first stage is gathering a dataset of pictures of plants that have different diseases. The dataset should be sufficiently large to provide a diverse set of examples for training the CNN model. Pre-processing is necessary to get rid of any noise, artifacts, or background distractions that can obstruct learning. To ensure that every image has a uniform size, resolution, and color depth, the images should be normalized.

Model Design: The second phase entails creating a CNN architecture that is most suited for identifying plant diseases. To extract features from the images, the model should comprise several convolutional layers with various kernel sizes and pooling layers. Additionally, the model needs to have fully interconnected layers for classification. To attain the greatest performance, experimentation should be used to determine the size and number of layers in the filters.

Model Training: Using the pre-processed dataset, the CNN model is trained in the third stage. A stochastic gradient descent approach should be used to train the model, and the learning rate should be modified as training progresses. It is important to keep an eye on the training process to make sure the model is not being overfit to the training set of data.

Model Evaluation: In the last step, the CNN model's performance is assessed using a validation dataset that wasn't used during training. The proposed approach's efficacy is evaluated using performance indicators including accuracy and precision. The model will also be evaluated on a different testing dataset to determine how well it generalizes. The outcomes will be contrasted with other cutting-edge methods for identifying plant diseases.

2.1 WorkFlow Diagram

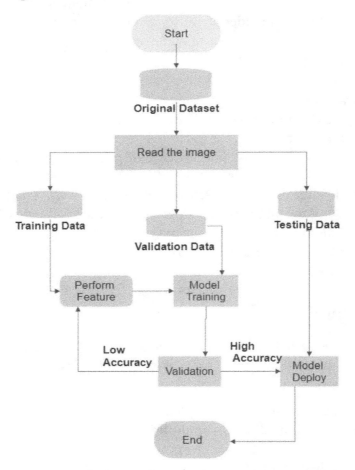

Fig. 27.1 The workflow diagram of the model describe in the thesis

2.2 CNN

A popular deep learning approach in computer vision applications is the CNN. It has evolved into a common method for image classification, object recognition, and other image-related tasks since it is particularly developed to identify patterns in images.

The picture or feature map is entered into the CNN's input layer, which then passes it through a sequence of convolutional, pooling, and activation layers to create the network's hidden layers. Through a sliding window called a kernel, the convolutional layers collect spatial characteristics from the input image. The generated feature maps' spatial dimensionality is decreased by the pooling layers. Additionally, the model's activation layers add nonlinearity.

The final classification or regression of the input image is often carried out by one or more fully linked layers in a CNN's output layer. The precise task at hand and the number of classes that need to be classified determine how many nodes are in the output layer. In conclusion, the hidden layers of a CNN perform convolution, pooling, and activation operations to extract features from the input after receiving the input image from the input layer. The ultimate classification or regression result is then generated by the output layer.

Convolutional Layer

In order to extract local information from the input image or feature map, convolutional layers apply a set of filters. The feature map is produced by the filters as they move through the entire picture or feature map and compute a dot product between the filter and the local patch of the image.

Through training on a sizable collection of labelled photos, each filter develops the ability to recognise a particular feature, such as edges, textures, or forms. The result of a convolutional layer is often a collection of feature maps that are integrated by a rectified linear unit (ReLU), a non-linear activation function that adds non-linearity to the network.

Pooling Layer

The spatial dimensionality of the feature maps is reduced through the use of pooling layers, which also introduce some degree of invariance to minute changes in the input image. When pooling, a small area of the feature map is typically chosen, and a single output value is computed using some aggregation function, such as the maximum or average value within the region.

By doing this, the network's requirements for parameters and computations are reduced while still maintaining the key elements of the input. Although average pooling and max pooling can also be employed, they are less frequently used in CNNs.

Max-pooling Layer

For image identification and other computer vision applications, CNNs frequently use the pooling layer type known as max pooling. The input feature map is partitioned into rectangular, non-overlapping pooling regions in a max pooling layer. The highest value within each pooling zone is calculated and transmitted to the following layer.

As a result, the input feature map's spatial resolution is decreased while the most prominent features are kept. Max pooling offers some degree of translation invariance while simultaneously minimising the number of parameters and calculations needed by the network.

The strongest feature in each pooling zone is chosen by max pooling, which is less susceptible to minute changes in the input image than other pooling techniques like average pooling.

Assume, for instance, that we apply a max pooling layer with a pool size of 2 × 2 and a stride of 2 (i.e., non-overlapping pooling zones) to an input feature map of size 8 × 8. Each value in the 4 × 4 output feature map would be the highest value found in its corresponding 2 × 2 input feature map pooling region.

Dense layer

In a dense layer, also referred to as a completely linked layer, every neuron in the layer is coupled to every neuron in the layer below it. In other words, each neuron in a dense layer receives input from every neuron in the layer below it, and their input and output are totally coupled. In neural networks, dense layers are frequently employed for a variety of tasks, such as feature extraction, regression, and classification. The weights and biases associated with each neuron in the layer are discovered during the training process, and the output of a dense layer is computed as a weighted sum of the input.

In order to reduce the error between the network's projected output and the actual target values, the backpropagation method is used to modify the weights and biases of each neuron in the dense layer during training. This procedure is repeated over a number of epochs until the network's performance on the training set is adequate.

Dense layers' capacity to recognise intricate data patterns is one of their key features. They might, however, be vulnerable to overfitting if the network is overly complicated or lacks insufficient regularisation. Because of this, it is frequently required to utilise methods like dropout or weight regularisation to stop over fitting in dense networks.

2.3 Model Building

As we discussed earlier we developed a convolutional neural netork model using convolutional, pooling and dense layer. The model summary is given in the below table.

Table 27.1 Model summary

Layer	Output Shape	Parameter
module_wrapper	(32, 256, 256, 3)	0
module_wrapper_1	(32, 256, 256, 3)	0
conv2d	(32, 254, 254, 32)	896
max_pooling2d	(32, 127, 127, 32)	0
conv2d_1	(32, 125, 125, 64)	18496
max_pooling2d_1	(32, 62, 62, 64)	0
conv2d_2	(32, 60, 60, 64)	36928

Layer	Output Shape	Parameter
max_pooling2d_2	(32, 30, 30, 64)	0
conv2d_3	(32, 28, 28, 64)	36928
max_pooling2d_3	(32, 14, 14, 64)	0
conv2d_4	(32, 12, 12, 64)	36928
Total Parameters	183,747	
Trainable Parameters	183,747	
Non-Trainable Parameters	0	

2.4 Filter and Feature Map

A filter (also known as a kernel) in a CNN is a compact matrix of weights that is applied to the input image to extract a specific feature. The filter weights that produce the best performance on the given task are determined by the network during training.

A convolutional layer's output, known as a feature map, shows the existence of a specific feature in the input image. A series of filters, each of which detects a different feature, are applied to the input image to create each feature map. The strength of each detected feature at each position in the image is indicated by the values in the feature map.

Consider a convolutional layer with three filters, each of which is 3 × 3 in size. Each filter would be applied to the image to extract a certain characteristic, and the input image to the layer may be a 32 × 32 pixel RGB image. A set of three feature maps reflecting the presence of various features in the input image would be the convolutional layer's output.

During the model-building process, a CNN's filters and feature maps are frequently visualized to aid in understanding how the network is learning to extract features from the input data. The network design can be improved with the use of this data in order to increase performance for the given task.

3. Result

3.1 Discussion

As we can see from the result above, the model's accuracy is quite good-nearly around 98%. These typically occurs as a result of the tiny and precise dataset we employ. These are our model's advantages and disadvantages. If the machine memory is enough, we can handle huge datasets. But at this time, we are unable to afford that. Therefore, we must reduce our dataset and train our model only on three specific classifications. Those being healthy potato leaves, potato late blight, and early potato blight. We tested tomato and apple leaves independently, and the results were nearly identical in both cases. This is an outstanding conclusion from the validation and testing accuracy. The accuracy of validation and testing is quite good, and this is an excellent take-away from the thesis that will aid us in future development.

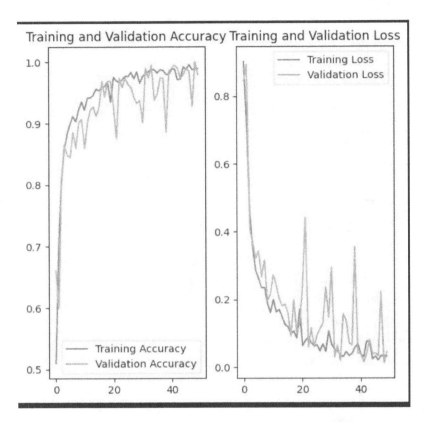

Fig. 27.2 The plot between validation and training accuracy and loss

4. Conclusion

We can see that the model has an overall accuracy of approximately 98%. It also performs satisfactorily in the test case circumstances. Because of this, we have stored the model and are attempting to use it on a platform that will enable us to create an app or website that can identify plant diseases. Even if the concept has some limits, we expect to get through them all through calm, steady advancement.

REFERENCES

1. [1] Akshitha M, Siddesh G M, S R Mani Sekhar, Parameshachari B D, "Paddy Crop Disease Detection using Deep Learning Techniques", *2022 IEEE 2nd Mysore Sub Section International Conference (MysuruCon)*, pp.1-6, 2022.
2. [2] Rohan Vora, Siddharth Nayak, Sarakshi Phate, Devanshi Shah, S.C. Shrawne, "Node Embeddings in Graph Convolutional Networks for Plant Disease Detection", *2022 4th International Conference on Advances in Computing, Communication Control and Networking (ICAC3N)*, pp.506-511, 2022.
3. [3] A.Sehgal, S.Mathur, "Plant Disease Classification Using SOFT COMPUTING Supervised Machine Learning", 3rd International Conference on Electronics, Communication, and Aerospace Technology (ICECA), Coimbatore, India, 12-14June 2019
4. [4] N.Petrillis, "Plant Disease Diagnosis With color Normalization", 8th International Conference on Modern Circuits and Systems Technologies(MOCAST), Thessaloniki, Greece,13-15 May 2019.
5. [5]GeeksforGeeks-https://www.geeksforgeeks.org/cnn-introduction-topooling-layer/

Smart Hospital and Healthcare: Hospital Management System

Taniya Baral*

Electronics and Communication Engineering, Institute of Technical Education and Research, Siksha 'O' Anusandhan University (Deemed to be University)

Anish Mohanty

Computer Science and Engineering, Institute of Technical Education and Research, Siksha 'O' Anusandhan University (Deemed to be University)

Abstract The smart hospital system operation starts from the patient using their medical to schedule an appointment on our app. This step involves doctor's unique ID and patient's unique ID which can be stored and separated by the hybrid cloud system. Hybrid cloud system will help us to store the patient's information in the private cloud and contain BFF and ERF module in the private cloud but ERF files in the form of microservice are transferred to the public cloud containing mobile BFF for the smooth flow of transmission between private and the public cloud. The doctors can store patient's history in the private cloud and can't get accessed by anyone else meanwhile the patient can also upload the history of their previous and present report in it through their ID. Doctor can easily access for further reference this will help in easy diagnosis. NFC tag will help to send any patient to the doctor of other department for diagnosis and can help the doctor to get the detail of the patient through one tap if this sticker is stuck on the wall, clipboard near their bed. Without disturbing the patient, the Doctor will get to know the history of the patient that will make it very hassle free and saves a lot of time. NFC tag will be coded earlier. ML and AI has a major role to play if the patient needs to be hospitalised or wants to book appointment, he/she can check for the availability of the bed and its price on our app online. AI and ML can obtain the calendar or schedule date from the patient's phone and crosscheck with Doctor's schedule. Once this gets done it would be able to identify a slot which would be comfortable for both patient and in sync with the availability of the Doctor. The patient can upload their prescription in our app under medical inventory section through their unique ID which will also link the physician. AI will check if the medicine prescribed by the physician is available or in the dispenser, if not it will immediately send a notification to prescribe alternate medicine. AI will also inform the Inventory management team about

---

*Corresponding author: taniyabaral2002@gmail.com

DOI: 10.1201/9781003489443-28

its shortage. This would help many old people who must go to the hospital alone and don't have to run to the market for medicine. To make the record keeping procedure easy and simple both for the patient and Doctor. To make the appointment and medicine buying procedure easier. This would help in the early detection of the disease and an affordable and accurate treatment can be given to the patient. In our system we help the patients to use their ID to schedule appointment and store their records in safe and secure way. The Doctor's can store patient's information too and even their research work through their ID. This would be done through hybrid cloud server which contains the public cloud with AI and ML based algorithm and hospital database contained in private cloud. NFC tags can also help in storing information of the patient. AI and ML also helps in inventory management. I would like to thank the department for giving us this opportunity to do a research work because this helped us a lot to explore about technology and gave us the idea of finding solution to real life problem.

Keywords   Hybrid cloud, ML, AI, NFC tag

1. Introduction

The healthcare sector has undergone a technological transformation in recent years, opening new possibilities for patient diagnosis and treatment that is quicker and more precise. As an illustration of how technology can be utilised to enhance healthcare services, consider the introduction of smart hospital systems. Patients can book appointments, store and manage their medical information, and even purchase medication through an app as part of this system, which seeks to offer hassle-free and economical healthcare services.

The hybrid cloud server that powers the smart hospital system enables secure patient data storage and smooth data transfer between private and public clouds. The hospital database, which is maintained in the private cloud and uses a unique ID for every patient, contains the patient's medical history, diagnosis, and therapy. Contrarily, the public cloud leverages AI and ML-based algorithms to support inventory management and appointment scheduling, ensuring that patients receive prompt and correct care.

In addition, the smart hospital system uses NFC tags to make it easier for clinicians to store and access patient information. Doctors may quickly and readily access a patient's medical history and other pertinent information with the use of NFC tags, which helps to speed up diagnosis and increase diagnostic accuracy.

In this paper, we'll go into detail on the implementation and advantages of the smart hospital system, emphasising the contribution that hybrid cloud technology, AI, ML, and NFC tags have made to raising the standard and effectiveness of healthcare. This system not only offers consumers a practical and economical option, but it also facilitates doctors' workflow, ensuring that patients get the finest care possible.

2. Proposed Model

The smart hospital system operation starts from the patient using their medical to schedule an appointment on our app. This step involves doctor's unique ID and patient's unique ID which can be stored and separated by the hybrid cloud system. Hybrid cloud system will help us to store the patient's information in the private cloud and contain BFF and ERF module in the private cloud but ERF files in the form of microservice are transferred to the public cloud containing mobile BFF for the smooth flow of transmission between private and the public cloud. The doctors can store patient's history in the private cloud and can't get accessed by anyone else meanwhile the patient can also upload the history of their previous and present report in it through their ID, this feature will help the patients a lot because they won't have to carry their report every time, there is no tension of forgetting the report at home or getting misplaced. Doctor can easily access for further reference this will help in easy diagnosis. Further doctor can also share their research work (not accessible by the patient) they will help them to form a community and help them know more about their achievements which will help them to send any patient to the doctor of other department for diagnosis.

2.1 NFC Tag

NFC tags can be used to store the patients ID no. which can help the doctor to get the detail of the patient through one tap if this sticker is stuck on the wall, clipboard near their bed. Without disturbing the patient, the Doctor will get to know the history of the patient that will make it very hassle free and saves a lot of time. NFC tag will be coded earlier

2.2 Appointment and Medical Inventory

ML and AI has a major role to play if the patient needs to be hospitalised or wants to book appointment, he/she can check for the availability of the bed and its price on our app online. AI and ML can obtain the calendar or schedule date from the patient's phone and crosscheck with Doctor's schedule. Once this gets done it would be able to identify a slot which would be comfortable for both patient and in sync with the availability of the Doctor. The patient can upload their prescription in our app under medical inventory section through their unique ID which will also link the physician. AI will check if the medicine prescribed by the physician is available or in the dispenser, if not it will immediately send a notification to prescribe alternate medicine. AI will also inform the Inventory management team about its shortage. This would help many old people who must go to the hospital alone and don't have to run to the market for medicine.

2.3 Detection and Treatment of Disease

AI will assist in detecting the problem in human body by enabling faster and more accurate analysis of the medical data. Detection and treatment of diseases are two of the most important applications of AI and machine learning in the healthcare industry. With the increasing amount of medical data being generated every day, machine learning algorithms can be used to analyse and interpret this data to detect diseases early and develop effective treatment plans.

One of the key areas where AI is being used for disease detection is in medical imaging. Machine learning algorithms can analyse large amounts of medical images like X-rays, CT scans, and MRI scans to detect signs of disease that may be difficult for human healthcare professionals to detect. For example, deep learning algorithms can detect subtle changes in medical images that may indicate the presence of cancer, helping healthcare professionals make a more accurate diagnosis.

3. Working

3.1 Cloud Computing

One of the main benefits of using hybrid cloud technology in hospitals is the ability to store and backup vast amounts of patient data securely. Hospitals generate large amounts of data every day, such as patient records, medical images, and test results. The hybrid cloud can provide a scalable and secure infrastructure for storing and backing up this data, allowing hospitals to access and manage it more efficiently. In addition, the hybrid cloud can provide hospitals with a reliable disaster recovery solution that ensures critical patient data and healthcare services are always available. Community cloud computing can also be used to link hospital, laboratory and blood banks. Multi cloud can be used collect and store report from all over the world which the doctor can go through the history of other patients. We will use Joukuu, Mozy, and Box.com for data storage and back up service application.

3.2 NFC Healthcare System Model

Typically, during the registration period, many hospitals throughout the world employ a paper-based flow chart to collect patient information that is eventually passed on to numerous other people over the course of various shifts of time. Even while hospital staff members attempt to update the paperwork every time a patient arrives, because it is handwritten, it is not always accurate.

One of the most significant new technologies that offers a better solution to this is NFC because it can help automate processes and also enable precise tracking of patient identity. Each NFC tag or wristband has a Unique Identification Number (UIN) that may be programmed and is also password-protected.

At the time of registration at the hospital, each patient may receive an NFC smart tag or wristband. This can be used to keep track of a patient's information while they are in the hospital as well as to save some crucial information about them, such as their blood type, allergies, the tests they need to have done, the medications they need to take, etc. Without the patient experiencing any inconvenience, this information can be read from an NFC encoded device at any moment. NFC can be used to send and receive data from patient smart tags to the systems of healthcare professionals.

Each NFC technology primarily comprises of a reader and a low power smart tag (smart tag). A microprocessor and a tiny antenna make up this tiny tag. By transforming the radio waves reflected by the smart tag into digital information, the reader can identify the data in the smart tag by transmitting electromagnetic rays that are picked up by the antenna in the smart tag. This data can be kept on any device, although it is primarily kept in the hospital patient management system. We can also access databases from other hospitals online. These days, NFC smart tags known as pokes are available, allowing users to exchange data like business cards and other items. When the reader and smart tag are brought together, the token will display some sign that a reader is there. If the data transmission is successful, it will display a green light; otherwise, it will display a red light. In this approach, we can even prevent many trees from being cut down for paper because the patient can use it at each hospital he visits.

3.3 AI and Machine Learning

AI and machine learning (ML) are transforming the healthcare industry, and smart hospitals are leveraging these technologies to improve patient care and operational efficiency. One of the ways in which supervised learning algorithms like regression can be used is to predict bed availability. By training the learning algorithm on historical data, the hospital staff can optimize bed allocation and ensure that patients receive the care they need. Unsupervised learning algorithms like clustering can be used to classify patients based on prescription information and prioritize appointments based on urgency. This can help allocate appointment slots more efficiently, reducing wait times and improving patient experience. Narrow AI algorithms can also be used to optimize staff scheduling and resource allocation in the hospital.

AI-powered chatbots can help patients communicate with the hospital in multiple languages and run basic diagnostics. These chatbots are powered by natural language processing (NLP) and can help reduce the burden on hospital staff while improving patient experience.

AI can also help patients with speech problems by using advanced text-to-speech technology. This can help them communicate with healthcare professionals more effectively and ensure they receive the care they need. Computer vision algorithms can also be used to analyse medical images like X-rays and MRI scans more accurately, helping healthcare professionals diagnose conditions more quickly and accurately.

The hospital can use machine learning to collect data on rare diseases from different sources and store them in a centralized cloud. This can help healthcare professionals diagnose rare diseases more accurately and provide better care for their patients. Cognitive computing solutions can also help healthcare professionals analyse large volumes of data more accurately and quickly, helping them make more informed decisions about patient care and treatment.

Supervised learning algorithms can be used to prioritize emergency admissions based on the severity of the patient's condition. This can help ensure that patients receive the care they need as quickly as possible. Finally, combining image analysis with natural language understanding and text analysis can provide healthcare professionals with a wealth of information and knowledge to help them make the best diagnosis possible.

AI and machine learning have the potential to revolutionize the healthcare industry, and smart hospitals are leveraging these technologies to improve patient care and operational efficiency. From predicting bed availability to prioritizing emergency admissions, these technologies are helping healthcare professionals provide better care for their patient.

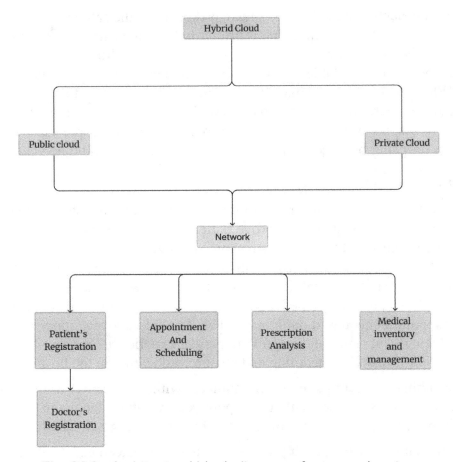

Fig. 28.1 Architectural block diagram of proposed system

3.4 Example

Tania, an 18-year-old girl, experienced swelling, pain, and discoloration in her left calf, along with a headache and fever. She used the hospital's app to schedule an appointment and register her symptoms and medical history. Using AI, the app linked Tania to an appropriate doctor and booked an appointment that was convenient for both. During the appointment, the doctor used AI and ML to analyse Tania's symptoms and narrow down the possible causes. The AI suggested that Tania may have Deep Vein Thrombosis (DVT), a serious condition that requires prompt treatment. To confirm the diagnosis, the doctor scheduled a series of tests using AI and ML to analyse Tania's medical images, lab results, and genetic information. Once DVT was confirmed, Tania was hospitalized, and an NFC tag was placed on her bed. The tag contained all the details of Tania's treatment, including the medicine given, tests run, and injections administered. The AI helped the hospital staff manage Tania's inventory of medicine and equipment, ensuring that she received the appropriate care. After Tania was discharged, she continued to use the hospital's app for flow-up appointments. The app stored all of her medical history and allowed the doctor to access it easily during each appointment. The AI helped the doctor prescribe the appropriate medicine and dosages based on Tania's medical history, improving her chances of a successful recovery. Overall, this scenario illustrates how technology can improve patient outcomes by streamlining hospital management, improving diagnosis and treatment, and providing personalized care. With the help of AI and ML, hospitals can become smarter, more efficient, and more effective at delivering healthcare services.

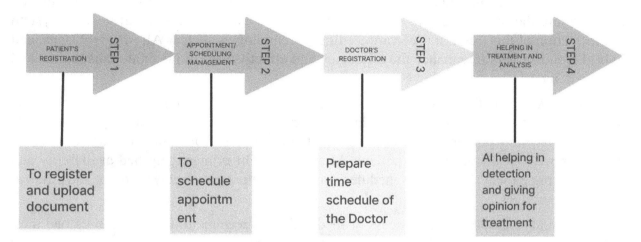

Fig. 28.2 Smart hospital flow diagram

4. Future Works

In a world of digital change, companies are looking to AI to really help them shape the future of work. AI can and inform future results. It enables people to do higher value work and businesses to imagine new models in the medical and health sector. It can automate decisions, processes, experiences. The truth is many organizations can't start because 80% of their data is locked in silos and not business ready. So, how do you turn our expectations into results? Through a (related to giving rules and instructions) set of steps we call the ladder AI, and it starts with (updating with the latest stuff) all your data on a single (raised, flat supporting surface) that runs on any cloud. Then on the ladder itself, there are four steps First, collect the data to make it simple and (easy to get to, use, or understand) Really think about the models you need to train

Second, organize your data to create a business-ready (information-giving's numbers) foundation for those AI models. Third, carefully study our data both for trust and clearness/open honesty Because there's no use in applying and scaling AI if we can't explain the result, detect bias or prove its (quality of being very close to the truth or true number) Fourth, once we can really trust our data and the AI that we end out and use Then we can (understand/make real/achieve) its full value Inside of the apps and processes that control our everyday work. In other words, the last step is to begin to operationalize AI throughout our whole business We help thousands of businesses/projects put AI to work by unlocking the value of their data and this AI and multi-cloud world. By delivering the right set of skills for their people and by building trust and clearness/open honesty into AI. That's the ladder to AI in a nutshell. Let's start climbing.

5. Conclusion

The emergence of the COVID-19 pandemic has brought significant changes in the healthcare industry. The pandemic has accelerated the adoption of digital technology in healthcare, transforming the way healthcare services are delivered. The proposed smart hospital system is a reflection of this trend. It uses various digital technologies such as AI, ML, NFC tags, and hybrid cloud servers to enhance the efficiency and effectiveness of hospital management and service delivery.

One of the key benefits of this system is its ability to simplify the process of scheduling appointments and managing patient information. Patients can use the app to schedule appointments, and doctors can retrieve patient information from other departments using NFC tags without disturbing the patient. AI and ML help in managing bed availability and crosschecking patient and doctor schedules, ensuring that patients receive prompt and efficient healthcare services.

The system also simplifies the process of medication management. Patients can upload their prescriptions on the app, and AI checks for medicine availability and notifies the inventory management team of any shortages. This eliminates the need for patients to run to the market for medicine, saving them time and effort.

Furthermore, the hybrid cloud server ensures the security and privacy of patient and doctor information. The system provides an easy and simple record-keeping procedure for both patients and doctors, enabling early detection of diseases and affordable and accurate treatment.

In conclusion, the use of technology in healthcare is a significant step towards solving real-life problems. The proposed smart hospital system is an example of this trend, providing patients with convenient, efficient, and hassle-free healthcare services. The system also helps hospital management to improve their services, enabling them to deliver quality healthcare services to patients. Overall, this system can revolutionize the healthcare industry and improve the quality of life of patients.

REFERENCE

1. Atluri Venkata Gopi Krishna, Cheerla Sreevardhan, S. Karun, S. Pranava Kumar. "NFC - based Hospital Real - time Patient Management System". International Journal of Engineering Trends and Technology (IJETT). V4(4):626-629 Apr 2013. ISSN:2231-5381. www.ijettjournal.org. published by seventh sense research group
2. www.nfc - forum.org
3. Devadass, Lingkiswaran & Sekaran, Sugalia & Thinakaran, Rajermani. (2017). CLOUD COMPUTING IN HEALTHCARE. International Journal of Students' Research in Technology & Management. 5. 25-31. 10.18510/ijsrtm.2017.516.
4. J. Naveen Ananda Kumar and S. Suresh, "A Proposal of smart hospital management using hybrid Cloud, IoT, ML, and AI," 2019 International Conference on Communication and Electronics Systems (ICCES), Coimbatore, India, 2019, pp. 1082-1085, doi: 10.1109/ICCES45898.2019.9002098.

5. N. Mahmood, A. Shah, A. Waqas, Z. Bhatti, A. Abubakar and H. A. M. Malik, "RFID based smart hospital management system: A conceptual framework," The 5th International Conference on Information and Communication Technology for The Muslim World (ICT4M), Kuching, Malaysia, 2014, pp. 1-6, doi: 10.1109/ICT4M.2014.7020594.
6. https://www.intel.com/content/www/us/en/healthcare-it/smart-hospital.html
7. Ngai, E.W.T., Poon, J.K.L., Suk, F.F.C. et al. Design of an RFID-based Healthcare Management System using an Information System Design Theory. Inf Syst Front 11, 405–417 (2009). https://doi.org/10.1007/s10796-009-9154-3
8. M. Saputra, I. Hermawan, W. Puspitasari and A. Almaarif, "How to Integrate Enterprise Asset Management System for Smart Hospital: a Case Study," 2020 International Conference on ICT for Smart Society (ICISS), Bandung, Indonesia, 2020, pp. 1-8, doi: 10.1109/ICISS50791.2020.9307535.

From Pollution to Power: Harnessing the Potential of Non-Biodegradable Wastes for Electricity Generation and Wireless Air Pollution Monitoring

Aditya Kumar Lenka, Amlan Adarsh and Biswaranjan Swain*
Siksha 'O' Anusandhan Deemed to be University, Bhubaneswar, India

Abstract The Non-biodegradable wastes pose a significant environmental crisis, causing soil and air pollution. Addressing this issue while fulfilling the fundamental human needs of safe drinking water and electricity is crucial. This study presents an innovative approach to tackle these challenges through the utilization of non-biodegradable wastes for electricity generation and wireless air pollution monitoring.

The primary objective of the project is to harness the potential of plastic and non-biodegradable wastes to produce efficient electricity, thereby mitigating soil and air pollution. Additionally, the project aims to address the problem of chemically infected saline or river water, rendering it unsuitable for human consumption. To achieve this, collected saline water from the sea or freshwater from the river undergoes a comprehensive process of dissolution of harmful particles and gases, cooling, filtration, and pH balancing to obtain drinkable water. Moreover, the project demonstrates the integration of wireless air pollution monitoring using ESP32, ThingSpeak and Blynq platform, enabling real-time data collection and analysis. This IoT-based approach facilitates remote monitoring of air pollution levels, providing valuable insights for decision-making.

This research showcases a sustainable solution to environmental problems, encompassing electricity generation from non-biodegradable wastes and wireless air pollution monitoring. The innovative utilization of waste materials, simultaneous treatment of chemically infected water, and real-time data monitoring make this project valuable for communities worldwide. The potential cost savings in environmental maintenance further emphasize the importance and impact of this research.

Keywords Non-biodegradable wastes, Plastic waste, Electricity generation, Wireless air pollution monitoring

*Corresponding author: biswaranjanswain@soa.ac.in

DOI: 10.1201/9781003489443-29

1. Introduction

The environmental crisis caused by non-biodegradable wastes has become a pressing global concern. The rapid accumulation of plastic and other non-biodegradable materials has resulted in severe soil and air pollution, posing significant challenges to environmental sustainability and human well-being. In response to this crisis, our study aims to address these environmental issues while simultaneously fulfilling two fundamental human needs: safe drinking water and electricity.

The primary objective of our study is to harness the untapped potential of nonbiodegradable wastes for electricity generation, thereby reducing the reliance on traditional energy sources that contribute to pollution and climate change. We envision a future where these wastes, which are often viewed as a burden, can be transformed into a valuable resource for sustainable energy production.

In addition to electricity generation, our study also integrates wireless air pollution monitoring to tackle the problem of air pollution caused by non-biodegradable wastes. We utilize the ESP32 microcontroller along with the ThingSpeak and Blynq IoT platforms to enable real-time data collection and analysis of air pollution levels. This innovative approach facilitates remote monitoring and provides valuable insights for informed decision-making and environmental management.

The significance of our study lies in its holistic approach, combining the utilization of non-biodegradable wastes for electricity generation and the integration of wireless air pollution monitoring. By adopting this approach, we aim to contribute to the development of sustainable solutions for waste management, environmental preservation, and the provision of essential services such as electricity and safe drinking water.

Through this research, we aspire to demonstrate the feasibility and effectiveness of this innovative approach, highlighting its potential for scalability and widespread adoption. By transforming non-biodegradable wastes into a valuable resource and implementing wireless air pollution monitoring, we can mitigate the environmental impact of these wastes and work towards a cleaner and more sustainable future.

In the subsequent sections of this paper, we will delve into the literature review, methodology, results and discussion, showcasing the outcomes of our study and their implications for addressing the environmental crisis caused by non-biodegradable wastes, while simultaneously fulfilling the needs for safe drinking water and electricity.

2. Literature Review

The following studies have contributed valuable insights into wireless air pollution monitoring systems based on IoT technology, as well as the conversion of plastic waste to energy:

Lu et al. propose a wireless air pollution monitoring system based on IoT technology. The system utilizes various sensors to collect air quality data, which is then transmitted to a central server for analysis and visualization. The study demonstrates the effectiveness of IoT-based solutions in real-time air pollution monitoring [1].

Li et al. present a comprehensive review of the conversion of plastic waste to energy. The study discusses various technologies such as pyrolysis, gasification, and thermal depolymerization, highlighting their advantages, challenges, and future prospects. The review provides a foundation for understanding the potential of plastic waste as a valuable energy resource [2].

Chand and Bhaskar propose an IoT-based air pollution monitoring system that incorporates sensors, wireless communication, and data analytics. The system collects real-time air quality data and provides insights into pollution levels, enabling effective environmental management. The study highlights the importance of IoT technologies in monitoring and addressing air pollution [3].

Ahmed et al. propose a wireless air pollution monitoring system based on IoT, incorporating sensors, microcontrollers, and cloud-based data analysis. The study demonstrates the feasibility and effectiveness of using IoT technology for real-time air quality monitoring, emphasizing the potential for enhanced environmental monitoring and management [4].

Sharma et al. present an IoT-based approach for air pollution monitoring, utilizing sensors, microcontrollers, and wireless communication. The study focuses on the realtime collection of air quality data and highlights the potential of IoT technologies in addressing air pollution challenges [5].

Ali et al. propose a smart air quality monitoring system based on IoT, incorporating sensors, microcontrollers, and cloud-based data analysis. The study demonstrates the feasibility and effectiveness of using IoT technology for air pollution monitoring and management [6].

Kaur et al. present a real-time air pollution monitoring system based on IoT technology. The study focuses on the integration of sensors, microcontrollers, and wireless communication to collect and analyze air quality data. The research emphasizes the significance of real-time monitoring in addressing air pollution concerns [7].

Zhu et al. propose an air quality monitoring system based on IoT technology, incorporating sensors, microcontrollers, and cloud-based data processing. The study highlights the effectiveness of IoT-based solutions in collecting and analyzing air quality data for environmental monitoring purposes [8].

Gupta et al. present an IoT-based air pollution monitoring system utilizing the ESP32 microcontroller. The study focuses on the integration of sensors, data processing algorithms, and wireless communication to monitor and analyze air pollution levels. The research emphasizes the potential of IoT technology in addressing air pollution challenges [9].

Paurush et al. propose a novel approach for air pollution monitoring using IoT technology. The study presents a system architecture that integrates sensors, microcontrollers, and cloud-based data analysis for real-time monitoring and management of air pollution. The research highlights the potential of IoT-based solutions in addressing air pollution concerns [10].

These studies contribute to the understanding of wireless air pollution monitoring systems based on IoT technology, as well as the conversion of plastic waste to energy. Our research builds upon these works by presenting an innovative approach that utilizes non-biodegradable wastes for electricity generation while integrating wireless air pollution monitoring using ESP32, ThingSpeak, and Blynq IoT platform.

3. Materials and Methods

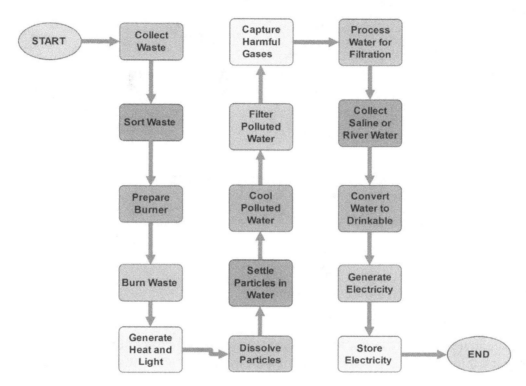

Fig. 29.1 Flowchart of the proposed system for electricity generation and wireless air pollution monitoring

The flowchart of the proposed system in Fig. 29.1, illustrates the sequential steps and processes involved in the proposed system, which combines the generation of electricity from non-biodegradable wastes and wireless air pollution monitoring. The flowchart provides a visual representation of the interconnected components, data flow, and decision points within the system. Each step in the flowchart corresponds to a specific action or process, including the collection and processing of nonbiodegradable wastes, the combustion of plastic waste for electricity generation, the integration of air pollution sensors and wireless communication, data transmission to the cloud-based platform, and the monitoring and analysis of air pollution levels. The flowchart serves as a guide for understanding the overall operation and functionality of the proposed system, highlighting the innovative approach towards addressing environmental challenges and achieving sustainable energy generation.

3.1 Experimental Setup

To investigate the feasibility of harnessing the potential of non-biodegradable wastes for electricity generation and integrating wireless air pollution monitoring, an experimental setup was established as illustrated in Fig. 29.2.

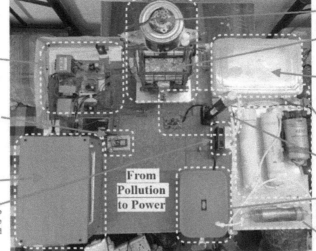

Fig. 29.2 Experimental setup for harnessing non-biodegradable wastes for electricity generation and air pollution monitoring using ESP32

The setup consisted of the following components:

Non-Biodegradable Wastes Collection and Processing

Plastic waste materials were collected from local waste management facilities and sorted based on their types. The collected plastics were then cleaned, shredded, and sorted further to remove any contaminants or non-plastic materials.

Electricity Generation System

The electricity generation system utilized the burned plastic waste as a fuel source. The shredded plastic waste was fed into a burner, where it was subjected to controlled combustion. The heat and light generated from the burning process were harnessed to drive a generator, producing electrical energy.

Air Pollution Monitoring System

To monitor air pollution levels, an IoT-based system was implemented using an ESP32 microcontroller, ThingSpeak, and Blynq IoT platform. The ESP32 microcontroller was integrated with an MQ135 air pollution sensor to measure various pollutants such as particulate matter (PM), volatile organic compounds (VOCs), and carbon monoxide (CO). The MQ135 sensor provided accurate and real-time data on air quality parameters.

4. Results and Discussion

4.1 Data Collection and Analysis

Throughout the experimental process, data on electricity generation efficiency, air pollution levels, and water purification efficiency were systematically collected for analysis. The electricity generation efficiency was evaluated by measuring the voltage and current output from the generator. This data provided insights into the performance and effectiveness of using non-biodegradable wastes for electricity generation.

Simultaneously, the air pollution sensors connected to the ESP32 microcontroller, including the MQ135 sensor, recorded data on PM, VOCs, and CO levels as depicted in Fig. 29.3. The MQ135 sensor played a crucial role in capturing the air pollution data accurately and reliably. These data points were vital in understanding the environmental impact of non-biodegradable waste combustion and assessing the effectiveness of the air pollution control measures implemented. The air pollution data collected from the MQ135 sensor was wirelessly transmitted to ThingSpeak, a cloud-based IoT platform. ThingSpeak served as a centralized database for storing the air pollution data and provided real-time monitoring and analysis capabilities. Through Thing-Speak, the air pollution data could be accessed remotely and visualized using graphs, charts, and other visualization tools.

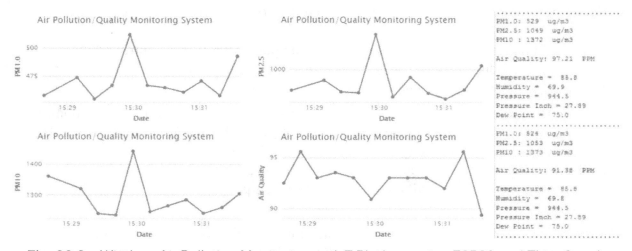

Fig. 29.3 Wireless Air Pollution Monitoring on IoT Platform using ESP32 and ThingSpeak

Furthermore, water purification efficiency was assessed by measuring key water quality parameters, including pH level, dissolved solids, and chemical contaminants, both before and after the purification process. These measurements provided valuable insights into the effectiveness of the water purification system integrated into the project.

To facilitate easy access and visualization of the air pollution data, Blynq was utilized as an interface. Blynq offered a user-friendly mobile application that allowed users to monitor the air pollution levels in real-time, view historical data, and receive notifications or alerts when pollution levels exceeded predefined thresholds. The integration of Blynq enhanced the usability and accessibility of the air pollution monitoring system illustrated in Fig. 29.4.

5. Conclusion

This paper presented a comprehensive approach for addressing the environmental crisis caused by non-biodegradable wastes through the generation of electricity and wireless air pollution monitoring. The primary objective of the project was to harness the potential of non-biodegradable wastes for electricity generation, thereby mitigating soil and air pollution

272 Prospects of Science, Technology and Applications

Fig. 29.4 Integration of Blynq as an interface for air pollution data visualization and access

associated with their improper disposal. By burning plastic waste in a controlled manner, heat and light were produced, which were then utilized to generate electrical energy using solar panels and thermodynamic plates.

Moreover, the project incorporated an IoT-based system for wireless air pollution monitoring. The integration of an ESP32 microcontroller, MQ135 air pollution sensor, and the use of ThingSpeak and Blynq IoT platforms facilitated real-time monitoring, analysis, and visualization of air pollution data. This enabled users to monitor air pollution levels, access historical data, and receive notifications or alerts when pollution levels exceeded predefined thresholds.

The experimental results demonstrated the feasibility and effectiveness of the proposed system. The electricity generation efficiency from non-biodegradable wastes was evaluated, and the air pollution levels were monitored in real-time. The system successfully demonstrated its potential in utilizing non-biodegradable wastes for electricity generation while simultaneously addressing air pollution concerns.

The project's innovative approach of integrating wireless air pollution monitoring using ESP32, ThingSpeak, and Blynq IoT platform adds significant value to environmental management efforts. It allows for effective monitoring and control of air pollution levels, enabling timely interventions and decision-making to mitigate the adverse effects of pollution on human health and the environment.

By providing a sustainable solution for electricity generation and real-time air pollution monitoring, this project contributes to environmental preservation, human wellbeing, and resource optimization. Furthermore, it offers potential cost savings by eliminating the need for costly disposal processes of non-biodegradable wastes.

REFERENCES

1. Lu, X., Zhang, Y., Zhang, L., Xu, X., & Wang, J. (2020). Wireless Air Pollution Monitoring System Based on IoT Technology. Journal of Physics: Conference Series, 1461, 012053.

2. Li, Y., Ma, F., & Zhang, X. (2019). Conversion of plastic waste to energy: A state-of-theart review and future perspectives. Waste Management, 92, 407–422.
3. Chand, A., & Bhaskar, V. V. (2018). IoT Based Air Pollution Monitoring System. In 2018 9th International Conference on Computing, Communication and Networking Technologies (ICCCNT) (pp. 1–5). IEEE.
4. Ahmed, A., Siddique, A. M., Wali, S., & Ansari, A. Q. (2019). Wireless Air Pollution Monitoring System Based on IoT. In 2019 10th International Conference on Computing, Communication and Networking Technologies (ICCCNT) (pp. 1–6). IEEE.
5. Sharma, A., Verma, N. K., & Gupta, S. K. (2017). Air Pollution Monitoring Using IoT Based Approach. In 2017 International Conference on Computing, Communication and Automation (ICCCA) (pp. 1–6). IEEE.
6. Ali, A., Imran, M., Javaid, N., & Ahmad, I. (2019). Smart Air Quality Monitoring System Based on IoT. In 2019 4th International Conference on Computing, Communication and Automation (ICCCA) (pp. 1–5). IEEE.
7. Kaur, A., Bhatia, R., Sood, A., & Kumar, M. (2019). Real-Time Air Pollution Monitoring Using IoT. In 2019 IEEE 7th International Conference on Reliability, Infocom Technologies and Optimization (Trends and Future Directions) (ICRITO) (pp. 586–591). IEEE.
8. Zhu, T., Wang, Q., Jin, X., Zhang, Y., & Jin, H. (2018). Air Quality Monitoring System Based on IoT. In 2018 5th International Conference on Systems and Informatics (ICSAI) (pp. 1050–1055). IEEE.
9. Gupta, G., Kumar, A., & Mishra, S. (2020). IoT-Based Air Pollution Monitoring System Using ESP32. In Proceedings of International Conference on Electrical, Communication, and Computing (ICECC) (pp. 1–5).
10. Paurush, A. K., Singh, S. K., & Kumar, S. (2019). A Novel Approach for Air Pollution Monitoring Using IoT. In 2019 4th International Conference on Internet of Things: Smart Innovation and Usages (IoT-SIU) (pp. 1–5). IEEE.

# A Formulation for Maximizing Solar Irradiance Based on Adjustment of Optimum Inclination Angle

Bibekananda Jena[1]

Department of Electrical and Electronics Engineering, ITER,
Siksha 'O' Anusandhan Deemed to be University, Bhubaneswar, India

Sonali Goel[2], **Renu Sharma**[3]

Department of Electrical Engineering, ITER,
Siksha 'O' Anusandhan Deemed to be University, Bhubaneswar, India

Abstract One of the most important variables affecting the PV system's ability to produce power is the tilt angle. This research provides a method for maximizing photovoltaic (PV) panel production throughout the year by determining the most appropriate tilt angle. The ideal tilt angles and orientation for the solar photovoltaic system on a daily, monthly, seasonal, and annual basis are identified for the smart city of Bhubaneswar, India, using a mathematical approach in MATLAB programming. The study is performed for a real-time 11 kWp solar photovoltaic system installed on the rooftop of a constituent institute in eastern India. Furthermore, the ideal tilt angle significantly influences overall power generation, which is investigated in this study. The results show that altering the tilt angle 12 times a year could increase PV panel power output by 4.5 percent.

Keywords Solar photovoltaic system, Tilt angle, MATLAB programming

1. Introduction

Due to the increased demand for energy in the residential and industrial sectors, conventional generating systems are operating at or over their environmental pollution increases impact on the environment. The world's need for traditional energy sources like coal might be reduced by solar energy, which might be the best way to reduce the pollutants that contribute to the greenhouse effect in the environment. The world's energy dilemma may be solved by solar

---

[1]bibekanandajena@soa.ac.in, [2]sonali19881@gmail.com, [3]renusharma@soa.ac.in

DOI: 10.1201/9781003489443-30

energy, which is widely available and cost-free on the globe's surface. Many researchers have proposed various methods for calculating optimal tilt inclination (β_{opt}) [1]. To maximize its performance, a PV system's tilt angle can be changed on an hourly basis, monthly, seasonally, and annually. The value of the optimal tilt angle (β_{opt}) has been determined to be dependent on the latitude (ϕ) and azimuth (β) [3, 4].

According to Ahmed and Tiwari, they proposed the optimal tilt angles estimation for India where the ideal annual tilt angle has been determined to be nearly equal to the location's latitude. The author suggested that the ideal tilt angle should be (ϕ-16°) in summer (ϕ+19°) in winter [2]. The optimum tilt angle for various countries has been investigated in [5-8].

There are various techniques for improving the amount of energy supplied by solar panels. When increasing the amount of energy generated, the PV panel's orientation is crucial; if they are fixed, they must face south and be inclined at an appropriate angle. Many programs exist that help in estimating this angle [9]. Sun-tracking technology is used in photovoltaic panels to trace the sun's movement across the sky. Sun tracker systems come in a variety of shapes and sizes. These can use a single axis, two equatorial or azimuth axes, and sensors, to orient themselves [10-12].

This study uses the sun's position with respect to the Earth to calculate the ideal tilt angle for the Solar panel. By employing a mathematical approach in MATLAB programming, the ideal tilt for Bhubaneswar, India, is found to vary on a daily, monthly, and annual basis. Throughout the year, keeping the panel tilted at the right degree is crucial to ensure optimal energy output.

2. Methodology

This study consist of a 11.2 kWp grid-connected solar photovoltaic system mounted on the roof of a constituent institute in ITER,SOADU, Bhubaneswar. The system consists of 40 solar photovoltaic panels built with a total surface area of 77.6 m². The arrays are connected in a series-parallel combination.

The modules are fixed with a tilt angle 22.5°, pointing south at an azimuth angle of 0°, with no shading effect. The basic guideline for solar panel placement in India, like in the northern hemisphere, is that solar panels should face true south. The complete PV system is supported by concrete pillars and installed on a metal frame. The detailed specification of the PV module is mentioned in Table 30.1.

Table 30.1 Specification of PV module at STC (1000W/m² and 25°C)

Parameters	Rating
Rated Peak Power (P_{max})	280 W
Open Circuit Voltage (V_{OC})	43 V
Short Circuit Current (I_{SC})	8.68 A
Voltage at MPP (V_{MP})	35 V
Current at MPP (I_{MP})	8 A
Number of series connected PV modules	20
Number of parallel connected strings	2

The dimension representation of the 11.2 kW roof top PV system is shown in Fig. 30.1.

Fig. 30.1 Dimension representation of the 11.2 kW roof top PV system

3. Results and discussions

The PV system's performance is primarily influenced by temperature and solar radiation. PV power production decreases with reduced irradiance and increases with temperatures beyond 25°C. Using MATLAB, graphs were produced for each month. Adjusting various tilt angles entails:

- Tilt angle adjustment on a daily basis
- Tilt angle adjustment monthly
- Tilt angle adjustment according to the season
- Tilt angle adjustment annually

3.1 Tilt angle adjustment on a daily basis

Daily adjusting of optimal tilt can be done by adjusting the solar panel by considering the sun's position. However, this adjustment can be a challenging option as this option are not economical and costs a lot. Therefore, seasonal adjustment is much simpler as the daily adjustment is not always possible. Figure 30.2 depicts the daily irradiance of each month.

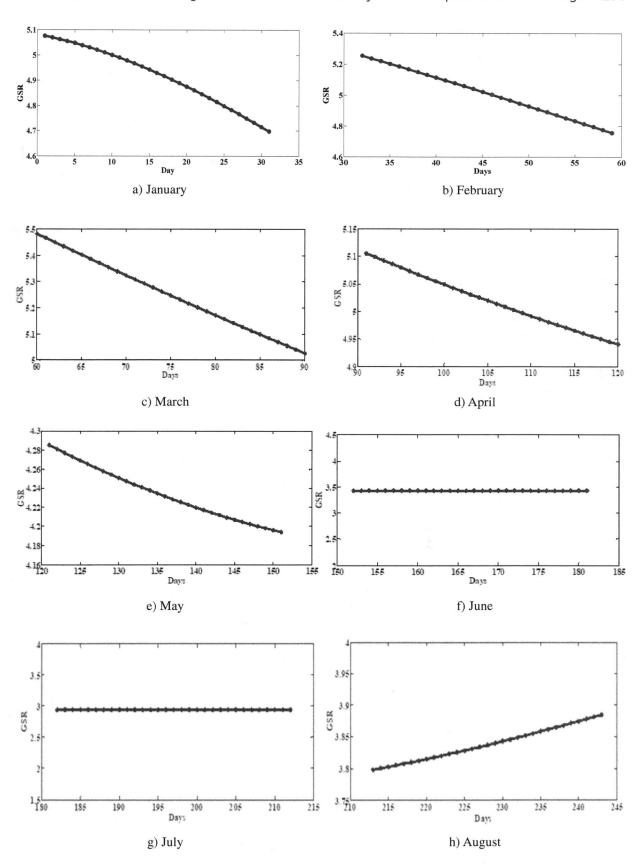

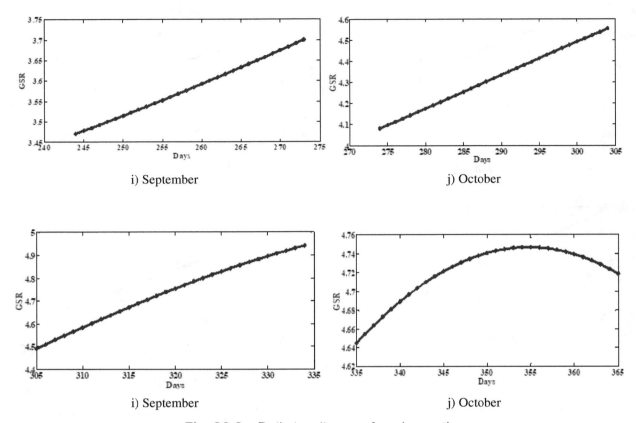

i) September j) October

i) September j) October

Fig. 30.2 Daily irradiance of each month

3.2 Tilt Angle Adjustment Every Month

For angles between 0^0 and 45^0, the ideal tilt variations for various months of the year were calculated. The relationship between the ideal tilt and the monthly average irradiance on a tilted plane is shown in Fig. 30.3.

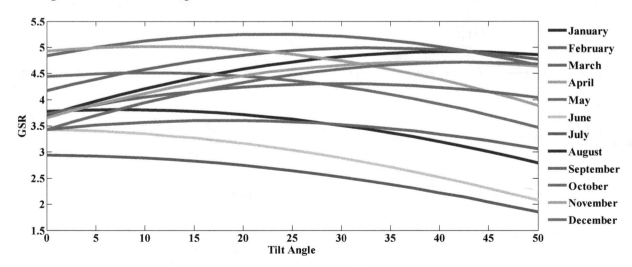

Fig. 30.3 Monthly adjustment of tilt for different months of the year

3.3 Tilt Angle Adjustment According to Season

To avoid the situation of changing the tilt angle every month and reduce the complicacy and operational cost, the tilt angle adjustment can be done seasonally. The tilt angle for different seasons such as summer, winter and equinox has been computed.

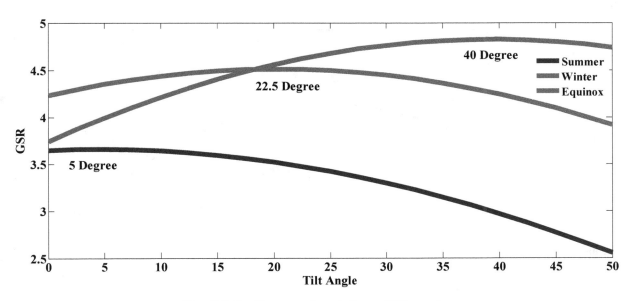

Fig. 30.4 Seasonal variation of tilt angle

Figure 30.4 shows that during the summer season i.e. for May, June, July, and August, when β is kept at 5^0, the solar panel receives a maximum GSR of 4.51 kWh/m^2. So, the panel can be kept at an optimum tilt of 50 for these four months. During the winter season i.e., for November, December, January, and February, it has been found that the solar panel received a maximum GSR of 4.92 kWh/m^2 when tilted at an angle of 40^0. So, the panel can be kept at an optimum tilt of 40^0 for these four months. During equinox, i.e., in March, April, September, and October, it has been found that the solar panel received a maximum GSR of 5.25 kWh/m^2 when tilted at an angle of 22.5^0. So, the panel can be kept at an optimum tilt of 22.50 for these four months.

The simulation results obtained from MATLAB are presented in Table 30.2.

Table 30.2 Summary of seasonal variation of optimum tilt angle from MATLAB Simulation

Season	Summer				Winter				Equinox			
Month	May	Jun	Jul	Aug	Nov	Dec	Jan	Feb	Mar	Apr	Sept	Oct
Degree	5	3	1	7	40	44	40	33	22.5	11	19	29
GSR (kWh/m^2)	4.51	3.43	2.94	3.81	4.72	4.05	4.92	4.89	5.25	5.02	3.61	4.30
Mean of GSR Gain (kWh/m^2)	3.67				4.64				4.54			

3.4 Tilt Angle Adjustment Annually

To reduce the complexity and operational cost of the PV system, the system can be mounted at a fixed tilt angle throughout the year at an angle of 22.5^0 to receive a maximum mean GSR of $4.65 kW/m^2$ on a tilted plane. Fig. 30.5 indicates the plot between tilt angle and GSR on tilt angle using simulated data.

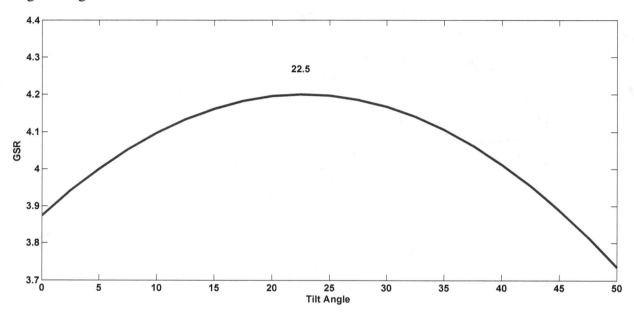

Fig. 30.5 Annual adjustment of tilt angle

4. Conclusion

PV arrays must be positioned with the proper tilt to receive the maximum irradiance. For a prototype site installation of an 11.2 kWp solar photovoltaic system at SOA, Bhubaneswar, India, the ideal tilt angle was determined, along with the daily, monthly, seasonal, and yearly optimal tilts for this site. Bhubaneswar's yearly optimum tilt angle is almost equal to the location's latitude of 22.5^0. Fixed tilt arrays don't provide as much energy as seasonal and monthly adjustments do. However, compared to the monthly adjustment, the seasonal adjustment is simpler. Seasonal adjustments may thus be applied in areas where monthly adjustments are impractical.

REFERENCES

1. Yakup, MohdAzmi bin HjMohd, and A. Q. Malik. "Optimum tilt angle and orientation for solar collector in Brunei Darussalam." *Renewable Energy* 24.2 (2001):223-234.
2. Jamil Ahmad, M., and G. N Tiwari. "Optimization of tilt angle for solar collector to receive maximum radiation." *The Open Renewable Energy Journal* 2.1 (2009).
3. Bari, Saiful. "Optimum orientation of domestic solar water heaters for the low latitude countries." *Energy Conversion and Management* 42.10 (2001):1205-1214.

4. Ahmad, M. Jamil, and G. N. Tiwari. "Optimum tilt angle for solar collectors used in India." *International Journal of Ambient Energy* 30.2 (2009): 73-78.
5. Bakirci, Kadir. "General models for optimum tilt angles of solar panels: Turkey case study." *Renewable and Sustainable Energy Reviews* 16.8 (2012):6149-6159.
6. Ertekin, Can, FatihEvrendilek, and RecepKulcu. "Modeling spatio-temporal dynamics of optimum tilt angles for solar collectors in turkey." *Sensors* 8.5 (2008):2913-2931.
7. Stanciu, Dorin, CameliaStanciu, and IoanaParaschiv. "Mathematical links between optimum solar collector tilts in isotropic sky for intercepting maximum solar irradiance." *Journal of Atmospheric and Solar-Terrestrial Physics* 137 (2016): 58-65.
8. Hartner, Michael, et al. "East to west–The optimal tilt angle and orientation of photovoltaic panels from an electricity system perspective." *Applied Energy* 160 (2015):94-107.
9. Photovoltaic Geographical Information System, http://re.jrc.ec.europa.eu/pvgis/apps4/pvest.php.
10. D. Cotfas, P. Cotfas, S. Kaplanis, D. Ursutiu, and C. Samoila, "Sun tracker system vs fixed system," *Bulletin of the Transilvania University of Brasov*, vol. 1, no. 50, pp. 545–552, 2008.
11. M. Serhan and L. El-Chaar, "Two axes sun tracking system: comparsion with a fixed system," in *Proceedings of the International Conference on Renewable Energies and Power Quality (ICREPQ '10)*, Granada, Spain, March, 2010.
12. H. Mousazadeh, A. Keyhani, A. Javadi, H. Mobli, K. Abrinia, and A. Sharifi, "A review of principle and sun-tracking methods for maximizing solar systems output," *Renewable and Sustainable Energy Reviews*, vol. 13, no. 8, pp. 1800–1818, 2009.

Complimentary Filter-Based Technique for Identification of Unmanned Aerial System Parameters

Dhrutidhara Behera[1]
STPI Bhubaneswar

Sobhit Panda[2]
CET Bhubaneswar

Subhranshu Sekhar Puhan[3], Sonali Goel[4], Renu Sharma[5]
Department of Electrical Engineering, ITER,
Siksha 'O' Anusandhan Deemed to be University, Bhubaneswar, India

Abstract This manuscript presents a complimentary based technique for stability and control of multi rotor unmanned aerial vehicle (UAV). The output of the inertial sensors used in UAV environments, like accelerometer and gyroscope is noisy. The evaluated sensor data are used to get the orientation and altitude of the UAV. In this manuscript, we have used the Kalman Filter algorithm enabled with complementary filter-based techniques to get precise values and sensor fusion analysis from the Inertial Measurement Unit. Robust PID algorithms are used to stabilize the UAV and for maneuver. In this manuscript, we have assumed that the rotational speed of the four motors used in this UAV can be increased or decreased independently to correct the orientation or position. The UAV enabled with Kalman filter is simulated in the ARDUINO environment, and the results are illustrated in the result section.

Keywords Stabilization, Multirotor, Kalman filter, PID, Flight controller, Unmanned aerial vehicle (UAV)

1. Introduction

An Unmanned Aerial Vehicle (UAV) an airplane without a human pilot on board, or one that doesn't use direct operator handling. UAVs come in two different flavors: autonomous and

---

[1]dhruti1357@gmail.com, [2]sobhitpanda25@gmail.com, [3]subhranshusekharpuhan@soa.ac.in, [4]sonali19881@gmail.com, [5]renusharma@soa.ac.in

DOI: 10.1201/9781003489443-31

computer-controlled. UAVs have primarily been utilized for military objectives in the past. However, as time has gone on, it has become increasingly used in civic applications, such as security and firefighting operations. UAVs can be effectively employed in search and rescue operations to locate missing or trapped people in locations that are out of human reach. Multi-rotor, fixed-wing, single-rotor, and hybrid VTOL UAVs are the different types of UAVs. The number of rotors on the platform can be used to further categorize multi-rotor UAVs. They are the quadcopter (four rotors), hexacopter (six rotors), and octocopter (eight rotors). Here, we have considered quadcopter for analysis purpose. The most difficult part of quadcopter control is to get angular position of the rotor because of absence of any direct data. The most effective sensors in the UAVs are of Micro electro mechanical sensors (MEMS) type, and sensors like gyroscope and accelerometer are generally used. But the advancement of digital filtering techniques opens up the door for UAV control. The Kalman and complementary filters are the most popular for determining the angular position of unmanned aerial vehicles (UAVs). The Kalman filter has been studied in [1-3]. Publications are available in which the authors have undertaken modifications to the structure of the Kalman filter. Author in [4] has developed an optimal Kalman filter position estimation and in [5] an unscented Kalman filtering techniques for position estimation of UAVs. In both the manuscript [4] and [5], the simulator object position specification is addressed. The most used digital filtering techniques for position estimation of UAVs are Kalman filter and complementary filter. In [6], authors have compared the availability of both these filtering techniques and it was come to notice that both the filter have some advantages in their own area of operation. The inertial sensor required for the development of inertial measurement unit (IMU) can be developed by using low cost and low weight MEMS technology as compared to conventional components. In general the measurement unit of IMU consists of gyroscope, accelerometer and equipment's like magnetometer. The sensor components are greatly affected by input variation and noise quantities in the output of sensors like temperature drift problem variation in the gyroscope unit, vibration in the accelerometer unit during UAV takeoff and external magnetic field effect in magnetometer [8]. Therefore, fusing the data from the IMU multi-sensor and filtering out the external interference to obtain the attitude with high reliability and high precision is a challenging work. The complementary filter is simple and reliable, and it has lower requirements from the IMU multi-sensor and filtering out the external interference to obtain the attitude with high reliability and high precision is a challenging work.

Variation in the gyroscope unit, vibration in the accelerometer unit during UAV takeoff, and external magnetic field effect in magnetometer [8]. Therefore, fusing the data from the IMU multi-sensor and filtering out the external interference to obtain the attitude with high reliability and high precision is a challenging work. The complementary filter is simple and reliable, and it has lower requirements for the accuracy of IMU. The application of a complementary filter (CF) and the Kalman Filter proposes accurate control of the inclination of roll and pitch [9-12]. There are some problems mentioned in other articles as follows [13-15]:

- Gyroscope measures angular change, not angular position directly.
- Accelerometer measures more than linear acceleration, such as gravitational acceleration and coriolis terms.
- Measurements are noisy and biased.

- Body-frame states need to be transformed to the inertial reference frame (Euler angles)
- The IMU gyro is less sensitive to vibration and drifts a little over time
- Vibration make the IMU accelerometer Inoperable
- Only the average of the IMU accelerometer is usable

Here in this manuscript we have proposed optimal complementary filter-based Kalman filtering techniques for the better feedback control of UAV along with better onboard attitude, roll, and pitch control by combining the advantages of Kalman and complementary filter. A complementary filter is used only for pitch angle calculation and that data is again used for Kalman filter controller design purpose.

The rest of the paper is organized as follows. Section II deals with the basics of complementary filter using Quaternion and Section III deals with basics of complementary filter and Section IV with Kalman filter modeling. The mostadvanced part of UAV detection using sensor fusion is mentioned in Section V. Section VI deals with Hardware results and in last Section VII depicts the conclusion of the manuscript.

1.1 Hardware Used

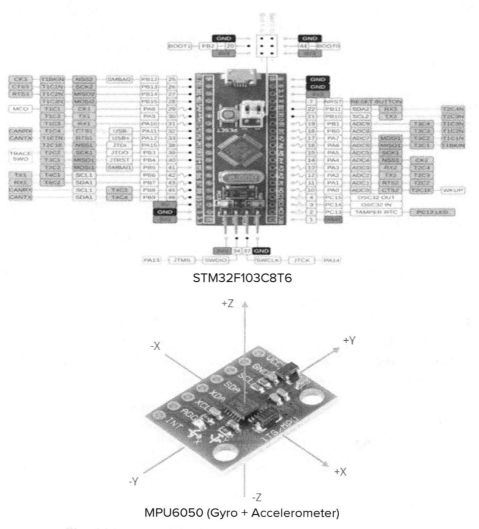

Fig. 31.1 Hardware used in the system representation

The IMU contains a 3-axis gyroscope to measure angular velocities and accelerometer and also barometer altimeter which is full of noisy inputs. The magnetometer data is eliminated here for simplification purposes (magnetic field intensity) and the barometer altimeter is taken in another dimension along with velocity and acceleration. The sensor fusion data sensitivities are illustrated in Table 31.1.

Table 31.1 MPU 6050 sensitivities

	Gyroscope	Accelerometer	Altimeter
Range (magnitude)	312(deg/s)	16 (g)	18(g)
Sensitivity	0.05deg/LSB	3.73(mag/LSB)	5.46(mag/LSB)

1.2 Connection Diagram

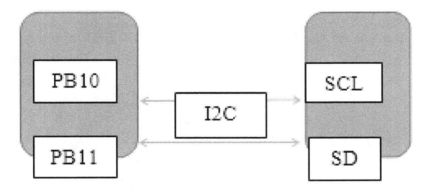

Fig. 31.2 Connection diagram in arduino

The connection diagram in Arduino platform is shown in Fig. 31.2. The 16MHZ crystal of the Arduino Board is already connected with Pin-10(XTAL1/PB7) and Pin-11(XTAL2/PB8) is connected with SD (Serial Data Signal) and SCL (Serial Clock Signal). SD and SCL are used for data transfer and synchronous clock signals between master and slave device. Master device always initiates the process of communication along with slave devices, and for communication purpose it needs the address of slave device.

2. Basics of UAV Modeling Using Quarternion

In order to describe UAV Quarternion a variable state x can be represented by considering the variable from gyroscope, accelerometer and magnetometer. Basic schematic diagram of Quarternion based UAV modeling is shown in Fig. 31.3(ii). The gyroscope data is fed to the accelerometer and then to the barometer altimeter as shown in Fig. 31.3.(ii). The measurement based on quarternion is described using equation no.1-8. Here authors have used quarternion based modeling than as compared to Euler modeling because former has ability to describe any dimension of 3 dimensional references in a holistic approach as compared to Euler rotating angle approach.

In a normal Kalman filter approach the data from gyroscope, accelerometer and magnetometer has fitted to Kalman filter and the Euler angle is calculated accordingly. But the basic Kalman filter approach lacks significantly when more noisy environment is there in the input parameter. The complementary filter process the Gyro data, accelerometer data, barometer altimeter data and the rotation data based on complementary filter approach as elaborated in next section.

The barometer altimeter uses a conditioning unit along with meter because of more noisy input in the gyroscope and accelerometer data. The complementary filter takes input from the signal conditioning unit and the rotation angle unit. Moreover, the Complementary filter based method uses a very simple quarternion method for posterior estimates.

The state of UAV for quarternion can be described as follows:

$$\omega = \begin{bmatrix} \omega_x & \omega_y & \omega_z \end{bmatrix} \tag{1}$$

$$a = \begin{bmatrix} a_x & a_y & a_z \end{bmatrix} \tag{2}$$

The angular rate and rate of acceleration of UAV for quartenion representation is describe as follows

$$q_0 = \begin{bmatrix} 0 & a_x a_y a_z \end{bmatrix} \tag{3}$$

$$q_\omega = \begin{bmatrix} 0 & \omega_x \omega_y \omega_z \end{bmatrix} \tag{4}$$

$$q_\omega = \begin{pmatrix} q_0 \\ q_\omega \end{pmatrix} \tag{5}$$

Where, $\omega_x, \omega_y, \omega_z$, and a_x, a_y, a_z are the angular data from gyroscope measurement and the acceleration from accelerometer respectively. The state of quarternion is assumed with two parameters, one with initial stage detection and any stage with actual angle from reference frame and angular acceleration as described in equation 5.

Let us consider the gyroscope rotates at some angle and in a time frame of 't', than quarternion representation can be represented as follows:

$$q_\omega = \begin{bmatrix} q_0 q_x q_y q_z \end{bmatrix} \tag{6}$$

$$q_\omega(\theta, \omega) = \begin{bmatrix} \cos\frac{\theta}{2} & \omega_x \sin\frac{\theta}{2} & \omega_y \sin\frac{\theta}{2} & \omega_z \sin\frac{\theta}{2} \end{bmatrix} \tag{7}$$

Complimentary Filter-Based Technique for Identification of Unmanned Aerial System Parameters 287

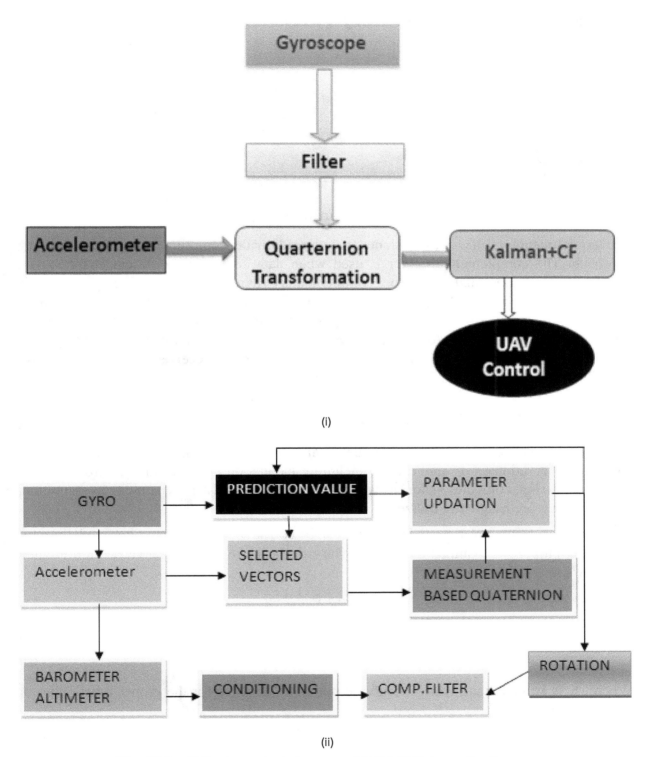

Fig. 31.3 (i) Basic control diagram of UAV, (ii) Schematic diagram

3. Basics of Complementary Filter

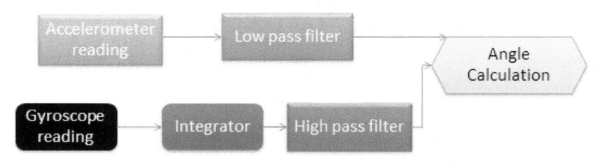

Fig. 31.4 Basics of complementary filter design

The complimentary filter was designed using readymade complimentary filter modeling approach. The complementary filter is used when data come from different channels and sensors. To obtain the final result from complementary filter a high pass and low pass filter is needed along with integrator application from some of the IMU data's (like Gyroscope reading). The complementary filter has main advantages is that it requires less processing power for effective and satisfactory performance analysis. For some variables, the complimentary filter relies quite heavily on the accelerometer data and therefore we receive a fairly noisy output. Complementary filter is used for controlling angle calculation.

4. Kalman Filter Modelling

An algorithm called the Kalman filter determines a system's states from ambiguous and sporadic observations. For applications including navigation and tracking, control systems, signal processing, computer vision, as well as econometrics and aerodynamics representation, Kalman filters are frequently utilized.

A causal Linear time invariant (LTI) system can be considered for kalman filter modelling for prediction purpose is assumed as follows. The state space representation can be represented as follows.

$$\vec{X}_{t+1} = A * \vec{x}_t + B * \vec{u}_t \tag{8}$$

$$\vec{Y}_{t+1} = \vec{z}_t - C * \vec{X}_t \tag{9}$$

$$P = (I - K * C)P \tag{10}$$

Where, K is kalman gain, which should be optimally adjusted for the proper functioning of Kalman filter.

$$S = C * P * C^T + R \tag{11}$$

$$K = P * C^T * S^{-1} \tag{12}$$

$$\vec{X}_{t+1} = A * \vec{x}_t + B * \vec{u}_t + \vec{w}_t \tag{13}$$

$$\vec{Y}_{t+1} = C * \vec{x}_t + \vec{v}_t \tag{14}$$

For simplicity purpose in kalman filtering modelling, we are processing without additive signal $\vec{w}_t$

$$P = A * P * A^T + Q \tag{15}$$

$$\vec{z}_{t+1} = \begin{bmatrix} \varphi t \\ \theta t \end{bmatrix} \tag{16}$$

$$\vec{x} = \begin{bmatrix} \Phi_t \\ b_{\Phi_t} \\ \theta_t \\ b_{\hat{\theta}_t} \end{bmatrix} \tag{17}$$

Let us consider the gyro bias is −Δt during time elapse of t second, the state space representation for kalman filter can be written as

$$\vec{X}_{t+1} = \begin{pmatrix} 1 & -\Delta t & 0 \\ 0 & 1 & 0 \\ 0 & 0 & 1 \end{pmatrix} \vec{x}_t + \begin{pmatrix} \Delta t & 0 \\ 0 & 0 \\ 0 & \Delta t \end{pmatrix} \vec{u}_t \tag{18}$$

The updated value of state equation can be find as mentioned already in equation 11.

Where, P, Q and R are error covariance matrix (already chosen), mean value of measured and noise value of IMU unit respectively.

P, Q and R are all diagonal matrices and the size of S depends on the optimal kalman gain parameter, what we have designed for better utilization of IMU data set. The matrix Q will tell the complementary filter about the uncertainty of model dynamics P and R will depict the presence of noise level in the IMU data set and $\vec{v}_t$ represents the dynamic velocity of UAV.

5. Sensor Fusion with Complementary Filter

Usually detection of UAV is done either by integration of more number of sensor with the help of machine learning algorithm or deep neural network algorithm. The sensor fusion stage gathers data set from multiple sensors and utilizes the data sets for more reliable sensor detection with mutual operation of sensors. In general the sensor fusion is nothing but an aggregation of input from multiple sensors and to form a single model or image environment of the drone /UAV platform. Basic steps involved in the sensor fusion is illustrated in a simple and comprehensive manner in the below paragraph.

The method is as follows:

1. Choose a variable α, such that 0 < α < 1. The larger α, the more the accelerometer measurements are 'trusted'. As α goes to zero, we base our estimate mainly on the IMU (e.g. gyroscope measurements). A good starting point is 0.001.
2. Initialize state estimate, e.g. $\hat{\phi}_{t=0} = 0$.
3. For each time step (sampling time Δt)
   (a) Retrieve raw accelerometer and gyroscope readings from IMU.
   (b) Calculate estimate of angle from accelerometer data

(c) Combine this estimate with equation 19.

$$\hat{\phi}_{t+1} = (1-\alpha).(\hat{\phi}_t + \phi_G * \Delta t) + \alpha * \phi_{Acc} \qquad (19)$$

The actual filtering is carried out in the final step of equation 18, a difference equation. In contrast to gyroscope measures, which are high-pass filtered, accelerometer measurements are low-pass filtered. Depending on the variable, these signals are then merged to get the final state estimate. So here the roll of complementary filter begins, which decreases the roll angle disturbance for some 30 to 35 second.

6. Simulation Results

For analyzing the proposed controller application we have used MATLAB/SIMULINK for modeling purpose. Figure 31.5 to 31.9 illustrates the various cases for modeling analysis of UAV behavior purposes. If we use low-pass filter the raw accelerometer data, for example by using a simple moving average filter, we get a much cleaner estimate. Figure 31.5 illustrates role estimate of UAV from raw acceleration data. The amplitude of the high-frequency motion however, from t = 30s to t = 35s, is attenuated by the low-pass filter this is not ideal, since this will not give us the correct roll angle at higher frequencies.

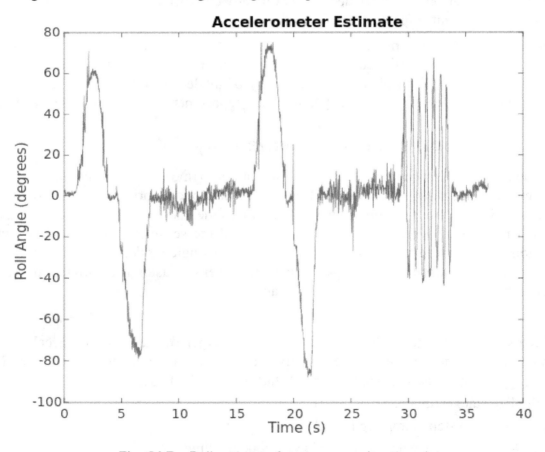

Fig. 31.5 Roll estimate from raw acceleration data

After getting estimation from acceleration data and processing it with Low pass filter, the filtered acceleration data based on role estimation is shown in Fig. 31.6.

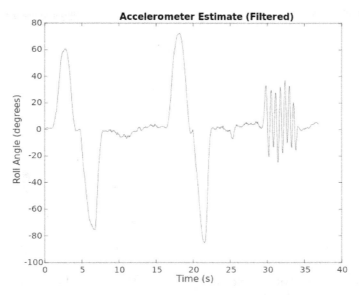

Fig. 31.6 Roll estimation from filtered acceleration data

The gyrodrift present in gyroscope datas from the IMU set is estimated using complementary filter on simple PID filter application. It is shown from below (Fig. 31.7) that gyrodrift true and estimated are exactly illustrated when the variable chosen is 0.1. The drift is obtained in an illustrative manner after short span of time about some 4-5 seconds.

As we know complementary filter need very low processing power, we have compared various level of variable data such as 0.4 and 0.1. When the variable data α is more, the complementary filter relies quite heavilyon accelerometer data of the IMU set and when the chosen variable is less as 0.1, it relies quite heavily on Gyroscope data.

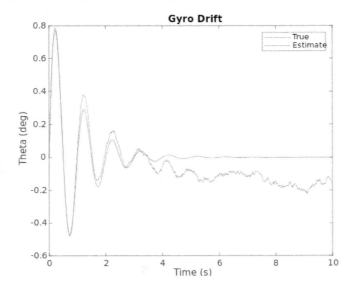

Fig. 31.7 Gyro drift present in gyroscope set of IMU

Figure 31.8 illustrates complimentary filter output for different values of IMU (e.g. Gyroscope measurement variable $\alpha = 0.4$ and 0.1 respectively).

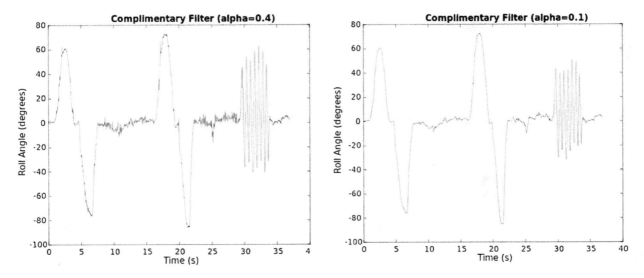

Fig. 31.8 Complementary filter output for roll angle in degrees ($\alpha = 0.4$, 0.1)

The below figure depicted in Fig. 31.9, shows the performance of the Kalman Filter compared to the complimentary filter. The obtained conclusion from Fig. 31.9 is that with general modelling of UAV, a very slight performance gain is obtained using Kalman filter but the main advantages of complementary filter i.e. low processing power and simpler in design perspective gives it a upper hand.

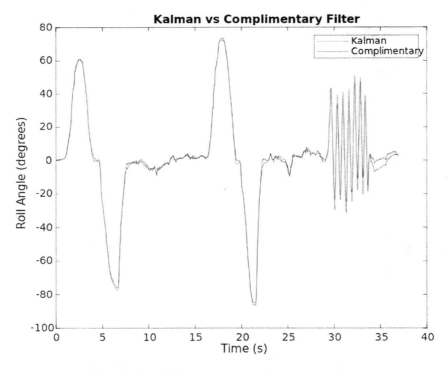

Fig. 31.9 Kalman vs. complementary filter

7. Conclusion

The complementary filter uses a linear estimator which is specifically based on posterior estimator. As complementary filter requires less processing power as compared to Kalman filter and it is simpler to design as compared to Kalman filter approach, it is generally used for UAV control. The Kalman filter approach of UAV also has certain merits like better estimates of roll and pitch and other states. So, we have proposed a combination of Kalman filter along with complimentary filter for better estimates of states and also lower processing power and simplicity in design purpose. At last the model dynamics is designed with the help of ARDUINO environment.

REFERENCES

1. Mao, G.; Drake, S.; Anderson, B.D.O. Design of an Extended Kalman Filter for UAV Localization. In Proceedings of the Information, Decision and Control, Adelaide, Australia, 12–14 February 2007; pp. 224–229.
2. Valade, A.; Acco, P.; Grabolosa, P.; Fourniol, J. A Study about Kalman Filters Applied to Embedded Sensors. Sensors **2017**,17, 2810.
3. Kim, Y.; Bang, H. Introduction to Kalman Filter and its applications. In Kalman Filter; Kordic, V., Ed.; Intechopen: London, UK, 2018.
4. Stephen P. Tseng, Wen-Lung Li, Chih-Yang Sheng, Jia-Wei Hsu, Chin- Sheng Chen Motion and Attitude Estimation Using Inertial Measurements with Complementary Filter" Institute of Mechatronic Engineering National Taipei University of Technology Taipei, Taiwan, ROC.
5. Xiong, J.J.; Zheng, E.H. Optimal Kalman filter for state estimation of a quadrotor UAV. Optik **2015**, 126, 2862–2868.
6. Di Corato F, Novi M, Pacini F, et al. "A nonlinear complementary filter for underwater navigation using inertial measurements", OCEANS 2015.
7. Hide, C., Moore, T. and Smith, M., "Adaptive Kalman filtering for low-cost INS/GPS," The Journal of Navigation, 56(1), pp.143-152, 2003.
8. Condomines, J.P. Nonlinear Kalman Filter for Multi-Sensor Navigation of Unmanned Aerial Vehicles; ISTE Press–Elsevier: London,UK, 2018.
9. Grewal, M.; Andrews, A. Kalman Filtering: Theory and Practice Using MATLAB; John Wiley and Sons: New York, NY, USA, 2001.
10. H. Xing, Z. Chen, C. Wang, M. Guo and R. Zhang, "Quaternion-based Complementary Filter for Aiding in the Self-Alignment of the MEMS IMU," 2019 IEEE International Symposium on Inertial Sensors and Systems (INERTIAL), Naples, FL, USA, 2019, pp. 1–4, doi: 10.1109/ISISS.2019.8739728.
11. A. Noordin, M. A. M. Basri, and Z. Mohamed, "Sensor fusion for attitude estimation and PID control of quadrotor UAV," International Journal of Electrical and Electronic Engineering and Telecommunications, vol. 7, no. 4, 2018.
12. L. Lasmadi, A. I. Cahyadi, R. Hidayat, and S. Herdjunanto, "Inertial navigation for quadrotor using Kalman filter with drift compensation," International Journal of Electrical and Computer Engineering, vol. 7, no. 5, p. 2596, 2017.
13. K. Bertan, M. Xhemajl, and C. Uran, "Noise reduction in Quadcopter Accelerometer and Gyroscope Measurements based on Kalman filter" (2015). UBT International Conference. 42.

14. Deibe, Á.; AntónNacimiento, J.A.; Cardenal, J.; López Peña, F. A Kalman Filter for Nonlinear Attitude Estimation Using Time Variable Matrices and Quaternions. Sensors 2020, 20, 6731.
15. Z. Tan, Y. Wu, and J. Zhang, "Fused attitude estimation algorithm based on explicit complementary filter and kalman filter for an indoor quadrotoruav," in 2018 Chinese Control And Decision Conference (CCDC), pp. 5813–5818, June 2018.

Innovative Methods and Greener Technology for Remediation of Microfiber Pollutants

Biswanath Naik, Lala Behari Sukla*

Biofuel and Bioprocessing Research Center (BBRC), ITER,
Siksha "O" Anusandhan (Deemed to be University), Bhubaneswar

Aditya Kishore Dash

Chemistry and BBRC, ITER, Siksha "O" Anusandhan
(Deemed to be University), Bhubaneswar

Abstract The industrial globalization of the appeal industry to meet the rising population demand shows enormous growth in the last decade. Today the global plastic market valuation is more than 457.73 billion USD and still rising. Due to the increase in demand and its durability nature synthetic polymer is used in every daily use essential of human life. Increasing production, mismanagement, and improper handling of these synthetic fibers create pollution and hazardous environmental effects. Today microfibers are minuscule synthetic fiber particles formed from larger fiber products, the major plastic pollutants of the environment, which gain attention in the research community. These tiny particles (size ranges from ≤100nm to ≤5mm) are released into the environment by dumping & landfill municipal waste, household laundering and waste disposal, sewage mismanagement, textile, and cosmetic industrial effluent, etc., found everywhere from the deep ocean to inaccessible mountains and enter the food system of living organisms by indirect ingestion. Due to the minuscule size, it was not easy to manage these fiber particles, sometimes they cannot be visible properly, and they are inert and non-disposable in nature. In this paper, we review some innovative methods & technologies to counter microfiber pollution using biological decomposition methods such as Biodegradation using microorganisms, algal absorption, and membrane technology in a sustainable manner. The release of energy during the processes will be reported.

Keywords Microfibers, Environmental pollution, Greener technology, Bioremediation, Microorganisms, Algal technology

---

*Corresponding author: lalabeharisukla@soa.ac.in

1. Introduction

Synthetic polymers make human life easier as it is required every day & everywhere as part of human life. Many plastic products are produced each and every year in the world from the day of development on synthetic polymer. Plastic production increased significantly on an industrial scale due to the increasing demand in various sectors such as agriculture, medicine & food packaging, textile industry, transportation industry, construction industry, retail, energy industry, petroleum industry, fishing industry, etc., as it is economically cheap, great mechanical and physiochemical properties, resilience nature, and multipurpose uses (Abomohra and Hanlet, 2022, Priya et. al., 2022). Synthetic fiber products come in various shapes and sizes, used in the textile and pharmaceutical industries, from large containers to small chips everywhere these synthetic fibers are used. Due to its enormous use in various sectors, and attentive mismanage, improper disposal of this chemically nondegradable plastic debris found in deep sea waters to inaccessible mountains causes pollution which imbalance the ecosystem, and environment of living organisms (Barnes et al., 2009, Weis and De Falco 2022). The report estimated that the annual global plastic production in 2017 was 380 Mt tons and in 2018 global plastic production increased to more than 450 Mt, and the following growing scenario of plastic production in 2025 will be estimated to be doubled (Geyer et al., 2017, Macleod et al., 2021, Suzuki et al. 2022,). A recent report estimated; due to the increase in demand in plastic production after the post covid (https://www.fortunebusinessinsights.com/plastics-market-102176) the market valuation of total global plastic was USD457.73 billion in 2022 and reached up to 643.37 billion by 2029. Microplastics or microfibers are usually referred to as minuscule plastic particles (solid) of regular and irregular shape and size nondegradable in soil and water including fragments, beads, spheres, films, fibers, and industrial pallets, and the breakdown of substantial plastic products, the size ranges from ≤0.5 mm to ≤100 nm (Nano plastics) i.e., the very much smaller than the microalgal cell diameter (Ferreira et al., 2019, Yin et al 2021). Most of these microfibers are reached the environment straight away or incidentally through anthropogenic misconduct such as domestic discharge, industrial discharge, cleaning products, cosmetics, laundering, fishing, dumping, etc. Most microfibers are in thread form or polymer resin and are not biodegradable or insoluble in water which is apparently made up of polyethylene, polystyrene, polyvinylchloride (PVC), polyurethane, polypropylene (PP), polyester, polyamide, polyethylene terephthalate (PET) (Anderson et al., 2016). HDPE (High-density polyethylene) including PP and PVC constitute major plastic form representing 59% of the world's plastic production, whereas 41% of low-density polyethylene, such as polystyrene, polyurethane, PET (polyethylene terephthalate), etc. (PlasticsEurope., 2022). A recent estimation reported that nearly ~0.28 Mn tons of worlds microfiber debris reached to the ocean each year, and Asian countries contribution is 65% (Belzagui et al., 2020). From Indian point of view, it is reported as the second-highest global synthetic fiber producer and 5th highest microfiber pollution creator in the world producing approximately ~2 Mt in the world (Mishra et al. 2019). MFs are tiny synthetic, artificial, and natural fabric particles of thread-like structure, with size ranges of more than one denier (<1D) (Henry et al., 2019). Microplastic and microfibers are major pollutants of the environment containing 70-90% sources originate from the land where as 10-25% originates from oceanic sources of global plastic waste and till date no proper action taken to control this microplastic pollution (Andrady, 2011). There are many reports and studies observing how these tiny microfiber particles enter the food system of many organisms, such as fishes (Wang et al., 2017), seabirds (Amelineau et al., 2016), mollusks (Su et al., 2018), echinoderms (Graham and Thompson, 2009), Zooplankton (Sun et al., 2017). In aquatic ecosystems, these microfibers act as potential habitats for several microorganisms such as bacteria, microalgae, and macroalgae, corals, etc. (Yokota et al., 2017, Amaral-Zettler et al., 2020, Peller et al., 2021). Algae can serve as the "plastisphere" as these microfibers are eventually entering the aquatic ecosystem (Borrelle and Law, 2020) because algae are known for their interaction with microfibers in the aquatic ecosystem (Yokota et al., 2017). Now, these microfibers are found suspended in water bodies or found as sediments which create delusion as food and ingested by aquatic organisms have the chance to infiltrate the human body by the food chain (Ramasamy and Subramanian, 2021). There are various studies reported that these synthetic microfibers are found in the human anatomy (Ragusa et al., 2021).

For that reason, there's a demand for banning single-use plastic items which are widely used in various packaging sectors. Similarly, proper management of these plastic products and their waste management need a proper action plan to combat plastic pollution, government level policy coupled with public awareness and its promotion at the local level is necessary. Again, in both national and international forums, G-7, G-20, UNEP (United Nations Environment Programme), and APEC (Asia Pacific Economic Cooperation) are the major global policy forums that take action to ban single-use plastics and international transportation ban on the trade of these types of plastic products to limit the plastic pollution (Law and Narayan, 2022). There are plenty of reports and studies obtained to contain large plastic debris in a sustainable way such as through bioremediation, biodegradation, and recycling models and procedures, but there are some reports rise concerns about microplastic or microfiber pollution (Ramasamy and Subramanian 2021, Weis and De Falco 2022) as there are no specific methods and technologies available for detection and it was not easy to collect these tiny particles.

The intention of this review study is not for the global concern on microfiber pollution including its effect on the environment which is already reported in various studies by researchers, but to how can we tackle microfiber pollution using the algae as a potential sustainable sinker of microfibers in the aquatic ecosystem. The absorption capacity of algae and the gradual disintegration of microfibers in contact with algae makes it an excellent bio-degrader of synthetic fibers which can be used on large scale for the dismantling of these tiny inert microfibers in aquatic conditions. We further reviewed some biodegradation methods using microorganisms and innovative techniques like membrane filtration which can be used in the filtration of synthetic fiber in the aquatic ecosystem in an eco-friendly manner.

Represents review assisted the information gathers from the search engine Sciencedirect (https://www.sciencedirect.com/), Nature (https://www.nature.com/) GoogleScholar (https://scholar.google.com/), ResearchGate.net (https://www.researchgate.net/), are searched through following search quarries: Microfiber Pollution and algal solution, Microfiber pollution and environmental effect, control of microfiber pollution and techniques. These publications are studied thoroughly and ultimately 35 literatures were summarized and the observed data was mentioned.

2. Identification of Microfiber in the Environment

Microfibers have Different types, and distinct physical and chemical properties, making it difficult to develop a universally precise technique of identification. Most common methods for quantifying and identifying microfibers/ MPs are gravimetric examination and visual inspection (Abomohra and Hanlet, 2022). The traditional methods like FTIR (Fourier transform infrared), Visual examination, and Raman spectroscopy are used for detecting plastic particles (Prata et al., 2019). Synthetic plastic particles may be classified visually based on their physical features, which can be observed directly with a microscope or using a fluorescence microscope. This strategy is being considered. It is now the most relevant and generally accessible method for identifying and quantifying plastic particles, and it is often utilized prior to additional chemical analysis (Prata et al., 2019). Identification of synthetic nano and microfibers is not easy from the mass of garbage and sand particles. Eriksen and his coauthors (2013) used SEM-EDS (Scanning Electron Microscopy coupled with Energy-Dispersive X-ray Spectroscopy) for identification of microfiber in their report. They observed that the numerous aluminum silicate, coal ash, and fly ash which looks similar to the microplastics. Several innovative approaches with low cost and great efficiency have been proposed, such as FTIR- FPA (FTIR combined with Focal Plane Array) detection, NR staining (Nile Red), and thermogravimetric analysis coupled with DSC (Differential Scanning Calorimetry) can be a possible method to identify these microfibers (Abomohra and Hanlet, 2022). The microplastic particles size between 20 to 100μm and 100μm to 5mm can be analyzed by IR spectroscopy and Raman spectroscopy method (Hanke et al., 2013). Some additional methods like composition of thermal extraction, and desorption-gas chromatography with mass spectrometry (TEDGC-MS), pyrolysis-gas chromatography composed with mass spectrometry (Pyr- GC- MS), and high temperature-gel permeation chromatography combined with IR detection (HT-GPC-IR) are used for detection and identification of microfibers simultaneously. Above studies indicating that identifying and detecting NPs is more difficult than detecting MPs as the minuscule size and can completely mixed with other garbage and sand it was difficult to identify and separate these micro particles by visualizing them. For example, Eriksen et al., (2013) in their study found that many particles which are visually identified as micro plastics, actually aluminum silicate of coal, coal ash, and coal fly ash. Therefore, there is a high chance of many aquatic and terrestrial organism consumes these particles unknowingly which creates imbalance in their food system as well as cause of many health hazards.

3. Environmental Impact of Microfiber Pollutions

Every day we use various pharmaceutical products where fibrous materials are used such as fairness cream, toothpaste, exfoliating scrub, shower gel, facewash, and other pharmaceutical products, etc., and wear products such as jackets, jeans, sweaters, t-shirts, shirts, socks, etc. Knowingly or unknowingly, we release these textile and daily use personal care products into the environment through laundering, washing, and mismanaging cause environment pollution. Fiber shed from garments or textiles and microfibers used in daily care products and cosmetics are the utmost known sources of microfibers (Lassen et al., 2015, Liu et al., 2020). Depending upon their source's MPs are reached to the environment either by primary discharge or secondary discharge. Primary discharge includes intentionally made micro-sized plastic products employ in cosmetics, personal care, and cleanser products (Lassen et al., 2015, Li et al., 2016, Napper and Thompson, 2016) similarly, in secondary discharge microplastics forge from pieces of large plastic products which are breaking out or disintegrating through anthropogenic such as fiber released from clothes during washing, waste management (plastic bags, bottles, electric wires, dismantled PVC products), and natural ways (earthquake, cyclone, flood, etc.) (Battelle Ocean Sciences, 1992) mechanical degradation, and UV radiation exposure, oceanic water current etc. (Priya et al., 2021). This plastic debris (MPs/MF) can be found everywhere, from the terrestrial ecosystem to the aquatic ecosystem (Fresh water and marine water) and also in the air (Bergmann et al., 2019). These plastic molecules circulate in the ecosystem due to their distinct features of nondegradable, chemical resonance, or stable structures with high molecular weight. Plastics are hydrophobic in nature having carbon backbone and the presence of stable functional groups like alkane & phenyl (Geyer et al., 2017). There are very few reports about the rate of disintegration of microplastics and emitting rate of pollution to the environment which provides a base to study microfibers/ microplastic pollution and its effect to the environment. Figure 32.1 shows different sources of microfiber pollution in the environment.

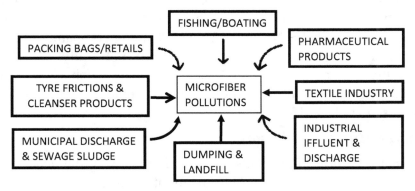

Fig. 32.1 Sources of microfiber pollution

3.1 Sources and Impact on the Soil

Although there are less of studies and research about microplastic effect on the terrestrial environment; reports estimate that microplastic pollution is found in a wide range of terrestrial ecosystems such as high to low altitudes of mountains, jungles, agricultural soil, urban & industrial lands, etc. (Zhang et al., 2020). According to UNEP (United Nations Environment Program), researchers found that in terrestrial ecosystem microplastic pollution is much higher from the marine ecosystem (i.e. 4 to 23 times higher) and 3 parts to 1 part of plastic waste integrated into the soil and eventually reached groundwater (https://www.unep.org/news-and-stories/story/plastic-planet-how-tiny-plastic-particles-are-polluting-our-soil). It is also reported that most of the terrestrial microfiber scarps comes from four main sources;

Dumping/Burying /Landfill of Plastic Debris

20-42% of global plastic wastes are dumped in landfill (Nizzetto et al., 2016) therefore the slow disintegration of larger plastic products into minuscule size (0.1-0.0.5 mm) thread-like fragmented structures (99.36%) or powdery bulky structures which have capacity to reach different ecosystem through the natural or anthropogenic passage (He et al., 2019). Industrial excretion of microplastic into the soil is caused by dumping/burying plastic wastes; due to the absorbance characteristics of MPs, it can absorb other organic pollutants and heavy metals like (Mn, Cu, Ni, Zn, and Pb etc.), and increase their concentration in soil, which causes sterile and unproductive land, therefore microplastic pollution in the soil is an emerging concern (Ferreira et al., 2019). This absorbance capacity of microplastic is due to the chemical properties of the polymer and their interaction with organic substances by using forces of interaction such as hydrophobic interaction, electrostatic interaction, and non-covalent interaction (π-π) (Torres et al., 2021).

Microfibers in Sewage Sludge

Sewage sludge is an important source where we found 80-90% of plastic particles, and sewage water used on agricultural land for water treatment or irrigation whereas the soil used as compost means a thousand & tons of microplastic ends up in soil every year (Rillig, 2012, ScienceDaily. 2018). According to the estimated data from countries in Europe, the United States of America, and Australia around 95% of microplastics pass through sewage sludge and enter the ocean in the form of biosolids (Kumari et al., 2022). It is estimated that biosolids in the agricultural ecosystem are up to 430,000 to 300,000 tons in Europe and North America respectively (Nizzetto et al., 2016, Ng et al., 2018). Cleaning companies used one of the most used microplastics Melamine or polyester (0.25-1.7mm), poly (allyl di-glycol carbonate) in scrubbers, and acrylic blast media is used in cleanser products such as industrial cleaning, car cleansers, paint cleansers, rust removal products, etc. without proper disposal system caused soil pollution (Eriksen et al., 2013).

Domestic Discharge/Municipal Discharge

The report says that 8 trillion microbeads are annually entering the ocean through domestic discharge. Microbeads are a powdery bulk form of microplastic having physiochemical properties employ in daily care products such as shampoo, toothpaste, exfoliating shower gel, makeup materials, etc., and comprise 90% of these products (Rochman et al., 2015). (Duis and Coors, 2016) reported that Particles like Polyolefin having size ranges from (74-420m), and an amorphous form were characterized as appropriate for use as exfoliants in the USA patent for skin cleansers comprising microparticles of plastic. Microplastic used in the medical sector is used as drug vector as it has the capacity to carry medicine to their target site, it may also be possible that in polluted soil they can carry disease-carrying microorganisms, apparently into our food (Patel et al., 2009, Science Daily, 2018).

Urban area and industrial activity are the reason for the heavy population size in an area and the reason for major anthropogenic source of soil pollution. IUCN (International Union for Conservation of Nature, 2017) reports every year ~0.8 to 2.5 Mt of primary microplastics makes their way into the ocean from which 35% of microfiber comes from textiles (Boucher and Firot, 2017). The textile industry grew rapidly by doubling clothing production from 2000 to 2014 because the garment purchase rate increasing 60% per annum. It is recorded that the clothing industry's growth rate increased, from 2013-2018 at a speed of 7.25% (Synthetic Fiber

Market Research Report- Global Forecast 2018). The textile industry mostly uses nylon, acrylic, polyester, recycled polyester, recycled nylon, rayon, cotton, viscous fiber, etc., as synthetic textiles making the clothes smooth and fancy and giving authentic looks. Textile industries also used some synthetic dyes such as cyanine, coumarin, xanthene, etc. for their water-propelling nature and antimicrobial and UV absorbent properties making the clothes attractive and feasible in any conditions (Seker, 2013). Nanotechnology is an emerging technology that is also used nowadays in textile industries for giving a better experience to the customer; this technology gives a finishing touch to clothes by using silver, copper, zinc oxide, titanium dioxide, etc., for their stain resistance, and flame retardancy properties (Gularjani, 2013). Now mismanage leaching of these types of industrial effluents gives undesirable microfiber pollutants which is a reason for soil pollution. Home laundering is also an important source for the release of microfiber as we use detergent to wash clothes these types of microplastics are released in fragmented sizes as microfiber shading and washed out through the water passage entering the soil causing soil pollution. Although many studies carried out to estimate the soil pollution that occurs by synthetic microfiber, we cannot find any reports that estimate the contamination rates of microplastic and the quantity of microplastic that ended up in the soil. There are many reports estimating the fate of microplastic released during laundering reaching the ocean passing by sewage sludge but it cannot estimate that when passing through the sewage what is the quantity of microplastic deposited in the soil.

Other Sources

Other sources of microplastic pollution of soil include (I) waste tires, paints, and (II) plastic mulching in soil. Loss of tires and tire waste released during transportation and paint and colours of shipments that directly or indirectly reach out to the terrestrial ecosystem a reason for soil pollution (Rilig, 2012, Bannick et al., 2015). Boucher and Friot, (2017) reported that the generation of microplastic by the friction of tires and released by spill caused road and soil pollution. Plastic mulching is another important source of microfiber pollution in the agricultural ecosystem. It will be in the agricultural sector for increasing temperature, so it retains water irrigation in the soil and protects crops. Although it has characteristics features to increase crop production it has a negative effect i.e., it can change the physiochemical property of soil such as changes in p^H, water absorbance capacity, carbon and nutrient content of the soil, etc., which cause soil de-fertilization (Zhang et al., 2015). These foils of plastics after burying in the ground release some harmful substances such as phthalates (50-120mg/kg) which after absorbed by crops enter the human digestive system and cause various health diseases (Wang et al., 2013). Moreover, synthetic polymer particles are used in horticulture to enhance the quality of the soil and as a composting material. Examples include PU (polyurethane) foam and expanded PS flakes of approximately 5 to 15 mm (Stoven et al., 2015).

3.2 Sources and Impact on Water

Aquatic ecosystems include (I) the Freshwater ecosystem (ponds, rivers, lakes, estuaries) and (II) the Marine water ecosystem (lakes, oceans) are polluted by microfibers released by anthropogenic plastic litter.

Fresh Water Ecosystem

Domestic discharge and municipal discharge of microfiber pollutants enter the river ecosystem by passing through the wastewater drainage system, which passes downstream, and gradually heavy microfibers settle down under the water and riverbanks, and sewage banks as sediments and rest of others entering into the ocean by the flow of water (Hartline et al., 2016). These microfiber sediments are gradually entering the ocean and are found to settle on the oceanic floor, and passes various region of the ocean by the oceanic current of water (Klein et al., 2015). Industrial effluents are also passing through the wastewater treatment plant and entering the river; reports estimate that in US and Canada, although many wastewater treatment plants retained most of the microfiber during the process, nearly 3.5 quadrillion of microfiber leaked into the fresh water and eventually entering to the marine water (Vasilenco et al., 2019). A report by Barnes, (2019) estimates that 9-23 million Mt of plastics are annually emitted to the rivers and lakes from which ≥0.3 million Mt enter the ocean and circulate by the oceanic water current (Borelle et al., 2020). It clearly indicates that a sizable amount of plastic litter entering to the terrestrial ecosystem, some are found on the river bed, some of found on shoreline sand near the beaches, some of these litters are ingested by water creatures, and some of the litters are degraded or sink, etc. (Barnes et al., 2019, Padervand et al., 2020). Estuaries and lakes are also polluted by these microfibers as extensive fishing and increasing visitors' day by day. Eventually, in these types of water bodies, microfibers are found as sediments and shoreline sand (Duis and Coors, 2016, Bruce et al., 2016). Freshwater environment (Report based on Trent River UK) textile fibers like wool & cotton constitute a great portion of microfiber pollution these textile fibers are degradable in nature (Stanton et al., 2019). Dying, coating and furnishing of natural fibers such as wool, silk, cotton, fax and regenerated cellulosic fibers such as, Tencel, lyocell, rayon and the blended products of these natural textiles are taking longer times to degrade and, in some cases, nondegradable due to their physiological changes cause water pollution (Liu et al., 2021).

Marine Water Ecosystem

According to the reports of UNEP (https://sdg.iisd.org/news/unep-publishes-scientific-assessment-of-plastic-pollution/), nearly ~85% of harmful plastic litter enter the ocean, from which 4.85 Tn microplastic flecks are found in oceans as sediment or floating (Eriksen et al., 2014). There are three major sources of water pollution caused by microfiber;

1. *Domestic/municipal sewage system:* According to the data the microplastic particle found in freshwater and marine water is equivalent i.e., 4.85 trillion and most of these microplastics are released from domestic discharge, municipal discharge, and sewage sludge (Vickers, 2017). Domestic discharge includes plastic bags, ripped clothes, home laundering wastes, daily use care products, etc. are released microfiber to the environment which are dragged by water to the municipal sewage system in urban areas. Sewage sludge is a dominant source of microplastic as much of domestic and municipal waste is released into it. Microbeads are the types of microplastics that are used in daily used products are entering through the sewage sludge in the form of biosolids enter into the ocean and create pollution in the coastal region as sediment soil.

2. Textile and industrial litter: Microfibers loose from textiles are the most prevalent kind of microplastic and around 35% of these fibers are found in the world's oceans. During home laundering microfiber shading and their untreated release into the water plays an important role in water pollution. A recent study estimates that home laundry in the USA and Canada produces 533 million tons of microfiber pollutants whereas a total estimation of 3.5 quadrillion of microfiber pollutants passes through municipal wastewater reached the ocean (Vasilenco et al., 2019), also their study reported that nearly 533 million tons of microfibers are emitted per year from home laundry in Canada and US, and in a single wash of synthetic garments, nearly 2000 microfibers are released into seawater by passing through municipal sewage water (Browne et al., 2011). Reports estimate that nearly 80% of the aquatic microfiber pollutants are comes from land sources and 10-25% comes from oceanic sources (Andrdy, 2011). Globalization is the cause of the increase in the appeal industry due to which both synthetic garments and wastes are produced. Despite the fact that microfibers are widely recognized as a growing environmental issue, a complete ban on synthetic fiber is not a solution for a sustainable economy. The worldwide annual water consumption by textile industries is estimated to be 93 billion Cm^3, which implies 4600 L/kg fiber, and synthetic fiber manufacturing requiring 38 L/kg fiber, and textile products requiring 88 L/kg fiber for dying (Liu et al., 2021). Synthetic microfibers which are mostly found as pollutants in marine water are Regenerated Cellulose, Nylon, PET (Polyethylene Terephthalate), polypropylene, etc.

3.3 Impact of Microfibers on Living Organisms

MFs (Microfibers) have been found in the body of a wide variety of aquatic animals, from minute invertebrates to huge predatory mammals. Intake of these tiny fibers cause negative effect on the organisms, and causing mechanical disruption and inflammatory reactions, and also provides a plausible pathway for the entry of some dangerous compounds (such as endogenous plastic additives, pollutants absorbed from the environment, and pathogenic microbes) into the aquatic food chain (Zelter et al., 2013). Additionally, widespread consumption of these microplastics by edible aquatic species poses a concern to food safety and human health. Microplastics have been found in the digestive systems or tissues of several marine creatures. Although recent research has revealed that microplastics can be hazardous to soil fauna in laboratory tests, few studies have looked at their influence on soil invertebrates in wild settings (Selonen et al., 2020). Many biological and non-biological factors can impact microplastic dissolution and fragmentation once they reach the soil. Physical soil characteristics, such as the existence of fissures, and macropores (which emerge in the summer season) that operates like a channel between the surface and deep layer soil, can improve microplastic dispersion (Rillig et al., 2017a). Agricultural processes in cultivation plowing (turn up the soil), tilling, and crop harvesting also enhance the movement of microplastics & microfibers from the soil surface into deep layers of soil. Species, such as earthworms, mites, and nematodes move microfibers across comparatively greater distances in laboratory circumstances, but smaller creatures move these particles ranges, around 3 cm for example collembola. Earthworms are effective transporters of microplastics with diameters less than 50 m and aid in the assimilation of plastic particles into the soil surface (Rillig et al., 2017a,). Microplastics are also ingested by humans through food; fish and shellfish, and also in salt, beer, and sugar. The researchers assume that buildup of plastics by terrestrial creatures is already widespread, even out of those species that do not "ingest" food. Plastic shards, can accumulate; example in yeasts and filamentous fungi. In a healthy terrestrial ecosystem, crucial functions of ecosystem like pollination, nutrient recycling, energy transfer, seed dispersal, trophic interaction, etc. are damaged if we do not properly intervein in these microfibers/ microplastics disposal pathways, we must adopt the biodiversity-ecosystem functioning approach (Brockerhoff et al., 2017).

4. Remediation Methods of Microfiber Pollutants

Microfiber possesses threat, to the environment and the human beings. Removal of microfibers and their degradation imposes a big challenge to us from the last decade as these are minuscule particles. There are many biotechnological applications and methods are introduced by many researchers, and articles from which bioremediation technologies using microorganisms, algae and biological membrane filtration methods are discussed in this paper.

4.1 Biodegradation of Microfibers by the Action of Microorganisms

Biodegradation using microorganisms is an efficient biological method for the degradation of microfibers. There are many types of microorganisms (bacteria & fungi) present in the environment which has the potential to hydrolyze the microfibers and decompose the plastic particles by releasing CH_4, CO_2, H_2O, and various inorganic compounds in anaerobic conditions, whereas in aerobic condition release of CO_2, and H_2O (Chandra and Enespa, 2020). Fig. 32.2 shows microfiber degradation by the action of microbes.

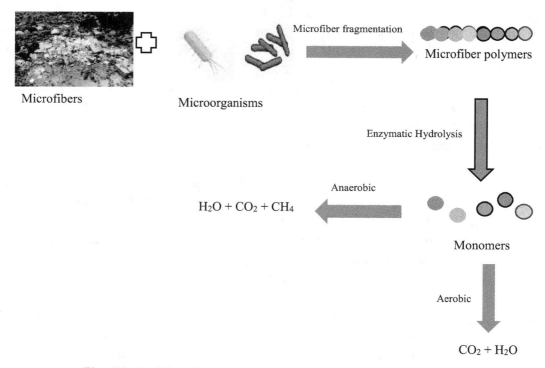

Fig. 32.2 Microfiber degradation by the action of microbes

There are several bacteria, fungi, and algae are observed in this process to have the capability to degrade the microfibers. Further using these types of microorganisms is economically viable and environmentally friendly as there is no use of chemical substances and high energy and these types of microorganisms are abundant in the environment (Badola et al., 2022,). Microorganisms can react with synthetic polymers and break down these complex forms into simpler forms such as *Zalerion maritimum* is a fungal species that degrade (PE) polyethylene by changing its physiochemical property. These fungal organisms secrete some mucilaginous substance which helps them to attach to microfiber as well as decomposition of the polymer (Paco et al., 2017). *Pestalotiopsis microspore* is a fungus that can hydrolyse PUR microplastics by releasing serine hydrolyse enzyme in the absence of O_2. *Penicillium simplicissimum* is another fungal example that can degrade polyethylene. Few marine fungal & bacterial species are discussed which has the capacity to degrade the microfibers in Table 32.1. Degradation of micro plastic using fungus is an emerging biological technique that can be widely applicable but their reaction rate with the plastic particles is slow and time-consuming. Therefore, pre-treatment of micro plastics is needed to increase the rate of reactions such as ozonolysis, photooxidation, solvolysis, etc. (Dey et al., 2020).

Table 32.1 Microorganisms (Fungal and Bacterial species) having degradation capacity

Organisms/ Species	Microfiber Type and Functions	Microfiber Degradation and its Impact	Reference
Fungus			
Aspergillus sp. *s45*	Degradation of polyester PUR film	15 to 20% decrease in weight	Osman et al., 2018
Penicillium simplicissimum	Polyethylene	4000 to 28000 Molecular Weight loss	Yamada et al., 2001
Aspergillus flavus *PEDX3*	Degradation of polyethylene	3.9025 ± 1.18%	Zhang et al.,2020
Pestolotiopsis microspora	PUR MPs Impranil DLN	Increase in 0.110± 0.031g of fungal biomass	Russel et al., 2011
Zalerion maritimum	Degradation of microplastics	43%	Paco et al., 2017
A. tubingensis *(VRKP1)*	Degradation of HDPE	Weight loss of 6.02 ± 0.2%	Devi et al., 2015
A.flavus *(VRKP2)*	Degradation of (High Density Polyethylene) HDPE	Weight loss of 8.51 ± 0.1%	Devi et al., 2015
Bacteria			
Comamonas testolteroni	Degradation of microplastic, PET	Particle diameter changes from 7.3 to 1.58, and 3.86 µm	Dey et al., 2020
Ideonella sakaiensis *201-F6*	Degradation of PET (Polyethylene Terephthalate)	0.13mg/cm^2 PET film degradation	Dey et al., 2020

Organisms/ Species	Microfiber Type and Functions	Microfiber Degradation and its Impact	Reference
Alcanivorax borkumensis	Degradation of LDPE (Low Density Polyethylene)	3.5% loss of weight	Delacuvellerie et al., 2019
Bacillus gottheilii	Degradation of PP, PET, PS, and PE	7.4% in vivo condition (40 days)	Auta et al., 2017
Bacillus cereus	Degradation of PP, PET, PS, and PE	1.6% in vivo condition (40 days)	Auta et al., 2017
Rhodoccus sp. *(strain 36)*	Degradation of Polypropylene	Increase in 6.4% bacterial biomass	Auta et al., 2018
Bacillus sp. *(strain 27)*	Degradation of polypropylene	Increase in 4.0% bacterial biomass	Auta et al., 2018
Bacillus substillis	Degradation of LDPE	Weight loss of 9.26%	Vimala and Mathew, 2016
Deinococcus thermus	Degradation of PP, PE, & PET	-	Kumari et al., 2022

4.2 Biodegradation of Microfibers by the Action of Algae

Algae is a lower-level fauna found in the aquatic ecosystem (both fresh and marine water), in both micro and macro forms, found that having the capability to degrade the microfiber in their habitat. Researchers found that algal treatment for microfiber pollution can be used as a biological tool for the removal of microfiber from wastewater as algae act as a natural sinker for microfiber (Cunha et al., 2020). Recent studies reported that several macro and microalgae present in the aquatic ecosystem have absorbance capacities of micro plastic causing degradation of that particle (Padervand et al., 2020). A marine microalga (*Phaeodactylum tricornutum*) might release a PETase enzyme to decompose PET plastics at a temperature of (21 °C), the enzyme remains active in a saline environment. Since this microalga has been transformed as a possible degradation chassis, it can fragmentize Polyethylene terephthalate (PET) into Mano-2-hydroxyethyl terephthalic acid (MHET), and Terephthalic acid (TPA). A brown algae *Fucus vesiculosus* can be adsorb microplastics/ microfibers (MPs/ MFs) because of presence of alginic acid, in its cell wall (Sundbaek et al., 2018). Absorption capacity of algae is proportional to the carboxylic acid (R- COOH) functional group present on the surface of Alginate polymer in brown algae. Hence plastic fiber particles are absorbed as adsorbents by the action of carboxylic functional group (R- COOH). The polystyrene micro plastics had a diameter of 20 m, and the algal cells having "sorbent" feature, very small micro channels which inhibit translocation of microfibers. Polystyrene micro plastics were found in the tissues, showed a significant absorption of micro plastics which can be explained by the action of alginate compounds released from the cell walls. Additionally, alginate can improve the attachment of polystyrene particles to the surface of seaweed because of its gelatinous qualities (Nolte et al., 2017). Again, report estimated that electropositive polystyrene particles adsorb to unicellular green algae, *Pseudokirchneriella subcapitata*, (20-500 nm polystyrene particles), so they conclude Micro plastics with electropositive nature are capable to absorbed on the algae's surface compared to those with electronegative nature. (Peller et al., 2021) studied on *Cladophora* a green macro alga and component of the submerged aquatic vegetation (SAV) community that often coexists with numerous other macro algae, microalgae, and microorganisms. The chemical composition, and broad surface area of algae seems to give the green algae an enhanced role for adsorption of synthetic microfibers. *Cladophora* was combined with micro plastic polyethylene particles in water like the microfibers, absorb, and intact with microfiber despite the adhered property of cell wall when agitated (Peller er al., 2021). There is a lack of papers which study the biodegradation of microfibers by algal absorption method from which it was mentioned that the algal degradation method is time consuming and the reaction rate is very slow but the overall process is eco-friendly and economically viable. In Table 32.2 some of the algal species which have the ability to degrade microfiber are mentioned.

Table 32.2 Algal species having microfiber degradation capacity

Algal species	MPs/MFs Type	Reaction/ Effect	Reference
Fucus vesiculosus	MPs	High sorption of MPs (~94.5%) by Alginic acids	Sundbeak et al., 2018
Scenedsmus dimorphus	Polyethylene	Polyethylene Polymer degradation	Kumar et al., 2017
Anabaena spiroides	Polyethylene	Polyethylene Polymer degradation	Kumar et al., 2017
Cladophora sp.	Microfibers	PET, PE, PA Microfiber Absorption method	Peller et al., 2020
C. vulgaris	PE, PVC, PA	Absorption method	Kiki et al., 2022
Dunaliella tertiolecta	PS	Absorption on the basis of charge	Davarpanah, and Guilhermino, 2015
Phaeodectylum tricornutum	PS	-	Davarpanah, and Guilhermino, 2015
Pyropia yezoensis	MPs	Absorption method	Li et al., 2020

In an experiment (Lagarde et al., 2016) fond that interaction of HDPE and PP with *Chlamydomonas reinhardtii* a model algal species showing negative growth rate at a concentration of 400mg/L. This is because trapping of micro plastics inhibits the photosynthetic reaction due to shading effect of micro plastics. But in case normal conditions LDPE interact with *Chlamydomonas reinhardtii* gives normal growth rate and enhance the production of desired compounds. Which conclude that micro algal and micro plastic interaction at long-term conditions give positive results.

4.3 Enzymatic Biodegradation by Forming Biofilm

In 2020 EPS (Extracellular Polymeric Substances) was first introduced by Sadaf Shabbir and his colleagues for decomposition of different synthetic microfibers such as PET, PE, and PP.

Algae and cyanobacterial species are interacted with microfibers and colonize in their surfaces by secreting EPS (Sarmah and Rout, 2018). The scavenging activity of EPS makes microalgae as a potential candidate for microfiber bioremediation in the environment. Several reports found that the potential of microalgal EPS is much high than the bacterial EPS i.e., metabolic activity of microalgae is 6.2-fold in light and 5.8-fold in dark conditions which is higher than the bacterial activity (Vital-j et al., 2021). Although biofilm development is a new concept but the potential of this method is highly recommendable and further studies and research needed for investigating new bacterial and algal species and their enzymatic biodegradation capacity by the action of EPS.

4.4 Removal of Microfibers Using Biological Membrane Filtration

Use of biological membrane technology is not a new concept, but rapid working in this concept is started from the last decade. There are many wastes water treatment plant in which can separate microfibers from the waste water but still there is many particles cannot be filtrated and entering to the aquatic ecosystem through the sewage sludge (Sun et al., 2019). Various report estimates that using membrane filtration technologies the separation of microfibers gives high amount of efficiency. Membrane bioreactor (MBR) technologies employ for elimination of microfibers from sewerage is the fusion of membrane filtration process and suspended growth biological reactors which is used for separation of effluent containing suspended, and dissolved particles (Xiao et al., 2019). The huge rise on both big (10,000 m^3/d) and small (1,000 m^3/d) scales is evidence of the technology's exponential appeal and extremely wide-reaching (over 100,000 m^3/d) plants everywhere. Beijing, China's first plant to be built with a super-large-scale capacity of 100,000 m^3 per day (Poerio et al., 2019). The function of MBR in MP treatment is to reduce the complexity of the solution by biodegrading the organic debris; this will enable MP purification and subsequent treatment of MP. Typically, the process begins when a pre-treated stream enters the bioreactor, where the biodegradation process takes place. For the separation operation, the created mixed liquor is subsequently pushed alongside a semi-crossflow filtering system (Poerio et al., 2019).

Microplastics were effectively removed from contaminated aquatic habitats using membrane technology. The effectiveness of removal over the membranes in particular entrenched the size, amount, and density of the microplastics/ microfibers, as well as their resilience property. Porous membranes and biological processes together might increase the elimination efficiency by up to 99.9% (Padervand et al., 2020). By using an MBR, for instance, the effectiveness on elimination of MPs/ MFs was increased up to 99.4% in comparison to the standard activated-sludge-based treatment (Pico et al., 2019).

5. Conclusion

Today microfibers possess immense threat to the mankind and the environment. Its reach to the human digestive tract is the proof that it can be invincible. Presence of microfibers in the food system and their circulation from lower to higher trophic level is even more dangerous as it causes organ damage and promote sterility of an organism in an ecosystem. Therefore, there is a need to explore various microbial species and their capabilities to degrade various types of synthetic microfibers in their natural habitat as greener technology. Utilization of microbes as a bio-degrader is often economic and eco-friendly alternative to disintegrate microfiber in aquatic environment. Interaction of microbes and microfiber have both negative and positive impact. As potential feedstock microfibers act as a bio plastisphere for microbial culture and growth. It can act as a nutrient source of microbes. The absorbance capacity of microbes and the binding property of microfiber coupled with enhance the gradual disintegration of microfibers. The present article suggests a new methodological approach that can be implemented for microfiber degradation in contaminated water through microbial & algal cultivation with biodegradation of microfiber in economically sustainable, approach. Removal of Microfibers by using biological methods is an optimistic method for which ground based frame work is needed to achieve the auspicious goal of microfiber free environment. A coupled mechanism of biological membrane filtration with algal absorption method can be a futuristic aspect which has the potential to make a large-scale impact on microfiber pollution which we have to focus on. Combination of both biological membrane filtration coupled with biodegradation using algal species and other microorganisms is can be a possible method of separation of microfibers and degradation in the environment. In addition to microfiber distribution in the environment serious action plans are needed to avoid the negative hazardous effect of microfibers.

Disclosure statement

The authors declare that there is no conflict of interest for it.

REFERENCES

1. Abomohra A, Hanelt D. Recent Advances in Micro-/Nanoplastic (MNPs) Removal by Microalgae and Possible Integrated Routes of Energy Recovery. Microorganisms. 2022 Dec;10(12):2400.
2. Priya AK, Jalil AA, Dutta K, Rajendran S, Vasseghian Y, Karimi-Maleh H, Soto-Moscoso M. Algal degradation of microplastic from the environment: Mechanism, challenges, and future prospects. Algal Research. 2022 Sep 10:102848.
3. Barnes DK, Galgani F, Thompson RC, Barlaz M. Accumulation and fragmentation of plastic debris in global environments. Philosophical transactions of the royal society B: biological sciences. 2009 Jul 27;364(1526):1985-98.
4. Weis JS, De Falco F. Microfibers: Environmental Problems and Textile Solutions. Microplastics. 2022 Nov 1;1(4):626-39.
5. Geyer R, Jambeck JR, Law KL. Production, use, and fate of all plastics ever made. Science advances. 2017 Jul 19;3(7): e1700782.
6. MacLeod M, Arp HP, Tekman MB, Jahnke A. The global threat from plastic pollution. Science. 2021 Jul 2;373(6550):61-5.
7. Suzuki G, Uchida N, Tanaka K, Matsukami H, Kunisue T, Takahashi S, Viet PH, Kuramochi H, Osako M. Mechanical recycling of plastic waste as a point source of microplastic pollution. Environmental Pollution. 2022 Jun 15; 303:119114.
8. Ferreira I, Venâncio C, Lopes I, Oliveira M. Nanoplastics and marine organisms: what has been studied? Environmental Toxicology and Pharmacology. 2019 Apr 1; 67:1-7.
9. Yin K, Wang Y, Zhao H, Wang D, Guo M, Mu M, Liu Y, Nie X, Li B, Li J, Xing M. A comparative review of microplastics and nanoplastics: Toxicity hazards on digestive, reproductive and nervous system. Science of The Total Environment. 2021 Jun 20; 774:145758.

10. Anderson JC, Park BJ, Palace VP. Microplastics in aquatic environments: Implications for Canadian ecosystems. Environmental Pollution. 2016 Nov 1; 218:269-80.
11. PlasticsEurope. Plastics–The Facts 2018: An Analysis of European Plastics Production, Demand and Waste Data; PlasticsEurope: Association of Plastic Manufacturers: Brussels, Belgium, 2018; Available online: https://plasticseurope.org/wp-content/ uploads/2021/10/2018-Plastics-the-facts.pdf
12. Belzagui F, Gutiérrez-Bouzán C, Álvarez-Sánchez A, Vilaseca M. Textile microfibers reaching aquatic environments: A new estimation approach. Environmental Pollution. 2020 Oct 1; 265:114889.
13. Mishra S, charan Rath C, Das AP. Marine microfiber pollution: a review on present status and future challenges. Marine pollution bulletin. 2019 Mar 1; 140:188-97.
14. Henry B, Laitala K, Klepp IG. Microfibres from apparel and home textiles: Prospects for including microplastics in environmental sustainability assessment. Science of the total environment. 2019 Feb 20; 652:483-94.
15. Andrady AL. Microplastics in the marine environment. Marine pollution bulletin. 2011 Aug 1;62(8):1596-605.
16. Wang J, Peng J, Tan Z, Gao Y, Zhan Z, Chen Q, Cai L. Microplastics in the surface sediments from the Beijiang River littoral zone: composition, abundance, surface textures and interaction with heavy metals. Chemosphere. 2017 Mar 1; 171:248-58.
17. Amélineau F, Bonnet D, Heitz O, Mortreux V, Harding AM, Karnovsky N, Walkusz W, Fort J, Grémillet D. Microplastic pollution in the Greenland Sea: Background levels and selective contamination of planktivorous diving seabirds. Environmental pollution. 2016 Dec 1; 219:1131-9.
18. Su L, Cai H, Kolandhasamy P, Wu C, Rochman CM, Shi H. Using the Asian clam as an indicator of microplastic pollution in freshwater ecosystems. Environmental pollution. 2018 Mar 1; 234:347-55.
19. Graham ER, Thompson JT. Deposit-and suspension-feeding sea cucumbers (Echinodermata) ingest plastic fragments. Journal of experimental marine biology and ecology. 2009 Jan 15;368(1):22-9.
20. Sun X, Li Q, Zhu M, Liang J, Zheng S, Zhao Y. Ingestion of microplastics by natural zooplankton groups in the northern South China Sea. Marine pollution bulletin. 2017 Feb 15;115(1-2):217-24.
21. Yokota K, Waterfield H, Hastings C, Davidson E, Kwietniewski E, Wells B. Finding the missing piece of the aquatic plastic pollution puzzle: interaction between primary producers and microplastics. Limnology and Oceanography Letters. 2017 Aug;2(4):91-104.
22. Amaral-Zettler LA, Zettler ER, Mincer TJ. Ecology of the plastisphere. Nature Reviews Microbiology. 2020 Mar;18(3):139-51.
23. Peller J, Nevers MB, Byappanahalli M, Nelson C, Babu BG, Evans MA, Kostelnik E, Keller M, Johnston J, Shidler S. Sequestration of microfibers and other microplastics by green algae, Cladophora, in the US Great Lakes. Environmental Pollution. 2021 May 1; 276:116695.
24. Borrelle SB, Ringma J, Law KL, Monnahan CC, Lebreton L, McGivern A, Murphy E, Jambeck J, Leonard GH, Hilleary MA, Eriksen M. Predicted growth in plastic waste exceeds efforts to mitigate plastic pollution. Science. 2020 Sep 18;369(6510):1515-8.
25. Ramasamy R, Subramanian RB. Synthetic textile and microfiber pollution: a review on mitigation strategies. Environmental Science and Pollution Research. 2021 Aug;28(31):41596-611.
26. Ragusa A, Svelato A, Santacroce C, Catalano P, Notarstefano V, Carnevali O, Papa F, Rongioletti MC, Baiocco F, Draghi S, D'Amore E. Plasticenta: First evidence of microplastics in human placenta. Environment international. 2021 Jan 1; 146:106274.
27. Law KL, Narayan R. Reducing environmental plastic pollution by designing polymer materials for managed end-of-life. Nature Reviews Materials. 2022 Feb;7(2):104-16.

28. Prata JC, da Costa JP, Duarte AC, Rocha-Santos T. Methods for sampling and detection of microplastics in water and sediment: a critical review. TrAC Trends in Analytical Chemistry. 2019 Jan 1; 110:150-9.
29. Eriksen M, Mason S, Wilson S, Box C, Zellers A, Edwards W et al (2013) Microplastic pollution in the surface waters of the Laurentian Great Lakes. Mar Pollut Bull 77:177–182
30. Hanke G, Galgani F, Werner S, Oosterbaan L, Nilsson P, Fleet D, Kinsey S, Thompson R, Palatinus A, Van Franeker J, Vlachogianni T. Guidance on Monitoring of Marine Litter in European Seas: a guidance document within the Common Implementation Strategy for the Marine Strategy Framework Directive. ISBN 978-92-79-32709-4.
31. Lassen C, Hansen SF, Magnusson K, Norén F, Hartmann NI, Jensen PR, Nielsen TG, Brinch A. Microplastics—occurrence, effects and sources of releases to the environment in Denmark, Environmental Project No. 1973. Danish Ministry of the Environment—Environmental Protection Agency, Denmark. 2015;204.
32. Liu F, Nord NB, Bester K, Vollertsen J. Microplastics removal from treated wastewater by a biofilter. Water. 2020 Apr 11;12(4):1085.
33. Li K, Ma D, Wu J, Chai C, Shi Y. Distribution of phthalate esters in agricultural soil with plastic film mulching in Shandong Peninsula, East China. Chemosphere. 2016 Dec 1; 164:314-21.
34. Nel HA, Froneman PW. A quantitative analysis of microplastic pollution along the south-eastern coastline of South Africa. Marine pollution bulletin. 2015 Dec 15;101(1):274-9.
35. Battelle Ocean Sciences (Organization), United States. Environmental Protection Agency. Oceans, Coastal Protection Division. Plastic pellets in the aquatic environment: Sources and recommendations. United States Environmental Protection Agency, Office of Water; 1992.
36. Bergmann M, Mützel S, Primpke S, Tekman MB, Trachsel J, Gerdts G. White and wonderful? Microplastics prevail in snow from the Alps to the Arctic. Science advances. 2019 Aug 14;5(8): eaax1157.
37. Zhang S, Liu X, Hao X, Wang J, Zhang Y. Distribution of low-density microplastics in the mollisol farmlands of northeast China. Science of the Total Environment. 2020 Mar 15; 708:135091.
38. https://www.unep.org/news-and-stories/story/plastic-planet-how-tiny-plastic-particles-are-polluting-our-soil
39. Nizzetto, L.; Futter, M.; Langaas, S. Are agricultural soils dumps for microplastics of urban origin? Environ. Sci. Technol. 2016, 50, 10777.
40. He P, Chen L, Shao L, Zhang H, Lü F. Municipal solid waste (MSW) landfill: A source of microplastics? -Evidence of microplastics in landfill leachate. Water research. 2019 Aug 1; 159:38-45.
41. Berlin F. An underestimated threat: Land-based pollution with microplastics. ScienceDaily. ScienceDaily. 2018 Feb; 5.
42. Rillig, M.C., 2012. Microplastic in terrestrial ecosystems and the soil? Environ. Sci. Technol. 46, 6453–6454
43. Kumari A, Rajput VD, Mandzhieva SS, Rajput S, Minkina T, Kaur R, Sushkova S, Kumari P, Ranjan A, Kalinitchenko VP, Glinushkin AP. Microplastic pollution: an emerging threat to terrestrial plants and insights into its remediation strategies. Plants. 2022 Jan 27; 11(3): 340.
44. Ng EL, Lwanga EH, Eldridge SM, Johnston P, Hu HW, Geissen V, Chen D. An overview of microplastic and nanoplastic pollution in agroecosystems. Science of the total environment. 2018 Jun 15; 627: 1377-88.
45. Rochman, Chelsea M., et al. "Scientific evidence supports a ban on microbeads." (2015): 10759-10761.

46. Duis K, Coors A. Microplastics in the aquatic and terrestrial environment: sources (with a specific focus on personal care products), fate and effects. Environmental Sciences Europe. 2016 Dec;28(1):1-25.
47. Patel MM, Goyal BR, Bhadada SV, Bhatt JS, Amin AF. Getting into the brain: approaches to enhance brain drug delivery. CNS drugs. 2009 Jan; 23:35-58.
48. Boucher J, Friot D. Primary microplastics in the oceans: a global evaluation of sources. Gland, Switzerland: Iucn; 2017 Feb.
49. Rathinamoorthy R, Raja Balasaraswathi S. A review of the current status of microfiber pollution research in textiles. International Journal of Clothing Science and Technology. 2021 Apr 29;33(3):364-87.
50. Sekar, N. "UV-absorbent, Antimicrobial, Water-repellent and Other Types of Functional Dye for Technical Textile Applications." Advances in the Dyeing and Finishing of Technical Textiles (2013) Feb: 47–77".
51. Gulrajani, M L. Advances in the Dyeing and Finishing of Technical Textiles. Cambridge, UK: Woodhead Publishing Ltd, 2013.
52. Bannick CG, Brand K, Jekel M, König F, Miklos D, Rechenberg B. Kunststoffe in der Umwelt-Ein Beitrag zur aktuellen Mikroplastikdiskussion. KA–Korrespondenz Abwasser, Abfall. 2015;62(1):36-41.
53. Zhang GS, Hu XB, Zhang XX, Li J. Effects of plastic mulch and crop rotation on soil physical properties in rain-fed vegetable production in the mid-Yunnan plateau, China. Soil and Tillage Research. 2015 Jan 1; 145:111-7.
54. Wang J, Luo Y, Teng Y, Ma W, Christie P, Li Z. Soil contamination by phthalate esters in Chinese intensive vegetable production systems with different modes of use of plastic film. Environmental Pollution. 2013 Sep 1; 180:265-73.
55. Stöven K, Jacobs F, Schnug E. Microplastic: a selfmade environmental problem in the plastic age. Journal für Kulturpflanzen. 2015 July;67(7): 241–50.
56. Hartline NL, Bruce NJ, Karba SN, Ruff EO, Sonar SU, Holden PA. Microfiber masses recovered from conventional machine washing of new or aged garments. Environmental science & technology. 2016 Nov 1;50(21):11532-8.
57. Klein S, Worch E, Knepper TP. Occurrence and spatial distribution of microplastics in river shore sediments of the Rhine-Main area in Germany. Environmental science & technology. 2015 May 19;49(10):6070-6.
58. Vassilenko K. Me, my clothes and the ocean: The role of textiles in microfiber pollution. University of British Columbia; 2019.
59. Barnes SJ. Understanding plastics pollution: The role of economic development and technological research. Environmental pollution. 2019 Jun 1; 249:812-21.
60. Borrelle SB, Ringma J, Law KL, Monnahan CC, Lebreton L, McGivern A, Murphy E, Jambeck J, Leonard GH, Hilleary MA, Eriksen M. Predicted growth in plastic waste exceeds efforts to mitigate plastic pollution. Science. 2020 Sep 18;369(6510):1515-8.
61. Padervand M, Lichtfouse E, Robert D, Wang C. Removal of microplastics from the environment. A review. Environmental Chemistry Letters. 2020 May; 18:807-28.
62. Bruce N, Hartline N, Karba S, Ruff B, Sonar S, Holden P. Microfiber pollution and the apparel industry. University of California Santa Barbara, Bren School of Environmental Science & Management (accessed 19 Aug 2016) http://brenmicroplastics. weebly. com/uploads/5/1/7/0/51702815/bren-patagonia_final_report. pdf. 2016.
63. Stanton T, Johnson M, Nathanail P, MacNaughtan W, Gomes RL. Freshwater and airborne textile fibre populations are dominated by 'natural', not microplastic, fibres. Science of the total environment. 2019 May 20; 666:377-89.

64. Liu J, Liang J, Ding J, Zhang G, Zeng X, Yang Q, Zhu B, Gao W. Microfiber pollution: an ongoing major environmental issue related to the sustainable development of textile and clothing industry. Environment, Development and Sustainability. 2021 Aug; 23:11240-56.
65. Eriksen M, Lebreton LC, Carson HS, Thiel M, Moore CJ, Borerro JC, Galgani F, Ryan PG, Reisser J. Plastic pollution in the world's oceans: more than 5 trillion plastic pieces weighing over 250,000 tons afloat at sea. PloS one. 2014 Dec 10;9(12): e111913.
66. Vickers NJ. Animal communication: when i'm calling you, will you answer too? Current biology. 2017 Jul 24;27(14): R713-5.
67. Browne, M.A., Crump, P., Nivens, S.J., Teuten, E., Tonkin, A., Galloway, T., Thompson, R., Accumulation of microplastics on shorelines worldwide: sources and sinks. Environ. Sci. Technol. 2011 Sept. 45 (21): 9175-79.
68. Zettler ER, Mincer TJ, Amaral-Zettler LA. Life in the "plastisphere": microbial communities on plastic marine debris. Environmental science & technology. 2013 Jul 2;47(13):7137-46.
69. Selonen S, Dolar A, Kokalj AJ, Skalar T, Dolcet LP, Hurley R, van Gestel CA. Exploring the impacts of plastics in soil–The effects of polyester textile fibers on soil invertebrates. Science of the Total Environment. 2020 Jan 15; 700:134451.
70. Rillig MC, Ingraffia R, de Souza Machado AA. Microplastic incorporation into soil in agroecosystems. Frontiers in plant science. 2017 Oct 18; 8: 1805.
71. Brockerhoff EG, Barbaro L, Castagneyrol B, Forrester DI, Gardiner B, González-Olabarria JR, Lyver PO, Meurisse N, Oxbrough A, Taki H, Thompson ID. Forest biodiversity, ecosystem functioning and the provision of ecosystem services. Biodiversity and Conservation. 2017 Dec; 26:3005-35.
72. Chandra P, Singh DP. Microplastic degradation by bacteria in aquatic ecosystem. InMicroorganisms for sustainable environment and health. Elsevier. 2020 Jan 1 (pp. 431–467).
73. Badola N, Bahuguna A, Sasson Y, Chauhan JS. Microplastics removal strategies: A step toward finding the solution. Frontiers of Environmental Science & Engineering. 2022 Jan; 16:1-8.
74. Paço A, Duarte K, da Costa JP, Santos PS, Pereira R, Pereira ME, Freitas AC, Duarte AC, Rocha-Santos TA. Biodegradation of polyethylene microplastics by the marine fungus Zalerion maritimum. Science of the Total Environment. 2017 May 15; 586:10-5.
75. Dey TK, Uddin ME, Jamal M. Detection and removal of microplastics in wastewater: evolution and impact. Environmental Science and Pollution Research. 2021 Apr; 28:16925-47.
76. Osman M, Satti SM, Luqman A, Hasan F, Shah Z, Shah AA. Degradation of polyester polyurethane by Aspergillus sp. strain S45 isolated from soil. Journal of Polymers and the Environment. 2018 Jan; 26:301-10.
77. Yamada-Onodera K, Mukumoto H, Katsuyaya Y, Saiganji A, Tani Y. Degradation of polyethylene by a fungus, Penicillium simplicissimum YK. Polymer degradation and stability. 2001 May 1;72(2):323-7.
78. Zhang J, Gao D, Li Q, Zhao Y, Li L, Lin H, Bi Q, Zhao Y. Biodegradation of polyethylene microplastic particles by the fungus Aspergillus flavus from the guts of wax moth Galleria mellonella. Science of the Total Environment. 2020 Feb 20; 704:135931.
79. Russell JR, Huang J, Anand P, Kucera K, Sandoval AG, Dantzler KW, Hickman D, Jee J, Kimovec FM, Koppstein D, Marks DH. Biodegradation of polyester polyurethane by endophytic fungi. Applied and environmental microbiology. 2011 Sep 1;77(17):6076-84.
80. Devi RS, Kannan VR, Nivas D, Kannan K, Chandru S, Antony AR. Biodegradation of HDPE by Aspergillus spp. from marine ecosystem of Gulf of Mannar, India. Marine pollution bulletin. 2015 Jul 15;96(1-2):32-40.

81. Delacuvellerie A, Cyriaque V, Gobert S, Benali S, Wattiez R. The plastisphere in marine ecosystem hosts potential specific microbial degraders including Alcanivorax borkumensis as a key player for the low-density polyethylene degradation. Journal of hazardous materials. 2019 Dec 15; 380:120899.
82. Auta HS, Emenike CU, Fauziah SH. Screening of Bacillus strains isolated from mangrove ecosystems in Peninsular Malaysia for microplastic degradation. Environmental Pollution. 2017 Dec 1; 231:1552-9.
83. Auta HS, Emenike CU, Jayanthi B, Fauziah SH. Growth kinetics and biodeterioration of polypropylene microplastics by Bacillus sp. and Rhodococcus sp. isolated from mangrove sediment. Marine Pollution Bulletin. 2018 Feb 1; 127:15-21.
84. Vimala PP, Mathew L. Biodegradation of polyethylene using Bacillus subtilis. Procedia Technology. 2016 Jan 1; 24:232-9.
85. Cunha C, Silva L, Paulo J, Faria M, Nogueira N, Cordeiro N. Microalgal-based biopolymer for nano-and microplastic removal: a possible biosolution for wastewater treatment. Environmental Pollution. 2020 Aug 1; 263:114385.
86. Sundbæk KB, Koch ID, Villaro CG, Rasmussen NS, Holdt SL, Hartmann NB. Sorption of fluorescent polystyrene microplastic particles to edible seaweed Fucus vesiculosus. Journal of applied phycology. 2018 Oct; 30:2923-7.
87. Nolte TM, Hartmann NB, Kleijn JM, Garnæs J, Van De Meent D, Hendriks AJ, Baun A. The toxicity of plastic nanoparticles to green algae as influenced by surface modification, medium hardness and cellular adsorption. Aquatic toxicology. 2017 Feb 1; 183:11-20.
88. Peller J, Nevers MB, Byappanahalli M, Nelson C, Babu BG, Evans MA, Kostelnik E, Keller M, Johnston J, Shidler S. Sequestration of microfibers and other microplastics by green algae, Cladophora, in the US Great Lakes. Environmental Pollution. 2021 May 1; 276:116695.
89. Kiki C, Rashid A, Zhang Y, Li X, Chen TY, Adéoye AB, Peter PO, Sun Q. Microalgal mediated antibiotic co-metabolism: Kinetics, transformation products and pathways. Chemosphere. 2022 Apr 1; 292:133438.
90. Davarpanah E, Guilhermino L. Single and combined effects of microplastics and copper on the population growth of the marine microalgae Tetraselmis chuii. Estuarine, Coastal and Shelf Science. 2015 Dec 20; 167:269-75.
91. Li H, Zhang L, Lu H, Ma J, Zhou X, Wang Z, Yi C. Macro-/nanoporous Al-doped ZnO/cellulose composites based on tunable cellulose fiber sizes for enhancing photocatalytic properties. Carbohydrate polymers. 2020 Dec 15; 250:116873.
92. Lagarde F, Olivier O, Zanella M, Daniel P, Hiard S, Caruso A. Microplastic interactions with freshwater microalgae: hetero-aggregation and changes in plastic density appear strongly dependent on polymer type. Environmental pollution. 2016 Aug 1; 215:331-9.
93. Sarmah P, Rout J. Efficient biodegradation of low-density polyethylene by cyanobacteria isolated from submerged polyethylene surface in domestic sewage water. Environmental Science and Pollution Research. 2018 Nov; 25:33508-20.
94. Sarmah P, Rout J. Cyanobacterial degradation of low-density polyethylene (LDPE) by Nostoc carneum isolated from submerged polyethylene surface in domestic sewage water. Energy, Ecology and Environment. 2019 Oct; 4:240-52.
95. Flores-Salgado G, Thalasso F, Buitrón G, Vital-Jácome M, Quijano G. Kinetic characterization of microalgal-bacterial systems: Contributions of microalgae and heterotrophic bacteria to the oxygen balance in wastewater treatment. Biochemical Engineering Journal. 2021 Jan 15; 165:107819.
96. Sun J, Dai X, Wang Q, Van Loosdrecht MC, Ni BJ. Microplastics in wastewater treatment plants: Detection, occurrence and removal. Water research. 2019 Apr 1; 152:21-37.

97. Xiao K, Liang S, Wang X, Chen C, Huang X. Current state and challenges of full-scale membrane bioreactor applications: A critical review. Bioresource technology. 2019 Jan 1; 271:473-81.
98. Poerio T, Piacentini E, Mazzei R. Membrane processes for microplastic removal. Molecules. 2019 Nov 15;24 (22):4148.
99. Pico Y, Alfarhan A, Barcelo D. Nano-and microplastic analysis: Focus on their occurrence in freshwater ecosystems and remediation technologies. TrAC Trends in Analytical Chemistry. 2019 Apr 1; 113:409-25.

Micro-Raman and ESR Studies for the confirmation of RTFM in $Zn_{1-x}Cu_xO$ ($0.00 \leq x \leq 0.1$)

Urmishree Routray

Department of Physics, Faculty of Engineering and Technology (ITER),
Siksha 'O' Anusandhan (Deemed to be University),
Khandagiri Square, Bhubaneswar 751030, Odisha, India

Jyoshnarani Mohapatra

Department of Physics, College of Basic Science and Humanities,
Orissa University of Agriculture and Technology,
Bhubaneswar 751003, Odisha, India

V. V. Srinivasu

Department of Physics, College of Science, Engineering and Technology,
University of South Africa, Johannesburg 1710, South Africa

Dilip Kumar Mishra*

Department of Physics, Faculty of Engineering and Technology (ITER),
Siksha 'O' Anusandhan (Deemed to be University), Khandagiri Square,
Bhubaneswar 751030, Odisha, India

Abstract Ball milling technique was adopted for the preparation of $Zn_{1-x}Cu_xO$ ($0.00 \leq x \leq 0.1$) powder samples by varying ball milling time period from 1 to 4h. Observation of impurity phases of CuO and Cu_2O is confirmed from micro-Raman studies. All the ball milled ZnO and Cu doped ZnO samples exhibit room temperature ferromagnetism. The presence of cationic interaction is confirmed from the electron spin resonance studies. The observation of same "g" value (2.12~2.14) and have peak-to-peak narrow line width of 260-290 G confirms the powders samples to be magnetically homogeneous system. The presence of oxygen vacancies and exchange interaction between the Cu^{1+} and Cu^{2+} ions are the prime factor for the occurrence of room temperature ferromagnetism.

Keywords Semiconductors, X-ray diffraction, Raman spectroscopy, Magnetic properties, ESR

---

*Corresponding author: dilipiuac@gmail.com; dilipmishra@soa.ac.in

DOI: 10.1201/9781003489443-33

1. Introduction

Among the well-known oxide semiconducting materials, ZnO has unique position which is being applied in optoelectronic, spintronics and space engineering [1-5]. The extensive progress has been made on undoped and doped ZnO based materials, motivated by, both basic sciences and potential advanced technologies [4, 5]. Due to its uniqueness, it is used for manufacturing UV lasers, luminophores, UV and IR radiation detectors, solar cells, transparent thin film transistors etc [1-5]. Again, the utilizing of both charge and spin characters of an electron to develop novel spintronics devices has led to an extensive search for ZnO based materials as the semiconducting properties of this materials can be integrated with the magnetic properties [4-6]. Earlier, it was assumed that the practical spintronics application of ZnO can be made only by doping or substituting magnetic ions. However, a number of reports suggest that in the nano form ZnO can exhibit ferromagnetic behaviour, without doping with magnetic ions or non-magnetic ions [7-14]. The observation of ferromagnetism in ZnO based materials has claimed as its inherent and intrinsic property whereas few groups claimed that the contribution of ferromagnetism is from magnetic impurities. To reach at a conclusion, some groups have attempted with Cu doped ZnO system as secondary phases of Cu, CuO and Cu_2O are nonmagnetic in nature [6, 15-22]. Theoretical and experimental predictions suggest that only p-type Cu doped ZnO is ferromagnetic nature and n-type Cu doped ZnO does not [18, 19]. But experimental observation supports the evidence of existence of room temperature ferromagnetism in n-type Cu doped ZnO [6, 20]. Even exchange interaction and point defects are considered as prime factor for the observation of ferromagnetism, but role of secondary phases is not discussed anywhere extensively.

To understand the role of point defects and presence of secondary phases, powder samples of $Zn_{1-x}Cu_xO$ ($0.00 \leq x \leq 0.1$) have been synthesized by ball milling technique by varying ball milling time from 1h to 4h. The role of secondary phases of Cu in the occurrence of ferromagnetism in Cu doped ZnO has been discussed through micro-Raman characterization and electron spin resonance studies.

2. Experimental Detail

Ball milling is a versatile route for the preparation of solid solution of any kind of materials. Hence, a series of undoped ZnO and Cu doped ZnO powder samples were prepared by varying the ball milling time period from 1h to 4h. $Zn_{1-x}Cu_xO$ powder samples were synthesized by changing the Cu concentration from 3 atomic weight % to 10 atomic weight %. For this purpose, 99.99 % spec-pure ZnO and CuO powders were used as per the stoichiometric ratio. The mixed powders were ground in a Retsch PM-100 planetary ball mill with 3 mm tungstate carbide balls to avoid the contamination of any magnetic materials. The grinding was carried out in dry medium at fixed rotations per minute (rpm) of 350 for varying time starting from one to four hours. A dwell time of 2 minutes was fixed after each 10 minutes of grinding. The grinded product was recovered and separated from the balls using a filtrated system.

The crystal structure, crystal perfection and structural defect have been studied using Micro-Raman Spectrometer (Renishaw inVia Raman Microscope), respectively. Room temperature magnetic field dependent magnetization measurements have been carried out using vibration sample magnetometer. Similarly, room temperature electron spin resonance (ESR) signals have been obtained by using a Bruker EMX spectrometer operating at a microwave frequency of 9.45 GHz with a field modulation frequency of 100Hz. The ESR signal was obtained for a sweeping magnetic field (DC) 0 G to 10 KG.

3. Results and Discussion

To understand on crystal structure and identification of secondary phases, micro-Raman analysis have been carried out. The presence of secondary phases can be identified properly from the Raman peak which is very difficult to identify from the XRD patterns. Micro-Raman spectra for 1h, 2h, 3h and 4h ball milled $Zn_{1-x}Cu_xO$ ($0 \leq x \leq 0.1$) have been shown in Fig. 33.1. The stable wurtzite crystal structure of ZnO exhibits C^4_{6v} space group with two formula units per primitive cell, where all the atoms occupy the C_{3v} symmetry [23]. According to group theory, the phonon modes of A_1, B_1, E_1 and E_2 are existing in ZnO structure near the centre of the Brillouin zone, given as a formula $\Gamma = A_1 + 2B_1 + E_1 + 2E_2$ in which B_1 modes are recognized as silence modes and A_1; E_1 and E_2 are Raman active modes.

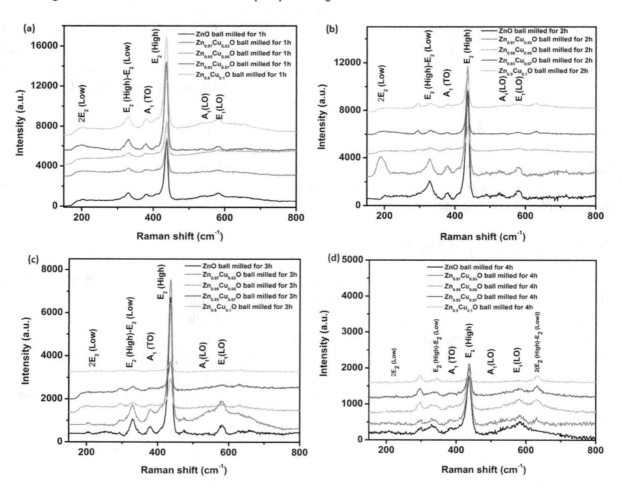

Fig. 33.1 Micro-Raman spectra of ball milled $Zn_{1-x}Cu_xO$ ($0 \leq x \leq 0.1$) powders

In all the ball milled samples, there is a prominent peak at around 435~437 cm^{-1}, which is indexed as E$_2$ high peak. This peak is known as the characterized peak of wurtzite structure of ZnO [11]. The spectra clearly represented that wurtzite structure of ZnO is not deteriorated with the increasing ball milled time. This peak strongly depends on the isotropic composition of ZnO. However, higher ball milled time creates amorphous nature in the sample. The characteristic nanocrystalline ZnO peak observed at 581 cm^{-1} is identified as the 1LO (A$_1$/E$_1$) mode. The peak is very distinct and sharp and treated as the characteristic feature of nanocrystalline ZnO [11, 23] which is appeared due to the association of structural disorder [23-25]. Hence, the presence of defects in ZnO based specimen is confirmed from the appearance of the sharp peak with its shoulder peak at 581 cm^{-1} as shown in Fig. 33.1. The appearance of this peak is expected to come from the vacancy clusters which arise from the interaction of vacancies on the surface of nanoparticles with nearby vacancies. The other peaks at 331 cm^{-1} and 512 cm^{-1} are assigned to possible multi phonon scattering modes and the peak at 276 cm^{-1} is attributed to the activation of the normally dormant B$_2$ modes. The peak at 385 cm^{-1} is related to the A$_1$ (TO) mode that are activated due to the presence of defects [11, 23]. Considering to all above peaks, few other peaks are also observed at around 200 cm^{-1}, 298 cm^{-1} and 630 cm^{-1}. It is expected that these peaks might be attributed to CuO/Cu$_2$O. Hence, we have attempted to analyse the Raman spectra inconsideration of the presence of oxide phases of Cu. It is reported that CuO belongs to the C$^6_{2h}$ space group with two molecules per primitive cell [24]. Near the center of the Brillouin zone, group theory predicts the existence of the following phonon modes as $\Gamma = 4A_u + 5B_u + A_g + 2B_g$. There are three acoustic modes (A$_u$+2B$_u$), six infrared actives modes (3A$_u$+3B$_u$), and three Raman active modes (A$_g$+2B$_g$) [24]. As per the group theory prediction, three peaks have been observed which are centred around 200 cm^{-1}, 298 cm^{-1} and 630 cm^{-1}. However, literature supports the presence of Raman peak at 278 cm^{-1}, 321 cm^{-1} and 608 cm^{-1} are attributed to CuO [24-26]. But in our case, there is the peak shift, and the peak is present at 298 cm^{-1}. As the observation of Raman peaks of CuO depends upon the synthesis technique, it is obvious to observe the Raman peaks 20 cm^{-1} away from the literature value. This peak is attributed clearly the presence of CuO. However, the peak at 331 cm^{-1} is also identified as CuO as it is difficult to distinguish from the ZnO modes of vibration as the intensity of the Raman scattering completely depends upon the number of scattering centers present in the volume illuminated by the laser beam. Again, the intensified peak at 200 cm^{-1} and 630 cm^{-1} are attributed to Cu$_2$O [24-26]. It indicates that both CuO and Cu$_2$O coexist in Cu doped ZnO samples.

Magnetic field dependent magnetization curves are shown in Fig. 33.2 for ball milled ZnO and Cu doped ZnO samples and determination of the values of remnant magnetization (M$_r$), coercivity (H$_c$) and saturation magnetization (M$_s$) of all Zn$_{1-x}$Cu$_x$O powders are given in Table 33.1. Figure 33.2a represents the M~H curve for ZnO samples ball milled from 1 to 4h. It is observed that the saturation magnetization increases with the increase in ball milled time. However, there is not much variation in the coercivity value. The coercivity value lies within the range of 95 to 110 Oe. With the increase in ball milled time, defect in the sample starts to develop and maximum in the higher ball milled time. This is one of the reasons for the observation of slightly higher saturation magnetization in ZnO ball milled samples with higher ball milled time. Low coercivity value indicates the ZnO as a soft magnetic materials and it can be achieved very low

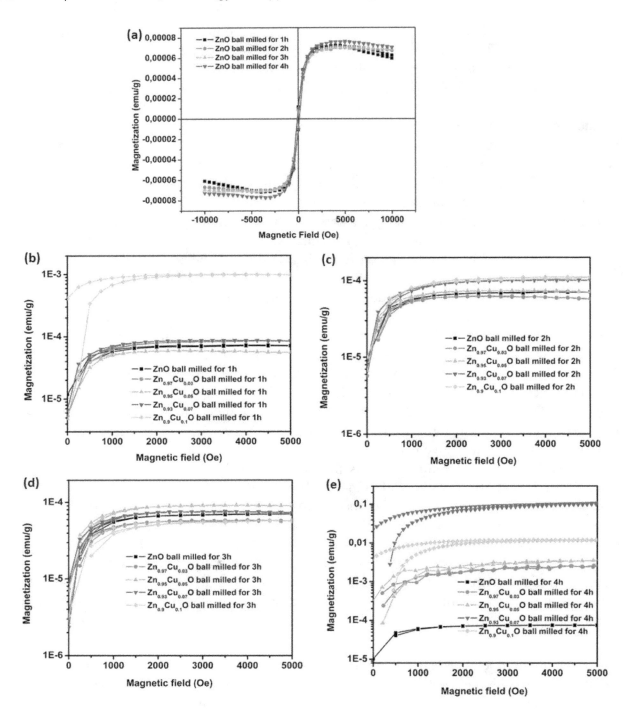

Fig. 33.2 Magnetic field dependent magnetization of ball milled $Zn_{1-x}Cu_xO$ ($0 \leq x \leq 0.1$) powders

coercivity value if optimization of control parameter will be carefully chosen. Magnetization vs magnetic field for 1h, 2h, 3h and 4h ball milled Cu doped ZnO are shown in Fig. 33.2b, c, d, e respectively. It is clearly observed that higher the percentage of Cu in ZnO and higher ball milled time gives higher saturation magnetization except leaving few cases. This case

Table 33.1 Determination of the values of remnant magnetization (M_r), coercivity (H_c) and saturation magnetization (M_s) of $Zn_{1-x}Cu_xO$ compounds ball milled for various hours

Ball milled time	Sample specifications	M_r (emu/g)	H_c (Oe)	M_s (emu/g)
1h	ZnO	1.25×10^{-5}	110	7.235×10^{-5}
	$Zn_{0.97}Cu_{0.03}O$	6.244×10^{-6}	85	8.5×10^{-5}
	$Zn_{0.95}Cu_{0.05}O$	6.11×10^{-6}	85	5.948×10^{-5}
	$Zn_{0.93}Cu_{0.07}O$	7.746×10^{-5}	65	8.80×10^{-5}
	$Zn_{0.9}Cu_{0.1}O$	4.248×10^{-4}	251	9.824×10^{-4}
2h	ZnO	1.009×10^{-5}	95	7.073×10^{-5}
	$Zn_{0.97}Cu_{0.03}O$	6.486×10^{-6}	80	6.217×10^{-5}
	$Zn_{0.95}Cu_{0.05}O$	6.696×10^{-6}	85	7.355×10^{-5}
	$Zn_{0.93}Cu_{0.07}O$	9.459×10^{-6}	76	1.003×10^{-4}
	$Zn_{0.9}Cu_{0.1}O$	1.003×10^{-5}	73	1.138×10^{-4}
3h	ZnO	1.009×10^{-5}	103	7.235×10^{-5}
	$Zn_{0.97}Cu_{0.03}O$	4.111×10^{-6}	72	5.939×10^{-5}
	$Zn_{0.95}Cu_{0.05}O$	9.520×10^{-6}	70	9.142×10^{-5}
	$Zn_{0.93}Cu_{0.07}O$	9.601×10^{-6}	90	7.589×10^{-5}
	$Zn_{0.9}Cu_{0.1}O$	3.942×10^{-6}	127	5.843×10^{-5}
4h	ZnO	1.108×10^{-5}	96	7.663×10^{-5}
	$Zn_{0.97}Cu_{0.03}O$	0.0002	92	0.002
	$Zn_{0.95}Cu_{0.05}O$	0.00017	71	0.004
	$Zn_{0.93}Cu_{0.07}O$	0.023	313	0.116
	$Zn_{0.9}Cu_{0.1}O$	0.004	381	0.012

is very well observed in 4 h ball milled samples. The saturation magnetization of Cu doped ZnO is much higher than the ZnO ball milled sample. The question arises that only oxygen vacancies creates in the ZnO during the synthesis is the only reason for the observation of higher magnetic moment or any other reasons is associated. In our previous literature [21], we reported that the Cu does not exist as only in single oxidation state. It is always existed as Cu^{1+} and Cu^{2+} state. The electronic structure of Cu^{1+} and Cu^{2+} state favours the feasible of exchange interactions between them through the oxygen vacancies. Micro-Raman observation clearly indicates the presence of oxygen vacancies clusters and as well the presence of both CuO and Cu_2O phase. Hence the percentage of Cu incorporates in the Zn site and CuO remains as Cu^{2+} state and the Cu_2O phase of Cu remains as Cu^{1+} state. This causes an exchange interaction between the Cu ions through the oxygen vacancies cluster and favours for the observation of ferromagnetism in Cu doped ZnO specimen.

ESR spectra of 1h and 4h ball milled ZnO and Cu doped ZnO is shown in Fig. 33.3 and Fig. 33.4. The ZnO and Cu mixed ZnO (commercially purchased) powders do not show any ESR signal which clearly indicated that the raw product is in pure form. It is observed that the

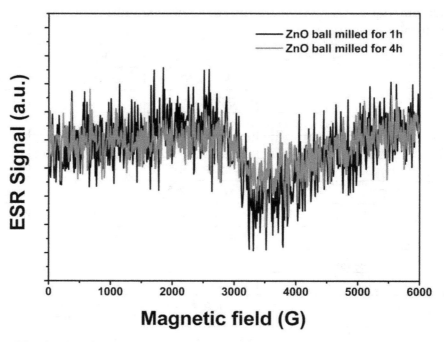

Fig. 33.3 ESR spectrum of 1h and 4h ball milled ZnO powders

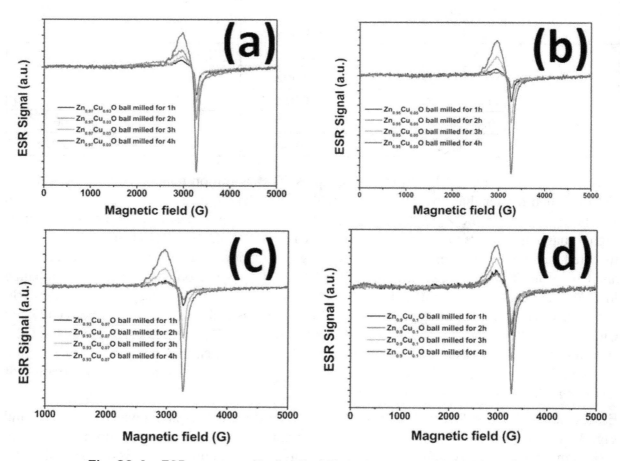

Fig. 33.4 ESR spectra of ball milled Zn1-xCuxO (0 ≤ x ≤ 0.1) powders

ESR signal is very weak in ball milled ZnO powder (Fig. 33.3) whereas the ESR signal is very strong in all Cu doped Ball milled powders (Fig. 33.4). All the samples show almost similar line shape with the centre of resonance at the same position. There is no observation of resonance peak shift in ball milled ZnO powder and as well as in Cu doped ZnO powders. It indicates that Cu has not magnetic effect on doping in ZnO. The resonant signal corresponding to an effective g-value of 2.12-2.14 is observed in all the samples. It is attributed to the existence of cationic type of interaction [27] in powder samples to provide ferromagnetic nature in ball milled undoped and Cu doped ZnO powders. All the samples have peak-to-peak narrow line width of 260-290 G which confirms the powders samples to be magnetically homogeneous system. The non-observation of low field signal in all the ball milled samples is a clear indication of the magnetically homogeneous system. However, the variation of saturation magnetization in all powder samples is due to the presence of oxygen vacancies and the interaction between the cations via oxygen vacancies. The confirmation of cationic interaction like Zn-Zn, Zn-Cu and Cu-Cu from the observation of g value and presence of oxygen vacancies from Micro-Raman analysis establish the occurrence of room temperature ferromagnetism in all ball milled samples. From all above studies, it can be concluded that the variation of magnetization value given in Table 1 depends on the presence of oxygen vacancies and on the exchange interaction. Again, the amorphization of powder sample reduces the ferromagnetic contribution of the powder samples. Hence, the variation is not systematic.

4. Conclusions

In summary, the series of synthesized $Zn_{1-x}Cu_xO$ (0.00 ≤ x ≤ 0.1) powder samples by ball milling technique exhibit room temperature ferromagnetism. The observation of +1 and +2 ionic state of Cu is confirmed from micro-Raman studies. The presence of cationic interaction is confirmed from the electron spin resonance studies. The observation of same "g" value (2.12~2.14) and have peak-to-peak narrow line width of 260-290 G confirms the powders samples to be magnetically homogeneous system. The presence of oxygen vacancies and exchange interaction between the Cu^{1+} and Cu^{2+} ions are the prime factor for the occurrence of room temperature ferromagnetism.

Acknowledgement

Authors JM, UR and DKM are thankful to Dr. S. K. Singh, CSIR-IMMT, Bhubaneswar, India for providing the synthesis facility of ball milling instrument. Authors are thankful to CIF, Pondicherry University for providing VSM measurement facility.

REFERENCES

1. Ozgur U, Alivov YL, Liu C, Teke A, Reshchikov MA, Dogan S, Avrutin V, Cho SJ, Morkoc HJ. A comprehensive review of ZnO materials and devices. *Journal of Applied Physics* 2005, **98**: 041301.
2. Hara K, Horiguci T, Kinoshita T, Sayama K, Sugihara H, Arakawa H. Highly efficient photon to electron conversion with mercurochrome-sensitized nanoporous oxide semiconductor solar cells. *Solar Energy Materials and Solar Cells* 2000, **64**: 115–134.

3. Heilnd G. Homogeneous semiconducting gas sensors. *Sensors and Actuators* 1981-1982, **2**: 343–361.
4. Prinz GA. Magnetoelectronics. *Science* 1998, **282**: 1660–1663.
5. Wolf SA, Awschalom DD, Buhrman RA, Daughton JM, Von Molnar S, Rukes ML, Chtchelkanova AY, Treger DM. Spintronics: A Spin-Based Electronics Vision for the Future. *Science* 2001, **294**: 1488–1495.
6. Shukla G. Magnetic and Optical properties of epitaxial n-type Cu doped ZnO thin films deposited on sapphire substrates. *Applied Physics A* 2009, **97**: 115–118.
7. Routray U, Dash R, Mohapatra JR, Das J, Srinivasu VV, Mishra DK. Temperature-dependent ferromagnetic behavior in nanocrystalline ZnO synthesized by pyrophoric technique. *Materials Letter* 2014, **137**: 29–31.
8. Mohapatra J, Mishra PK, Singh SK, Mishra DK. Room temperature ferromagnetism in ZnO nanoparticles synthesized by auto combustion technique. *Nanoscience and Nanotechnology Letters* 2010, **2**: 30-34.
9. Rainey K, Chess J, Eixenberger J, Tenne DA, Hanna CB, Punnoose A. Defect induced ferromagnetism in undoped ZnO nanoparticles. *Journal Applied Physics* 2014, **115**: 17D727.
10. Sundaresan A, Bhargavi R, Rangarajan N, Siddesh U, Rao CNR. Ferromagnetism as a universal feature of nanoparticles of the otherwise nonmagnetic oxides. Physical Review B 2006, **74**: 161306(R).
11. Mishra DK, Mohapatra J, Sharma MK, Chattarjee R, Singh SK, Varma S, Behera SN, Nayak SK, Entel P. Carbon doped ZnO: Synthesis, characterization and interpretation. *Journal of Magnetism and Magnetic Materials* 2013, **323**: 146–152.
12. Khan ZA. Ghosh S. Robust room temperature ferromagnetism in Cu doped ZnO thin films. *Applied Physics Letters* 2011, **99**: 042504.
13. Pan H, Yi JB, Shen L, Wu RQ, Yang JH, Lin JY, Feng YP, Ding J, Van LH, Yin JH. Room Temperature Ferromagnetism in Carbon doped ZnO. *Physical Review Letters* 2007, **99**: 127201.
14. Li XL, Guo JF, Z.-Y. Quan, X.-H. Xu, G.A. Gehring. Defects inducing Ferromagnetism in carbon doped ZnO films. *IEEE Transactions on Magnetics* 2010, **46**: 1382.
15. Mohapatra J, Mishra DK, Mishra PK, Bag BP, Singh SK. Enhancement of ferromagnetism in nanocrystalline $Zn_{1-x}Cu_xO$ ($0.03 \leq x \leq 0.07$) Systems. *NANO* 2011, **6**: 387–393.
16. Zhang Z, Yi JB, Wong LM, Seng HL, Wang SJ, Tao JG, Li GP, Xing GZ, Sum TC, Huan CHA, Wu, T. Cu doped ZnO nano needles and nano nails: morphological evolution and physical properties. *J. Phys. Chem. C* 2008, **112**: 9579–9585.
17. Owens FJ. Room temperature ferromagnetism in Cu doped ZnO synthesized from CuO and ZnO nanoparticles. *Journal of Magnetism and Magnetic Materials* 2009, **321**: 3734–3737.
18. Park MS, Min BJ. Ferromagnetism in ZnO codoped with transition metals: $Zn_{1-x}(FeCo)_xZn_{1-x}(FeCo)_xO$. *Physical Review B* 2003, **68**: 224436.
19. Buchholz DB, Chang RPH, Song JH, Ketterson JB. Room temperature ferromagnetism in Cu-doped ZnO thin films. *Applied Physics Letters* 2005, **87**: 82504.
20. Hou DL, Ye XJ, Meng HJ, Zhou HJ, Li XL, Zhen CM, Tang GD. Magnetic properties of n-type Cu- doped ZnO thin films. *Applied Physics Letters* 2007, **90**: 142502.
21. Mohapatra J, Routray U, Medicherla VRR, Mishra DK. Existence of Cu^{1+} and Cu^{2+} oxidation states for the occurrence of ferromagnetism in $Zn_{0.95}Cu_{0.05}O$. *Advanced Science Letters* 2016, **22**: 479–481.
22. Mohapatra J, Mishra DK, Berma V, Kamilla SK, Sakthivel R, Mohapatra BK, Singh SK. Structural and magnetic properties of $Zn_{1-x}Cu_xO$ ($0 \leq x \leq 0.1$) systems. *Advanced Science Letters* 2011, **4**: 458–462.

23. Das J, Pradhan SK, Sahu DR, Mishra DK, Sarangi S, Nayak BB, Verma S, Roul BK. Micro Raman and XPS studies of pure ZnO ceramics. *Physica B* 2010, **405**: 2492-2497 and *references therein*.
24. Fuentes S, Zarate RA, Munoz P, Diaz-droguett E. Formation of hierarchical CuO nanowires on a copper surface via a room temperature solution-immersion Process. *Journal of the Chilean Chemical Society (JCCS)* 2010, **55**: 147-149 and *references therein*.
25. Gong YS, Lee C, Yang CK. Atomic force microscopy and Raman spectroscopy studies in the oxidation of Cu thin films. *Journal of Applied Physics* 1995, **77**: 5422.
26. Balamurugan B, Mehta BR, Avasthi DK, Singh F, Arora AK, Rajalakshmi M, Raghavan G, Tyagi AK, Shivaprasad SM. Modifying the nanocrystalline characteristics – structure, size and surface states of copper oxide thin films by high energy heavy ion irradiation. *Journal of Applied Physics* 2002, **92**: 3304.
27. Motaung DE, Mhlongo GH, Nkosi SS, Malgas GF, Mwakikunga BW, Coetsee E, Swart HC, Abdallah HMI, Moyo T, Ray SS. Shape-Selective Dependence of Room Temperature Ferromagnetism Induced by Hierarchical ZnO Nanostructures. *ACS Applied Materials & Interfaces* 2014, **6**: 8981–8995.